The Combined
Finite-Discrete Element
Method

The Combined Finite-Discrete Element Method

Ante Munjiza
Queen Mary, University of London, London, UK

John Wiley & Sons, Ltd

Other Wiley Editorial Offices

John Wiley & Sons Inc., 111 River Street, Hoboken, NJ 07030, USA

Jossey-Bass, 989 Market Street, San Francisco, CA 94103-1741, USA

Wiley-VCH Verlag GmbH, Boschstr. 12, D-69469 Weinheim, Germany

John Wiley & Sons Australia Ltd, 33 Park Road, Milton, Queensland 4064, Australia

John Wiley & Sons (Asia) Pte Ltd, 2 Clementi Loop #02-01, Jin Xing Distripark, Singapore 129809

John Wiley & Sons Canada Ltd, 22 Worcester Road, Etobicoke, Ontario, Canada M9W 1L1

Wiley also publishes its books in a variety of electronic formats. Some content that appears
in print may not be available in electronic books.

Library of Congress Cataloging-in-Publication Data

Munjiza, Ante.
   The combined finite-discrete element method / Ante Munjiza.
      p. cm.
   ISBN 0-470-84199-0 (Cloth : alk. paper)
   1.   Deformations (Mechanics) – Mathematical models. 2.   Finite element
method. I.   Title.
TA417.6 M87 2004
620.1′123′015118 – dc22

                                                              2003025485

British Library Cataloguing in Publication Data

A catalogue record for this book is available from the British Library

ISBN 0-470-84199-0

Typeset in 10/12pt Times by Laserwords Private Limited, Chennai, India
Printed and bound in Great Britain by Antony Rowe Ltd, Chippenham, Wiltshire
This book is printed on acid-free paper responsibly manufactured from sustainable forestry
in which at least two trees are planted for each one used for paper production.

To Jasna and Boney

Contents

Preface

Computational mechanics of discontinua is a relatively new discipline of computational mechanics. It deals with numerical solutions to engineering problems and processes where constitutive laws are not available. Particle- based modelling of micro-structural elements of material is used instead. Thus, the interaction and individual behaviour of millions, even billions, of particles is considered to arrive at emergent physical properties of practical importance.

Methods of computational mechanics of discontinua include DEM (Discrete Element Methods), DDA (Discontinua Deformation Analysis Methods), Methods of Molecular Dynamics, etc.

In the last decade of the 20th century, the Discrete Element Method has been coupled with the Finite Element Method. The new method is termed the 'Combined Finite-Discrete Element Method'. Thanks to the relatively inexpensive high performance hardware rapidly becoming available, it is possible to consider combined finite-discrete element systems comprising millions of particles, and most recently even billions of particles. These, coupled with recent algorithmic developments, have resulted in the combined finite-discrete element method being applied to a diversity of engineering and scientific problems, ranging from powder technology, ceramics, composite materials, rock blasting, mining, demolition, blasts and impacts in a defence context, through to geological and environmental applications.

Broadly speaking, the combined finite-discrete element method is applicable to any problem of solid mechanics where failure, fracture, fragmentation, collapse or other type of extensive material damage is expected.

In the early 1990s, the combined finite-discrete element method was mostly an academic subject. In the last ten years, the first commercial codes have been developed, and many commercial finite element packages are increasingly adopting the combined finite-discrete element method. The same is valid for the commercial discrete element packages. The combined finite-discrete element method has become available to research students, but also to engineers and researchers in industry. It is also becoming an integral part of many undergraduate and postgraduate programs.

A book on the combined finite-discrete element method is long overdue. This book aims to help all those who need to learn more about the combined finite-discrete element method.

Acknowledgements

This book could not have happened without the support and encouragements of the publishers, Wiley, and many of my colleagues. I would like first to mention Professor J.R. Williams from MIT. He has always been an inspiration and a motivator, and ready to help throughout. I started my PhD under his guidance, and finished it while working in his laboratory at MIT. He was a great teacher then, just as he is today. He is one of those professors well remembered by students. I also owe a great deal to many others. This book could not have happened without the support from Dr. D.S. Preece, Prof. G. Mustoe, Dr. C. Thornton, Dr. J.P. Latham, Prof. D.R.J. Owen, Prof. B. Mohanty, Prof. E. Hinton, Prof. O. Zienkiewicz, Dr. K.R.F. Andrews, Prof. J. White, Dr. N. John, Prof. R. O'Connor, Prof. F. Aliabadi, Prof. Mihanovic, Prof. Jovic, and many others from whom I have learned. You have been real friends. Thank you.

1

Introduction

1.1 GENERAL FORMULATION OF CONTINUUM PROBLEMS

The microstructure of engineering materials is discontinuous. However, for a large proportion of engineering problems it is not necessary to take into account the discontinuous nature of the material. This is because engineering problems take material in quantities large enough so that the microstructure of the material can be described by averaged material properties, which are continuous. The continuous nature of such material properties is best illustrated by mass m, which is defined as a continuous function of volume through introduction of density ρ such that

$$\rho = \frac{dm}{dV} \tag{1.1}$$

where V is volume. The microscopic discontinuous distribution of mass in space is thus replaced by hypothetical macroscopic continuous mass distribution. In other words, microscopic discontinuous material is replaced by macroscopic continuum of density ρ.

The continuum hypothesis introduced is valid as long as the characteristic length of the particular engineering problem is, for instance, much greater than the mean free path of molecules. For engineering problems the characteristic length is defined by either the smallest dimension of the problem itself, or the smallest dimension of the part of the problem of practical interest. The hypothesis of continuum enables the definition of physical properties of the material as continuous functions of volume. These physical properties are very often called *physical equations*, or the *constitutive law*, and they are then combined with balance principles (balance equations). The result is a set of governing equations. The balance principles are *a priori* physical principles describing materials in sufficient bulk so that the effects of discontinuous microstructure can be neglected. Balance principles include conservation of mass, conservation of energy, preservation of momentum balance, preservation of moment of momentum balance, etc.

Governing equations are usually given as a set of partial differential equations (strong formulation) or integral equations (weak or variational formulation). Governing equations, when coupled with external actions in the form of boundary and initial conditions (such as loads, supports, initial velocity, etc.), make a boundary value problem or an initial boundary value problem. The solution of a particular boundary value problem is sometimes expressed in analytical form. More often, approximate numerical methods are employed.

The Combined Finite-Discrete Element Method   A. Munjiza
© 2004 John Wiley & Sons, Ltd   ISBN: 0-470-84199-0

These include the finite difference method, finite volume method, finite element method, mesh-less finite element methods, etc.

The most advanced and the most often used method is the finite element method. The finite element method is based on discretisation of the domain into finite sub-domains, also called finite elements. Finite elements share nodes, edges and surfaces, all of which comprise a finite element mesh. The solution over individual finite elements is sought in an approximate form using shape (base) functions. Balanced principles are imposed in averaged (integral or weak) form. These usually yield algebraic equations, for instance equilibrium of nodal forces, thus effectively replacing governing partial differential equations with a system of simultaneous algebraic equations, the solution of which gives results (e.g. displacements) at the nodes of finite elements.

1.2 GENERAL FORMULATION OF DISCONTINUUM PROBLEMS

Taking into account that the mean free path of molecules for most engineering materials is very small in comparison to the characteristic length of most of the engineering problems, one may arrive at the conclusion that most engineering materials are well represented by a hypothetical continuum model. That this is not the case is easily demonstrated by the following problem:

A glass container of square base is filled with particles of varying shape and size, as shown in Figure 1.1. The particulate is left to fall from the given height. During the fall under gravity, the particles interact with each other and with the walls of the container. In this process energy is dissipated, and finally, all the particles find state of rest. The

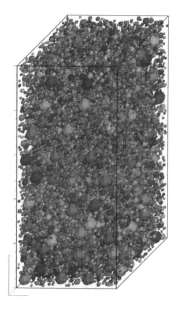

Figure 1.1 Letting particles of different shape and size pack under gravity inside the glass container. The question is how much space will they occupy?

question is, what is the total volume occupied by the particulate after all the particles have found the state of rest?

This problem is subsequently referred to as *container problem*. It is self-evident that the definition of density ρ given by

$$\rho = \frac{dm}{dV} \tag{1.2}$$

and the definition of mass m given by

$$m = \int_V \rho \, dV \tag{1.3}$$

are not valid for the container problem.

The total mass of the system is instead given by

$$m = \sum_{i=1}^{N} m_i \tag{1.4}$$

where N is the total number of particles in the container and m_i is the mass of the individual particles. In other words, the total mass is given as a sum of the masses of individual particles. It is worth mentioning that the size of the container is not much larger than the size of the individual particles. The particles pack in the container, and the mass of particles in the container is a function of the size of the container, the shape of individual particles, size of individual particles, deposition method, deposition sequence, etc.

Mathematical description of the container problem ought to take into account the shape, size and mass of individual particles, and also the interaction between the individual particles and interaction with the walls of the container. The mathematical model describing the problem has to state the interaction law for each couple of contacting particles. For each particle, the interaction law is combined with a momentum balance principle, resulting in a set of governing equations describing that particle. Sets of differential equations for different particles are coupled through inter-particle interaction. The resulting global set of coupled differential equations describes the behaviour of the particulate system as a whole. The solution of the global set of governing equations results in the final state of rest for each of the particles. In the case of hypothetical continuum the total number of governing partial differential equations does not depend on the size of the problem. In the container problem, each particle has a set of differential equations governing its motion, and the total number of governing partial differential equations is proportional to the total number of particles in the container.

The container problem, together with many similar problems pervading science and engineering, are by nature discontinuous. They are called problems of discontinuum, or *discontinuum* problems. Problems for which a hypothetical continuum model is valid are, in contrast, called problems of continuum, or *continuum problems*. The mathematical formulation of problems of continuum involves the constitutive law, balance principles, boundary conditions and/or initial conditions.

The mathematical formulation of problems of discontinua involves the interaction law between particles and balance principles. Analytical solutions of these equations are rarely available, and approximate numerical solutions are sought instead. The most advanced and most often used numerical methods are Discontinuous Deformation Analysis (DDA) and Discrete Element Methods (DEM). These methods are designed to handle contact situations for a large number of irregular particles. DDA is more suitable for static problems, while DEM is more suitable for problems involving transient dynamics until the state of rest or steady state is achieved.

A division of computational mechanics dealing with computational solutions to the problems of discontinua is called *Computational Mechanics of Discontinua*. Computational Mechanics of Discontinua is a relatively new discipline. Pioneering work in the late 1960s and early 1970s was done by researchers (Williams, Cundal, Gen-She, Musto, Preece, Thornton) from various disciplines. They handled complex problems of discontinua with very modest computing hardware resources available at the time. A second generation of researchers, such as Munjiza, Owen and O'Connor, benefited not only from more sophisticated computer hardware, available with RAM space measured in megabytes, but also from the UNIX operating system and graphics libraries combined with a new generation of computer languages, such as C and C++. This has enabled the key algorithmic solutions to be developed and/or improved. The third generation of researchers (late 1990s and the early years of this century) has benefited further from increased RAM space, now measured in gigabytes, relatively inexpensive CPU power, sophisticated visualisation tools, the internet and public domain software. As a result of this progress, discontinua methods have been applied to a wide range of engineering problems, which include both industrial and scientific applications.

1.3 A TYPICAL PROBLEM OF COMPUTATIONAL MECHANICS OF DISCONTINUA

The difference between problems of Computational Mechanics of Continua and problems of Computational Mechanics of Discontinua are best illustrated by the container problem. The container problem is about how many particles can be placed in a given volume, how they interact and the mechanics of the pack in general. To demonstrate the key elements of discontinua analysis, a numerical simulation of gravitational deposition of different packs inside a rigid box of size $250 \times 250 \times 540$ mm is shown in this section. The total solid volume deposited is constant for all simulations shown, and is equal to $V = 9.150e{-}03$ m^3.

The particles are deposited in three stages:

- In the first stage, a regular pattern is used to initially place all the particles inside the box, (Figure 1.2 (left)).

- In the second stage, random velocity field is applied to all the particles, making particles move inside the box until a near random distribution of particles inside the box is achieved. There is no gravity at this stage (Figure 1.2 (right)).

- In the third stage, the velocity of all particles is set to zero, and acceleration of gravity $g = 9.81$m/s^2 is applied in the z-direction. Under gravity the particles move towards the bottom of the box. Due to the interaction with the box and with each other, the particles closer to the bottom of the box slowly settle into the final

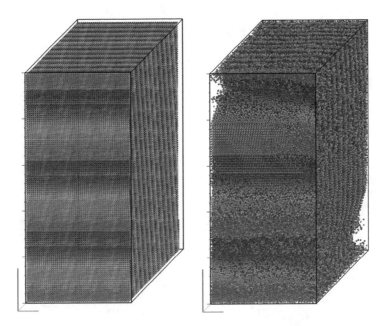

Figure 1.2 Gravitational deposition of identical spheres of diameter $d = 2.635\,$mm; Left: initial pack; Right: randomly perturbed pack before the start of deposition, i.e. at time $t = 0\,$s.

position. Due to energy dissipation, eventually the velocity of all particles is zero and the pack is at the state of rest – which is also the state of static equilibrium. Thus, through dynamic governing equations taking into account dynamic equilibrium and through contact and the motion of individual particles, static equilibrium of the pack is achieved.

In Figures 1.2–1.6, such deposition sequence of a pack comprising identical spheres of diameter $d = 2.635\,$mm is shown. As explained earlier, the total volume of the solid deposited is $V = 9.150\mathrm{e}{-03}\,$m3. Thus, the pack comprises 955,160 identical spheres.

The maximum theoretical density for packs comprised of identical spheres is given by

$$\rho = \frac{\pi\sqrt{18}}{18} \cong 0.7405 \tag{1.5}$$

A density profile obtained using gravitational deposition is normalised using this theoretical density. Density profiles for different packs are shown in Figure 1.7. Initially (solid line), the particles are almost uniformly distributed over the box; the density is only slightly larger at the bottom of the box than at the top.

The motion of the particles under gravity gradually increases density at the bottom of the pack and decreases density at the top of the pack, making the upper parts of the box empty. As the particles settle, the pack gets denser. At the state of rest (dashed line), almost uniform density over most of the pack is achieved, with rapid density decrease towards the top of the pack.

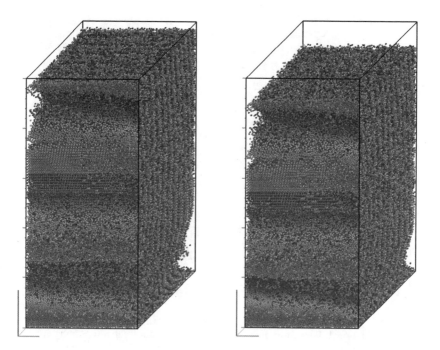

Figure 1.3 Deposition sequence at 0.55 s (left) and 0.105 s (right).

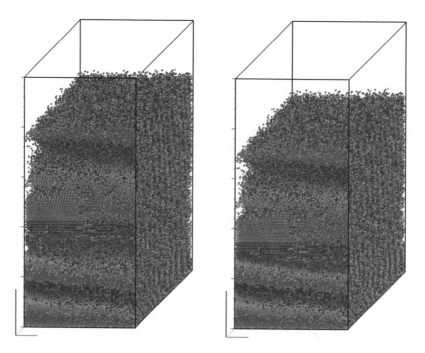

Figure 1.4 Deposition sequence at 0.155 s (left) and 0.180 s (right).

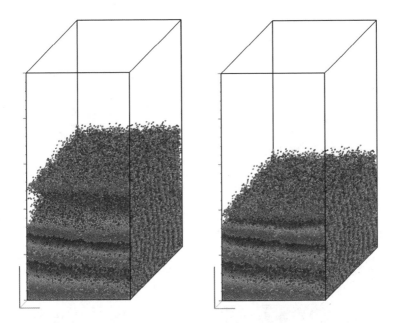

Figure 1.5   Deposition sequence at 0.230 s (left) and 0.255 s (right).

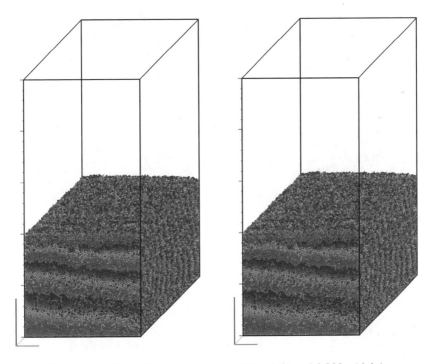

Figure 1.6   Deposition sequence at 0.605 s (left) and 0.830 s (right).

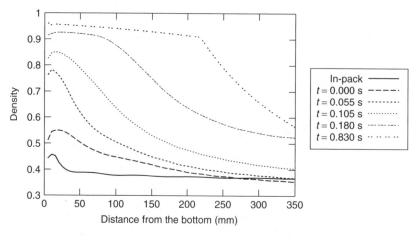

Figure 1.7 Averaged density over horizontal cross-section of the box as function of distance of the cross-section from the bottom of the box.

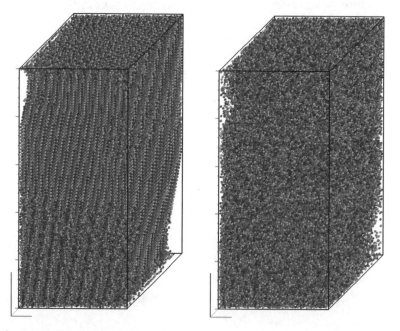

Figure 1.8 Initial disturbance (left) and spatial distribution of spheres at the start of gravity induced deposition, i.e. time 0.0 s (right).

The final packing density achieved is smaller than the theoretical density. The reasons behind this can be investigated if the same volume of solid is deposited using identical spheres of a larger diameter. Thus, in Figures 1.8–1.13, numerical simulation of gravitational deposition of 139,968 spheres all of the same diameter ($d = 4.998$ mm) is performed. The total solid volume of the spheres is the same as in the previous example, i.e. $V = 9.150\mathrm{e}{-03}\,\mathrm{m}^3$.

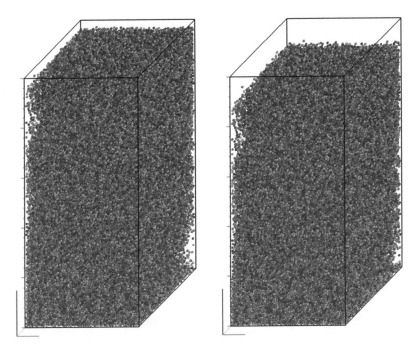

Figure 1.9   Gravity induced motion of the pack after 0.055 s and 0.105 s.

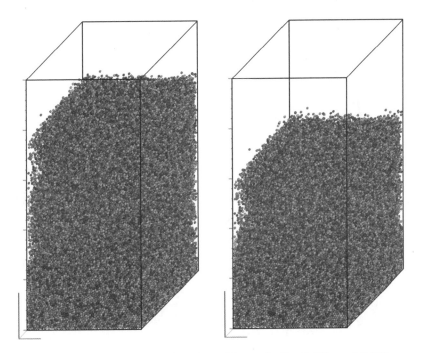

Figure 1.10   Gravity induced motion of the pack after 0.155 s and 0.205 s.

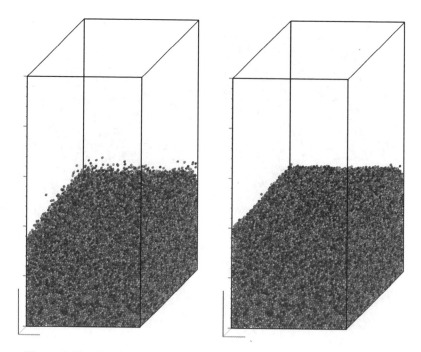

Figure 1.11   Gravity induced motion of the pack after 0.255 s and 0.305 s.

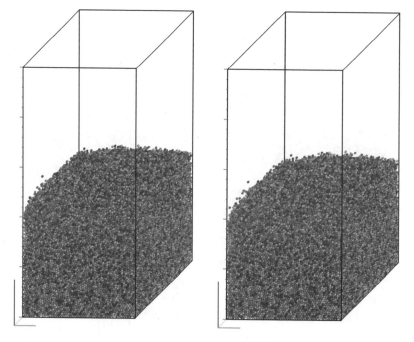

Figure 1.12   Gravity induced motion of the pack after 0.405 s and 0.430 s.

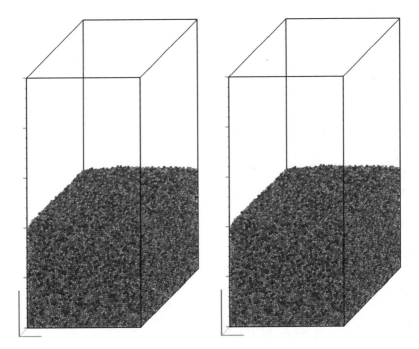

Figure 1.13   Gravity induced motion of the pack after 0.605 s and 0.830 s.

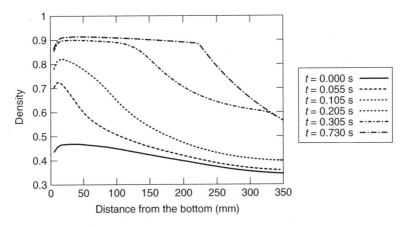

Figure 1.14   Averaged density over horizontal cross-section of the box as function of distance of the cross-section from the bottom of the box.

Density profiles for this pack are shown in Figure 1.14. The density profiles shown are normalised using the theoretical density. Initially (solid line), the particles are loosely packed inside the box. The motion of the particles under gravity gradually increases density at the bottom of the pack and decreases density at the top of the pack, finally making the upper parts of the box empty. As the particles settle, the pack gets denser. At the state of rest (top line), almost uniform density over most of the pack is achieved, with rapid density decrease towards the top of the pack. The final density is almost 10% smaller than the theoretical density for the same problem with periodic boundaries.

A further demonstration of the influence of sphere size on packing density is obtained through deposition of the same volume of solid, but comprised of larger spheres. Numerical experiments included:

- **41472 spheres of** $d = 7.497$ mm (Figure 1.15)
- **5184 spheres of** $d = 14.994$ mm (Figure 1.15)
- **648 spheres of** $d = 29.988$ mm (Figure 1.16)
- **192 spheres of** $d = 44.982\,9$ mm (Figure 1.16)
- **81 spheres of** $d = 59.976$ mm (Figure 1.17)
- **50 spheres of** $d = 70.439$ mm (Figure 1.17).

The final density profiles for each of the packs are given in Figure 1.18. The density of packs comprising very large spheres is very far from the theoretical density. This is due to the influence of the boundary conditions. For large spheres, the box is simply too small, and theoretical packing cannot be achieved.

As the spheres get smaller, the influence of the box diminishes and the density gets closer to the theoretical density, as shown in Figure 1.19, where packing density as a function of normalised sphere diameter (sphere diameter divided by the size of the edge of the box base – 250 mm) is plotted. For large spheres the density can go either up or down with a reduction in sphere size, i.e. discontinuum behaviour is strongly pronounced. As the spheres get smaller, more uniform convergence toward the theoretical result is achieved. Thus, at zero sphere diameter, theoretical density is achieved.

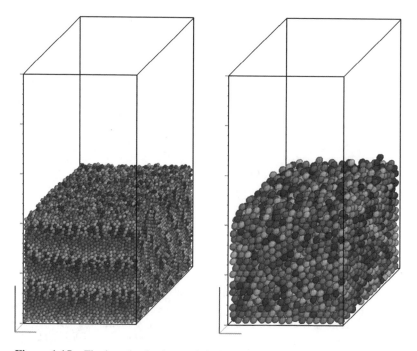

Figure 1.15   Final pack of spheres: left $d = 7.497$ mm, right $d = 14.994$ mm.

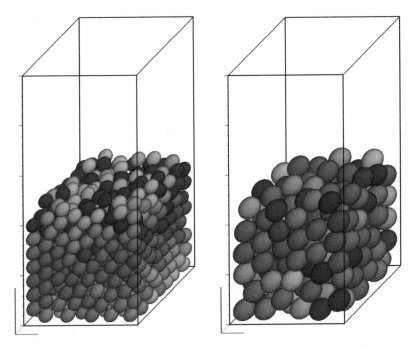

Figure 1.16   Final pack of spheres: left $d = 29.988$ mm, right $d = 44.982$ mm.

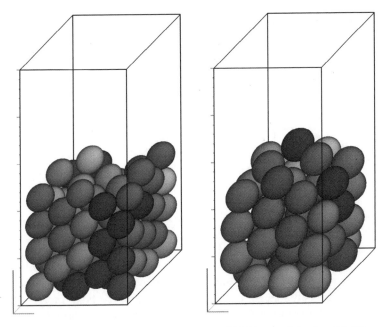

Figure 1.17   Final pack of spheres: left $d = 59.976$ mm, right $d = 70.439$ mm.

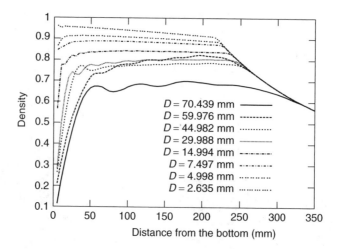

Figure 1.18   Final density profiles for selected diameters of spheres.

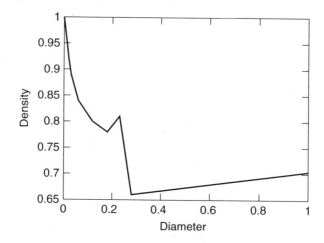

Figure 1.19   Maximum density as function of diameter of sphere and size of the box.

The container problem is a typical problem where continuum-based models cannot be applied. This problem also demonstrates that discontinuum-based simulations recover continuum formulation when the size of individual discrete elements (the diameter of the sphere in the problem described above) becomes small in comparison with the characteristic length of the problem being analysed. In the problem shown, the characteristic length of the problem is the length of the smallest edge of the box. The continuum-based models are simply a subset of more general discontinuum-based formulations; applicable when microstructural elements of the matter comprising the problem are very small in comparison to the characteristic length of the problem being analysed.

The behaviour of discontinuum systems is a function of the properties of microstructural elements (particles or discrete elements) making the system. The size of discrete elements is not the only parameter governing the behaviour of discontinuum system. This is demonstrated by using the same container problem as described above. This time size distribution of the spheres making the pack is changing. In Figure 1.20, the formation sequence of a pack comprising of 648 identical spheres is shown.

The normalised density averaged over a horizontal cross-section as a function of distance from the bottom of the box is shown in Figure 1.21.

The gravitational deposition of a pack comprising spheres of two different sizes is shown in Figure 1.22, while a packing density profile for the same pack is shown in Figure 1.23. The packing density has increased in comparison to the pack comprising mono-sized spheres.

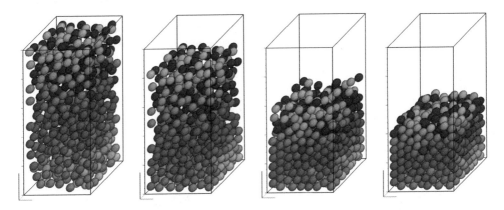

Figure 1.20   Gravitational deposition of a pack comprising 648 identical spheres of diameter $d = 29.988$ mm and total solid volume $V = 9.150\mathrm{e}{-03}\,\mathrm{m}^3$.

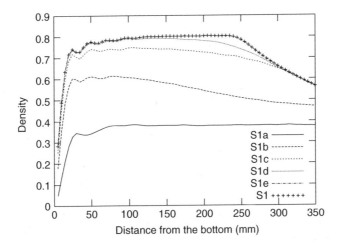

Figure 1.21   Normalised average density for a horizontal cross-section as a function of distance from the bottom of the box – pack of 648 identical spheres of diameter $d = 29.988$ mm and total solid volume $V = 9.150\mathrm{e}{-03}\,\mathrm{m}^3$; S1a is initial density profile, S1b–S1e are transient density profiles, and S1 is the final density profile, which corresponds to the sate of rest.

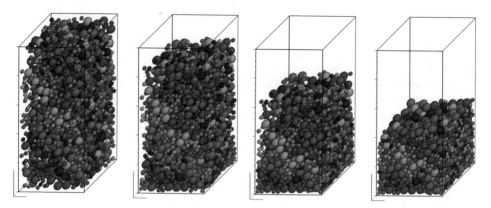

Figure 1.22   Gravitational deposition of a pack comprising spheres of two different sizes: 399 large spheres of diameter $d = 29.988$ mm, and 1992 smaller spheres of diameter $d = 14.994$ mm; the total solid volume is $V = 9.150e{-}03$ m^3.

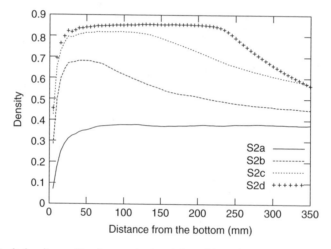

Figure 1.23   Pack density profiles for gravitational deposition of a pack comprising spheres of two different sizes: 399 large spheres of diameter $d = 29.988$ mm, and 1992 smaller spheres of diameter $d = 14.994$ mm; the total solid volume is $V = 9.150e{-}03$ m^3; S2a is initial density profile, S2b and S2c are transient density profiles, and S2 is the final density profile corresponding to the state of rest.

The gravitational deposition of a pack comprising spheres of three different sizes is shown in Figure 1.24. The transient and final packing density profile for the same pack is shown in Figure 1.25, which shows increased final packing density relative to the pack comprising spheres of two different sizes. It is also noticeable that large particles tend to segregate, and can be found on the top of the pack.

The gravitational deposition of a pack comprising spheres of four different sphere sizes is shown in Figure 1.26. Again, some increase in packing density (Figure 1.27), coupled with segregation of the largest particles, which can be found on the top of the pack, is observed.

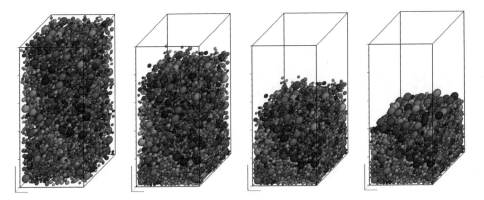

Figure 1.24 Gravitational deposition sequence of a pack comprising spheres of three different sizes: 319 spheres of diameter $d = 29.988$ mm, 1595 spheres of diameter $d = 14.994$ mm, 3506 spheres of diameter $d = 9.996$ mm; the total solid volume is $V = 9.150\mathrm{e}{-03}\,\mathrm{m}^3$.

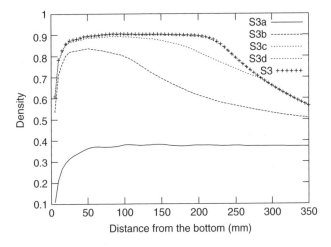

Figure 1.25 Density profile for gravitational deposition of a pack comprising spheres of three different sizes: 319 spheres of diameter $d = 29.988$ mm, 1595 spheres of diameter $d = 14.994$ mm, 3506 spheres of diameter $d = 9.996$ mm; the total solid volume is $V = 9.150\mathrm{e}{-03}$; S3a is initial density profile, S3 is the final density profile corresponding to the state of rest, and S3b, S3c and S3d are transient density profiles.

Further increase in the fraction of smallest particles is obtained by using five different particle sizes. The gravitational deposition of a pack comprising spheres of five different sphere sizes is shown in Figure 1.28, while the packing density profile for the same pack is shown in Figure 1.29. It is observed that the packing density has increased in comparison to the pack comprising spheres of four different sizes. However, it is still smaller than the maximum theoretical density for packs comprising identical spheres.

None of the size distributions employed has produced a pack of a density that would exceed the theoretical density of a pack comprising identical spheres. This is because the

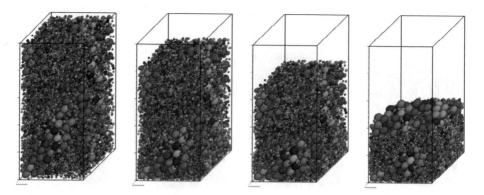

Figure 1.26 Gravitational deposition sequence of a pack comprising spheres of four different sizes: 277 spheres of diameter $d = 29.988$ mm, 1379 spheres of diameter $d = 29.988/2$ mm, 3036 spheres of diameter $d = 29.988/3$ mm and 5520 spheres of diameter $d = 29.988/4$ mm; the total solid volume is $V = 9.150e{-}03$ m^3.

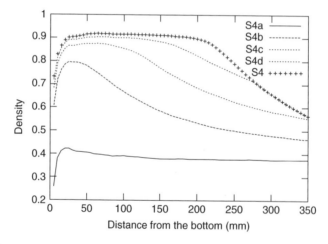

Figure 1.27 Density profile for gravitational deposition of a pack comprising spheres of four different sizes: 277 spheres of diameter $d = 29.988$ mm, 1379 spheres of diameter $d = 29.988/2$ mm, 3036 spheres of diameter $d = 29.988/3$ mm and 5520 spheres of diameter $d = 29.988/4$ mm; the total solid volume is $V = 9.150e{-}03$ m^3; S4a is the initial density profile, S4 is the final density profile corresponding to the state of rest, and S4b, S4c and S4d are transient density profiles.

packs shown are assembled using uniform size distribution given by the formula

$$y = \frac{\log(x/x_{min})}{\log(x_{max}/x_{min})} \tag{1.6}$$

where y is passing, x is the sieve size, x_{min} is the size of the smallest particle (corresponding to 0% passing) and x_{max} is the sieve size of the largest particle (corresponding to 100% passing) (Figure 1.30).

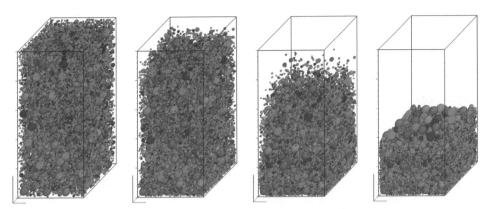

Figure 1.28 Gravitational deposition sequence of a pack comprising spheres of five different sizes: 249 spheres of diameter $d = 29.988$ mm, 1240 spheres of diameter $d = 29.988/2$ mm, 2728 spheres of diameter $d = 29.988/3$ mm, 4960 spheres of diameter $d = 29.988/4$ mm and 8184 spheres of diameter $d = 29.988/5$ mm; the total solid volume is $V = 9.150\mathrm{e}{-}03\,\mathrm{m}^3$.

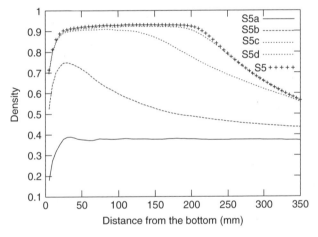

Figure 1.29 Density profile for gravitational deposition of a pack comprising spheres of five different sizes: 249 spheres of diameter $d = 29.988$ mm, 1240 spheres of diameter $d = 29.988/2$ mm, 2728 spheres of diameter $d = 29.988/3$ mm, 4960 spheres of diameter $d = 29.988/4$ mm, and 8184 spheres of diameter $d = 29.988/5$ mm; the total solid volume is $V = 9.150\mathrm{e}{-}03\,\mathrm{m}^3$; S5a is the initial density profile, S5 is the final density profile corresponding to the state of rest, and S5b, S5c and S5d are transient density profiles.

It is well known that materials with a uniform size distribution do not pack densely. Dense packs are known to be obtained using the power law size distribution, given by

$$y = \left[\frac{x}{x_{\max}} \right]^m \qquad (1.7)$$

where y is passing, x is the sieve size, $x_{\max}$ is the sieve size of the largest particle (corresponding to 100% passing), and m is the parameter (uniformity exponent) (Figure 1.31).

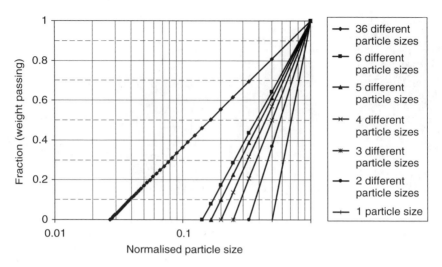

Figure 1.30 Uniform size distribution used to assemble the packs–points show different particle sizes comprising individual packs.

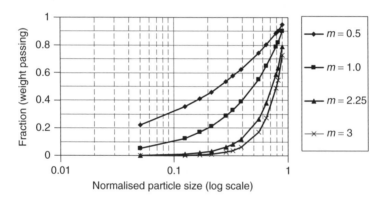

Figure 1.31 Power Law–for $m = 2.25$ points show different particle sizes comprising individual packs.

A pack comprising spheres of 13 different sizes, with a proportion of each size obtained using power law size distribution (137 spheres of diameter 29.988, 136 spheres of diameter 0.9·29.988, 57 spheres of diameter 0.8173·29.988, 274 spheres of diameter 0.789·29.988, 273 spheres of diameter 0.650·29.988, 547 spheres of diameter 0.553·29.988, 410 spheres of diameter 0.391·29.988, 408 spheres of diameter 0.331·29.988, 817 spheres of diameter 0.287·29.988, 829 spheres of diameter 0.212·29.988, 1211 spheres of diameter 0.169·29.988, 2690 spheres of diameter 0.125·29.988 and 6128 spheres of diameter 0.05·29.988) is shown in Figure 1.32.

By visual inspection of the deposition sequence and final state of rest, it is evident that small particles fill the space between the large particles, achieving a degree of locking that reduces segregation. While packs assembled using uniform size distribution have shown

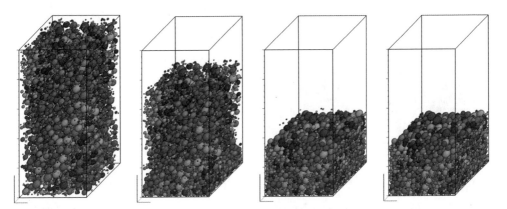

Figure 1.32 Deposition sequence of a pack of size distribution defined by the power law with uniformity exponent $m = 2.25$ and maximum sphere diameter 29.988 mm. The total volume of spheres is $V = 9.150e{-}03\,\text{m}^3$.

large particles 'floating' over small particles, and therefore ending up at the top of the pile, in this pack larger particles are distributed all over the pack. The density profile for this pack is shown in Figure 1.33. This time, the density is much larger than the maximum theoretical density of packs comprising mono-sized spheres. The pack is characterised by a packing density exceeding 85%. The inclusion of even smaller particles would result in this density being increased further.

A similar pack comprising smaller spheres of 13 different sizes, with a proportion of each size obtained using power law size distribution (1096 spheres of diameter 14.994, 1088 spheres of diameter 0.9·14.994, 456 spheres of diameter 0.8173·14.994, 2192 spheres of diameter 0.789·14.994, 2184 spheres of diameter 0.650·14.994, 4376

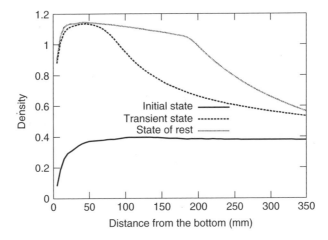

Figure 1.33 Density profile of a pack of size distribution defined by the power law with uniformity exponent $m = 2.25$ and maximum sphere diameter 29.988 mm. The total volume of spheres is $V = 9.150e{-}03\,\text{m}^3$.

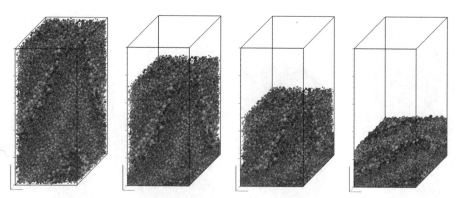

Figure 1.34 Deposition sequence of a pack of size distribution defined by the power law with uniformity exponent $m = 2.25$ and maximum sphere diameter 14.994 mm. The total volume of spheres is $V = 9.150\text{e}{-}03\,\text{m}^3$.

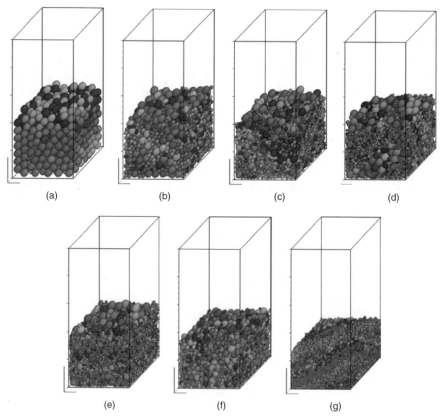

Figure 1.35 Visual comparison of different packs: (a) mono-sized spheres, (b) uniform size distribution using two different particle sizes, (c) uniform size distribution using three different particle sizes, (d) uniform size distribution using four different particle sizes, (e) uniform size distribution using five different particle sizes, (f) power law size distribution using 13 different particle sizes with maximum diameter 29.988 mm and, (g) power law size distribution using 13 different particle sizes with maximum diameter 14.994 mm; The total volume of spheres is the same in all packs $V = 9.150\text{e}{-}03\,\text{m}^3$.

spheres of diameter 0.553·14.994, 3280 spheres of diameter 0.391·14.994, 3264 spheres of diameter 0.331·14.994, 6536 spheres of diameter 0.287·14.994, 6632 spheres of diameter 0.212·14.994, 9688 spheres of diameter 0.169·14.994, 21520 spheres of diameter 0.125·14.994 and 49024 spheres of diameter 0.05·14.994) is shown in Figure 1.34. By reducing the size of the largest sphere, the influence of boundary on the pack is reduced. As a result, the smaller spheres fit the spaces between larger spheres even better.

It is evident that the size distribution plays an important role in pack formation, and has important impact on packing density. Visual comparison of packs obtained using uniform size distribution and packs obtained using power law size distribution is shown in Figure 1.35, while a comparison of density profiles is shown in Figure 1.36.

It can be seen that normalised density of packs obtained using uniform size distribution does not exceed 1.0, i.e. the density does not exceed the maximum theoretical density for mono-sized spheres

$$\rho = \frac{\pi \sqrt{18}}{18} \cong 0.7405 \tag{1.8}$$

It can be further observed that by varying the size distribution theoretically, normalised packing density much greater than 1 can be achieved. In addition, segregation phenomena is less pronounced with well graded packs (Figure 1.35).

The shape of individual particles also plays an important role in discontinua problems. The importance of the shape in the container problem described above is best illustrated by the results shown in Figure 1.37, where packs comprising particles of different shapes are shown. By visual inspection, it can be observed that the packing density varies considerably with the shape of individual particles comprising the pack.

Transient dynamics of the pack during gravitational deposition is also a function of the shape of the particles comprising the pack. This is illustrated by the deposition

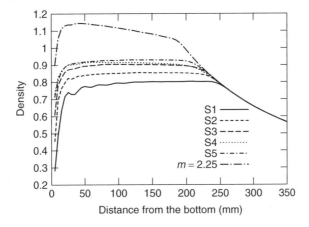

Figure 1.36   Density as function of size distribution. S1 mono-sized spheres; S2 uniform size distribution using two different particle sizes; S3 uniform size distribution using three different particle sizes; S4 uniform size distribution using four different particle sizes; S5 uniform size distribution using five different particle sizes; $m = 2.25$ power law size distribution using uniformity coefficient $m = 2.25$ and 13 different sphere sizes. Maximum sphere diameter in all cases is 29.988 mm, and the total volume of spheres is the same in all packs $V = 9.150e{-}03$ m3.

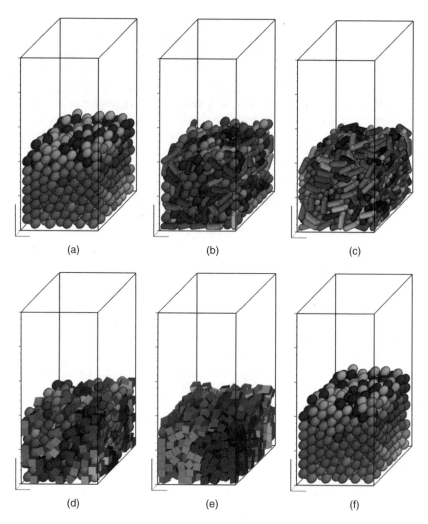

Figure 1.37 Visual comparison of packs comprised of mono-sized particles of different shape. (a) 648 spheres, (b) 324 spheres plus 324 pills, (c) 648 pills, (d) 324 spheres plus 324 cubes, (e) 648 cubes, (f) same as (a), i.e. 648 spheres. The total solid volume of particles is $V = 9.150e{-}03\,\mathrm{m}^3$, and is the same for all packs.

sequences shown in Figures 1.38–1.42. The most interesting motion sequence is obtained using cube-shaped particles, where a column type collapse of a pile is observed. On the other hand, pill-shaped particles show much interlocking and relatively quick settlement. Mixtures of cubes and spheres and pills and spheres show motion sequence, which is different from the corresponding packs comprised mono-shaped particles.

The discontinua simulations shown in this section best illustrate the difference between continuum and discontinuum problems. In continuum problems, the number of particles, the shape of particles and similar micromechanical characteristics are abstracted through the constitutive law, and they do not appear in computational simulations. In discontinua problems, the number of particles, size of particles, shape of particles, size distribution,

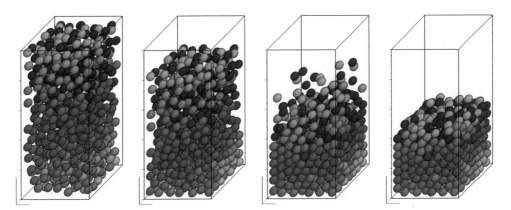

Figure 1.38 Deposition sequence of the pack comprising 648 mono-sized spheres of total volume $V = 9.150e-03 \, \text{m}^3$.

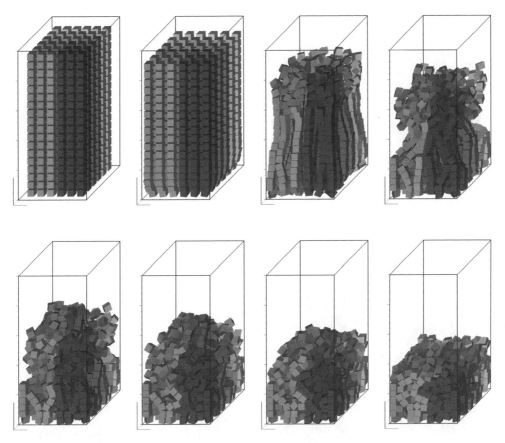

Figure 1.39 Deposition sequence of the pack comprising 648 mono-sized cubes of total volume $V = 9.150e-03 \, \text{m}^3$.

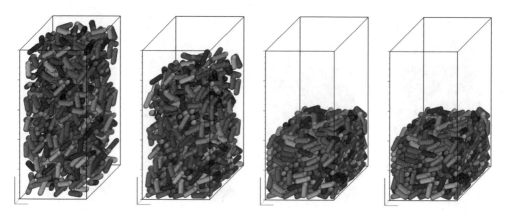

Figure 1.40 Deposition sequence of the pack comprising 648 mono-sized pills of total volume $V = 9.150\text{e}-03\,\text{m}^3$.

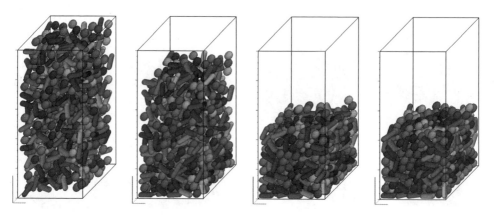

Figure 1.41 Deposition sequence of the pack comprising mixture of 324 spheres and 324 pills of equal volume with total volume being equal $V = 9.150\text{e}-03\,\text{m}^3$.

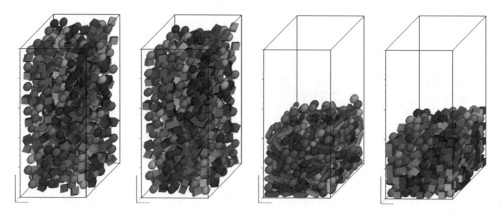

Figure 1.42 Deposition sequence of the pack comprising mixture of 324 spheres and 324 cubes of equal volume with total volume being equal $V = 9.150\text{e}-03\,\text{m}^3$.

shape distribution, size of the box, etc., are the key parameters governing the behaviour of the systems.

In this context, discontinua-based formulations are not just an alternative to continua-based formulations. On the contrary, discontinua-based formulations are the only available solution for problems where the continuum hypothesis is not valid. This fact has been recognised by researchers working in different disciplines from nano-scale and atomic-scale to terrestrial bodies.

1.4   COMBINED CONTINUA-DISCONTINUA PROBLEMS

Consider the container problem with particles being made of very soft rubber or jelly so that, in addition to interacting with each other, they deform as well. In addition, the walls of the container also deform. This problem is called the *flexible container problem*. Even in the case of less deformable particles, the deformation of the container and deformation of individual particles significantly influences the way particles move inside the container. Thus, the total mass of the particles deposited into the container is also influenced by deformability (elastic properties) of both particles and the container.

Each individual particle deforms under external forces and interaction with other particles already in the container, and also under interaction with container walls. Changes in the shape and size of individual particles is in essence a problem of finite strain elasticity (since finite rotations at least are present). The deformability of individual particles is therefore well represented by a hypothetical continuum-based model. Interaction among individual particles and interaction between particles and the container is best represented by discontinua-based model. The flexible container problem involves aspects of both continua and discontinua.

Problems such as the flexible container problem are a combination of continua and discontinua, and are therefore termed combined continua-discontinua problems. The deformability of individual particles is best described using the hypothetical continua formulation. Interaction and the motion of individual particles is best described using discontinua formulation. The set of governing equations obtained describes both the deformability of individual particles and interaction with container walls, and also between particles. The number of equations is a function of the total number of particles in the container. Analytical solutions of the governing equations obtained are rarely available, and numerical approaches have to be employed. These include DDA and DEM with added features to capture deformability.

However, the most advanced approach is to use the sate of the art method (the finite element method) to model continuum-based phenomena (in this case deformability) and the sate of the art method (the discrete element method) to model discontinuum-based phenomena (interaction and motion of individual particles). The new method is therefore a combination of both the finite element method and the discrete element method, and is termed *the combined finite–discrete element method (FEM/DEM)*.

In the combined finite-discrete element method, each particle (body) is represented by a single discrete element that interacts with discrete elements that are close to it. In addition, each discrete element is discretised into finite elements. Each discrete element has its own finite element mesh. The total number of finite element meshes employed is equal to the total number of discrete elements. Each finite element mesh employed

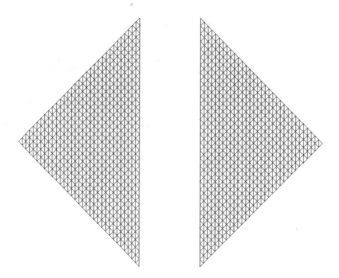

Figure 1.43 The combined finite-discrete element problem comprising two discrete elements–in the combined finite-discrete element method, each discrete element is discretised into finite elements.

captures the deformability of a single discrete element (particle, body). This is shown in Figure 1.43, where a combined finite-discrete element problem comprising two discrete elements is shown. Each discrete element is discretised into finite elements.

1.5 TRANSITION FROM CONTINUA TO DISCONTINUA

In flexible container type problems, it is often the case that individual particles can also fracture or fragment in addition to being deformable and interacting with each other. Fracture and fragmentation processes are in essence processes of transition from continua to discontinua. It is also in principle possible to imagine an inverse process of particles merging together.

There exists a whole class of engineering problems where transition from continua to discontinua plays a major role. Transition from continua to discontinua is in general a result of failure, fracture or fragmentation. Very often, the purpose of industrial operation itself is to induce failure, fracture or fragmentation of a solid. This is best demonstrated by a rock blasting operation, where an explosive charge is introduced into a borehole to break and fragment the rock.

Similar rock fracture and fragmentation processes can be observed in the case of rock crushers, where rock is broken by machinery to achieve a desired size distribution.

Failure, fracture and fragmentation is also observed in structural demolition. Very often, tall buildings are demolished by introducing carefully placed explosive charges to initiate the collapse using the potential energy of the building itself, together with inertia effects of individual parts of the building.

Structural collapse also occurs due to hazardous loading conditions such as impacts, earthquakes, blasts, explosions, etc. Structural collapse due to these loads can be progressive (e.g. Twin Towers, New York, September 2001). In progressive collapse, the potential energy of a building is transformed into kinetic energy, which in turn is transformed into strain energy at impact. Strain energy results in the material limits of structural elements being exceeded, and therefore the material is broken, bent or crushed, resulting in progressive collapse, which eventually transforms the building into a pile of rubble.

Similar situations have in the past occurred with a derailed train impacting against the bridge above the rail, resulting in the failure and collapse of the bridge (accident at Eschede in Germany, June 1998). Again, at the impact, the kinetic energy of the train together with the potential energy of the bridge is transferred into strain energy, resulting in the failure of the structural elements of the bridge and leading to the final collapse of the bridge. Yet another case of structural collapse can be seen in a situation where a house has collapsed due to the failure of ground under it (Walsall, West Midlands, May 1999). Again, what was a continuous structure has been transformed into a pile of separate particles. In this process kinetic energy, inertia forces and interaction between individual structural elements or falling and failing fragments has played an important role.

In the past, a common cause of structural collapse has been gas explosion. In modern days, blast loads are also becoming a frequent hazard due to wars, civil unrest and terrorism.

The common feature of all the above listed problems is a transition from continua to discontinua, i.e. failure, fracture, fragmentation and collapse. In the process, a transition from a static system to a dynamic system occurs. Subsequent contact-impact and energy dissipation mechanisms lead to the final state of rest, which often is just a pile of rubble. With predictive modelling capacity enabling simulations of various collapse scenarios, perhaps better design, structural, protective or procedural measures could be taken to at least reduce the risk to human life. The practical importance of knowing how a structure is going to collapse in the demolition process, or how a structure is going to behave under hazardous loads (for instance, will the evacuation routes be blocked), or what will be the size distribution of blasted or crushed material, cannot be overestimated.

1.6 THE COMBINED FINITE-DISCRETE ELEMENT METHOD

The only numerical tool currently available to a scientist or engineer that can properly take systems comprising millions of deformable discrete elements that simultaneously fracture and fragment is the combined finite-discrete element method. The combined finite-discrete element method merges finite element tools and techniques with discrete element algorithms. Finite element-based analysis of continua is merged with discrete element-based transient dynamics, contact detection and contact interaction solutions. Thus, transient dynamic analysis of systems comprising a large number (from a few thousands to more than a million) of deformable bodies which interact with each other and in this process can break, fracture or fragment, becomes possible.

A typical combined finite-discrete element simulation is shown in Figure 1.44. The right-hand support of a simply supported beam is suddenly released. Thus, the beam rotates about the left-hand support and breaks due to inertia forces, as shown in Figure 1.45.

Inertia forces play an important role in the failure of the beam. The beam breaks at the point of maximum bending moment due to the combined self-weight and inertia forces. The left-hand part of the beam subsequently breaks away from the right-hand support, resulting in a free fall of both parts of the beam as shown in Figure 1.45.

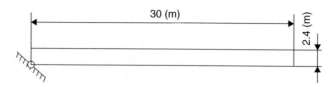

Figure 1.44 The simply supported beam after the release of the right-hand support.

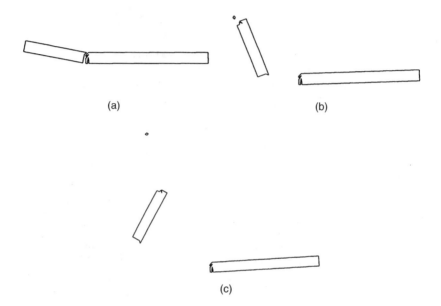

Figure 1.45 Failure (a) and subsequent motion (b and c) of the broken beam.

In Figure 1.46, the collapse sequence of a bridge due to the impact of a heavy vehicle is shown. The result of the combined finite-discrete element simulation clearly indicates the collapse sequence of the bridge together with the time intervals for each sequence and the total time necessary for the collapse of the bridge.

Similar failure of a retaining wall is shown in Figure 1.47. The pressure of water saturated clay results in overturning of the wall, and finally, its free fall towards the ground. The clay has disintegrated in the process. Despite the very coarse finite element mesh being used to model clay, a combination of the loss of its structural integrity and consequent increased pressure against the retaining wall has resulted in wall overturning and the clay further disintegrating until the wall reaches the state of free rotation about the foundation.

In Figure 1.48 the chimney stack is shown. After the base was blasted away, the combined finite-discrete element simulation shows a chimney stack leaning towards the right. Finally, due to the combined effect of inertia forces and self-weight, it breaks in the middle part and falls freely towards the ground until it crashes against the ground.

It is worth mentioning that the combined finite-discrete element simulation has correctly predicted all three stages of the demolition process:

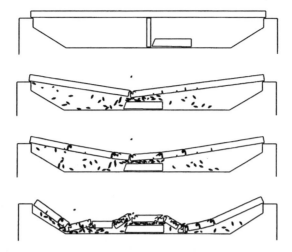

Figure 1.46   The combined finite-discrete element simulation of the collapse sequence of a bridge impacted by a heavy vehicle.

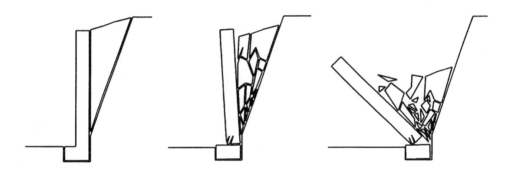

Figure 1.47   Initial conditions, failure and collapse of the retaining wall.

- disintegration of the base,
- break due to the inertia forces,
- final fragmentation due to the crash against the ground.

In all the examples shown, deformation of individual discrete elements has been taken into account, interaction among discrete elements has been resolved at every time instance, transient dynamics has been considered and finally, transition from continua to discontinua including failure, fracture and fragmentation have all been included.

The simulations shown above are some of the early simulations obtained by using the combined finite-discrete element method (early 1990s). Ever since Munjiza first proposed the combined finite-discrete method in 1989, the field of applications has widened and the complexity of the simulations has increased. The combined finite-discrete element method is now a fast developing area of computational mechanics of discontinua involving researchers and engineers from various disciplines. In recent years, 3D simulations

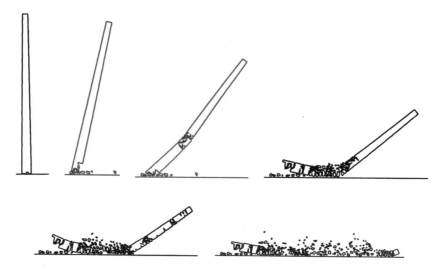

Figure 1.48 Chimney stack demolition – collapse sequence and crashing against the ground.

comprising fluid coupling have become widely available, and the first commercial codes have been produced.

1.7 ALGORITHMIC AND COMPUTATIONAL CHALLENGE OF THE COMBINED FINITE-DISCRETE ELEMENT METHOD

As explained before, the combined finite discrete element method combines finite elements with discrete elements. A typical combined finite-discrete element problem may contain thousands, even millions, of discrete elements. Each discrete element has a separate finite element mesh. Thus, the model may contain thousands to millions of separate finite element meshes.

The nature of the deformation of individual discrete elements involves at least finite rotations. Finite strains may be involved depending on the material that discrete elements are maid of. In addition, material non-linearity including fracture and fragmentation are considered. Thus, the transition from continua to discontinua results in ever changing geometry of individual discrete elements and/or changing the total number of discrete elements comprising the model.

Transient dynamics of each of discrete element, together with the possible state of rest, is considered. External loads on individual discrete elements often include interaction with fluid. Such is the case, for instance, in explosive induced fragmentation, where a detonation gas pushes against the walls of discrete elements causing further fracture and fragmentation, or increasing the kinetic energy of the system, i.e. increasing the velocity of individual discrete elements.

Energy dissipation mechanisms such as elastic hysteresis, plastic straining of the material, fracture of the material and friction between discrete elements, eventually lead to the state of rest being reached when all discrete elements have zero kinetic energy.

All these combined together result in a need for robust, CPU and RAM efficient algorithmic solutions. These include algorithms such as

- contact interaction algorithms
- contact detection algorithms
- finite strain elasticity-plasticity
- temporal discretisation and integration
- transition from continua to discontinua algorithms
- coupled problems such as gas-solid interaction
- application-specific algorithms, visualisation, key diagnostics, etc.

A number of different algorithmic solutions for each of the algorithmic aspects listed have been proposed in the scientific literature. Some of the solutions proposed have subsequently been replaced by more efficient and robust solutions, and are therefore only of historical interest.

To avoid confusion and make this book as short as possible, only algorithms that have found wider acceptance and are superior in terms of accuracy, CPU efficiency, RAM efficiency or robustness are given attention in this book. However, for those who want to find out more, an extensive bibliography is given at the end of the book.

This book is not meant to be a literature survey. Instead, it is a summary of best practices and state of the art algorithms available to researchers, students and practising engineers. The algorithms are therefore presented in sufficient detail that enables almost direct translation into a computer code. Very often, pseudo-codes and data structures, are given. When pseudo-codes, and especially data structures, are given an effort is made not to link the code to any computer language. Thus, for instance, pointers are not used, making the implementation in C (which has pointers) and in FORTRAN (which does not have pointers) possible.

Numerical examples in this chapter are meant to give the reader a general idea of what the combined finite-discrete element method is about. Numerical examples when given in the rest of the book are inserted only when these are necessary in an understanding of the particular algorithmic solution.

The combined finite-discrete element method is now being applied to a wide range of problems, with a large number of researchers and engineers working in the field. It is hoped that this book will help both researchers who are further developing the generic algorithmic solutions and also engineers and researchers applying the combined finite-discrete element method to a specific engineering or scientific problem. For this reason, a detailed version of implementation of the key algorithmic procedures is given in Chapter 10. These are accompanied with necessary explanations, guidelines, things to avoid, etc. These codes are intended to be used by any users of the book who want to assemble their own application-specific computer programs, but also by engineers who may wish to understand the implementation details behind commercial codes, such as the meaning of a particular input parameter.

2

Processing of Contact Interaction in the Combined Finite Discrete Element Method

2.1 INTRODUCTION

The combined finite-discrete element method is aimed at problems involving transient dynamics of systems comprising a large number of deformable bodies that interact with each other, and that may in general fracture and fragment, thus increasing the total number of discrete elements even further. Each individual discrete element is of a general shape and size, and is modelled by a single discrete element. Each discrete element is discretised into finite elements to analyse deformability, fracture and fragmentation. A typical combined finite-discrete element system comprises a few thousand to a few million separate interacting solids, each associated with separate finite element meshes. In this context, one of the key issues in the development of the combined finite-discrete element method is the treatment of contact, i.e. the enforcement of the constraint that no point in space is occupied by more than one body at the same time.

From an algorithmic point of view, there are two aspects to contact in the combined finite-discrete element method:

- contact detection,
- contact interaction.

Contact detection is aimed at detecting couples of discrete elements close to each other, i.e. eliminating couples of discrete elements that are far from each other and cannot possibly be in contact. In other words, contact detection is aimed at avoiding processing contact interaction when there is no contact. In that sense, contact detection is aimed at reducing CPU requirements, i.e. reducing processing (run) times.

Once couples of discrete elements in contact have been detected, a contact interaction algorithm is employed to evaluate contact forces between discrete elements in contact.

The Combined Finite-Discrete Element Method   A. Munjiza
© 2004 John Wiley & Sons, Ltd   ISBN: 0-470-84199-0

Contact interaction between neighbouring discrete elements occurs through solid surfaces which are in general irregular and, as a consequence, the contact pressure between two solids is actually transferred through a set of points at which surfaces touch. At small normal pressure, surfaces only touch at a few points, and with increasing normal pressure, elastic and plastic deformation of individual surface asperities occurs, resulting in an increase in the real contact area.

Theoretical and micro-mechanical models of contact take into account this complex phenomenon, and are usually based on assumptions such as shape, distribution and deformation of individual asperities, etc.

In the computational literature, theoretical assumptions about contact are simplified by employing variational formulation of contact combined with the most simple contact law that defines contact pressure as a function of approach, with tangential resistance to motion being a function of normal pressure and/or slip condition. This is usually done using variational formulation.

The variational formulation of a boundary value problem with contact is equivalent to the problem of making a functional Π stationary subject to the contact constraints over boundaries of the domain Γ:

$$\mathbf{C}(\mathbf{u}) = 0 \tag{2.1}$$

Variational formulation of contact problems involves an additional functional due to contact, through which no penetration conditions are enforced. Among the classic approaches employed are:

- *The least square method:* to enforce contact constraints on the boundary Γ, functional

$$\int_\Gamma \mathbf{C}^T(\mathbf{u})\mathbf{C}(\mathbf{u})\,d\Gamma \tag{2.2}$$

 is introduced. This functional is always positive except when contact constraints are satisfied exactly. Minimisation of this functional produces

$$\mathbf{C}(\mathbf{u}) = 0 \tag{2.3}$$

- *The Lagrangian multipliers method:* to enforce contact constraints, the additional functional

$$\int_\Gamma \boldsymbol{\lambda}^T \mathbf{C}(\mathbf{u})\,d\Gamma \tag{2.4}$$

 is added to the original functional

$$\tilde{\Pi}(\mathbf{u}, \boldsymbol{\lambda}) = \Pi(\mathbf{u}) + \int_\Gamma \boldsymbol{\lambda}^T \mathbf{C}(\mathbf{u})\,d\Gamma \tag{2.5}$$

 where $\boldsymbol{\lambda}$ is a set of independent functions defined on Γ. These functions are known as Lagrangian multipliers. A stationary point is found by employing the variation of this functional

$$\delta\tilde{\Pi}(\mathbf{u}, \boldsymbol{\lambda}) = \delta\Pi(\mathbf{u}) + \delta\int_\Gamma \boldsymbol{\lambda}^T \mathbf{C}(\mathbf{u})\,d\Gamma = 0 \tag{2.6}$$

which is zero, providing that

$$\delta\Pi(\mathbf{u}) = 0; \quad \mathbf{C}(\mathbf{u}) = 0; \quad \text{hence } \delta\mathbf{C}(\mathbf{u}) = 0 \tag{2.7}$$

In the finite element approximation, Lagrangian multipliers are approximated using the base functions. These result in additional unknown variables, the physical meaning of which is contact force. Thus, in the Lagrange multiplier method, contact is enforced by increasing the total number of unknowns. In static and implicit transient dynamics problems, the Lagrange multiplier method is implemented in a rigorous manner by solving the system of simultaneous algebraic equations. In the explicit transient dynamics problems, the system of algebraic equations is not solved and the term *Lagrange multiplier method* only implies that the impenetrability conditions are satisfied approximately (often through an iterative solution of coupled system of equations and sufficiently small time steps).

- *The penalty function method:* this is introduced with the aim of eliminating the drawbacks of the Lagrange multipliers method. To enforce contact constraints on the boundary Γ, the additional functional

$$p \int_{\Gamma} \mathbf{C}^T(\mathbf{u})\mathbf{C}(\mathbf{u}) \, d\Gamma \tag{2.8}$$

is added to the original functional

$$\tilde{\Pi}(\mathbf{u}, \boldsymbol{\lambda}) = \Pi(\mathbf{u}) + p \int_{\Gamma} \mathbf{C}^T(\mathbf{u})\mathbf{C}(\mathbf{u}) \, d\Gamma \tag{2.9}$$

where p is the penalty parameter. As

$$\int_{\Gamma} \mathbf{C}^T(\mathbf{u})\mathbf{C}(\mathbf{u}) \, d\Gamma \geq 0 \tag{2.10}$$

if Π is a minimum of the solution, then p must be a positive number. The solution obtained by minimising the modified functional satisfies the contact constraint only approximately. The larger the value of penalty, the better the contact constraints achieved. Only with an infinite penalty are the contact constraints satisfied exactly. The penalty function method is either used to impose an impenetrability condition in an iterative manner, or to violate this condition in such a way that the correct response of the physical system is still recovered. This is achieved by using a sufficiently large penalty term.

In an implementation of any of the above formulations in the combined finite-discrete element method, the handling of kinematics of contact is of major importance. In the finite element method, kinematics of contact is considered by employing slideline algorithms, where one surface is designated as the master (target) surface, while the other is designated as the slave (contactor) surface. In early algorithms, discretisation of master and slave surfaces was often performed in such a way that each slave node had a designated master node, thus only node-to-node contact was handled.

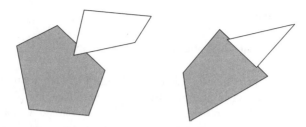

Figure 2.1   Node to edge (left) and edge to edge (right) contact.

More general situations are handled by allowing node-to-line contact and, equivalently, edge-to-edge, node-to-edge, node-to-node and node-to-surface contact (Figure 2.1). The common feature of all algorithms that handle contact kinematics in this way are concentrated contact forces, as opposed to distributed contact force approaches, which are usually based on the consideration of overlapping volumes.

Both the distributed and concentrated approaches involve relatively complicated kinematics of contact with many branches of code, in which case processing of kinematics of contact can be on the critical path of the otherwise efficient solution.

In 3D applications contact kinematics becomes so difficult that it is almost impossible to resolve contact due to the non-existence of surface normals at some corner points. The result is inconsistent contact kinematics that produces energy imbalance. An example of this is shown in Figure 2.2. As the potential energy is proportional to

$$\delta^2 \tag{2.11}$$

where δ is penetration, the total amount of kinetic energy transferred into potential energy is proportional to

$$\delta_B{}^2 \tag{2.12}$$

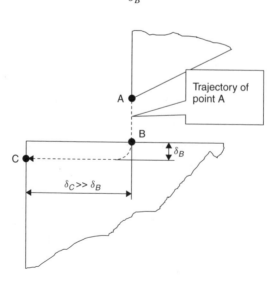

Figure 2.2   Node A penetrates through point B and exits through point C, thus numerically creating energy.

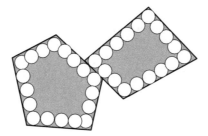

Figure 2.2a   Impenetrability condition imposed using pinballs placed on the boundary of discrete elements.

while the total amount of recovered kinetic energy after contact release is proportional to

$$\delta_C{}^2 \tag{2.13}$$

As

$$\delta_C{}^2 > \delta_B{}^2 \tag{2.14}$$

the final total energy is greater than the initial total energy.

An example of an attempt to simplify contact kinematics related procedures is a so-called pinball algorithm, which is among the simplest slideline algorithms. Its core idea is to embed pinballs into a surface, and therefore enforce the impenetrability condition only to pinballs. Difficulties can arise from the large number of pinballs needed to discretise the surface, or an unrealistic distribution of contact forces due to the discrete nature of pinballs, Figure 2.2a.

Pinballs can produce unrealistic behaviour of the system, as shown in Figure 2.3, where even with smooth surfaces and the absence of friction, sliding of contact surfaces over each other is difficult.

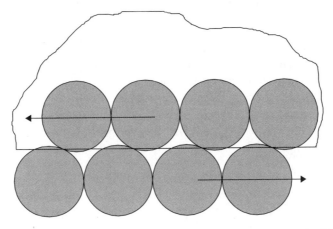

Figure 2.3   Sliding of surface A over smooth surface B is made numerically difficult due to the introduction of pinballs.

Problems of contact interaction in the context of the combined finite-discrete element method are even more important, due to the fact that in this method, the problem of contact interaction and handling of contacts also defines the 'constitutive' behaviour of the system, because of the presence of large numbers of separate bodies. Thus, algorithms employed must pay special attention to contact kinematics in terms of the realistic distribution of contact forces, energy balance and robustness.

One of the early contact interaction algorithms that took into account energy balance considerations is the contact interaction algorithm, based on the so-called concept of a contact element, in which the penalty parameters are given physical meaning through the concept of a contact layer, which is in a way a crude approximation of surface roughness. This early algorithm was originally developed to handle 2D problems, and was difficult to implement in 3D problems where, for instance, a surface normal may not be defined at all surface points. In addition, if overlap of discrete elements in contact exceeds the contact layer, the energy balance is not preserved, which is also the case if new surfaces are created due to a fracture process. The contact forces resulting from the algorithm are concentrated, which greatly influences stress and strain fields close to the boundary, and may considerably influence the results of fracture and fragmentation analysis. For this reason, this and many other contact algorithms of mainly historical importance are not considered in this book.

The latest generation of contact interaction algorithms makes use of finite element discretisations of discrete elements, and combines this with the so-called potential contact force concept. These algorithms assume discretisation of individual discrete elements into finite elements, thus imposing no additional database requirements on handling the geometry of individual discrete elements. They also yield realistic distribution of contact forces over finite contact areas resulting from the overlap of discrete elements that are in contact. Thus, numerical distortion of local strain fields close to the boundary due to contact is much reduced—an important aspect when, for instance, the fracture of brittle material is analysed.

2.2   THE PENALTY FUNCTION METHOD

The penalty function method in its classical form assumes that two bodies in contact penetrate each other, and this penetration results in a contact force. The standard contact functional for the penalty function method takes the form

$$U_c = \int_{\Gamma_c} \frac{1}{2} p (\mathbf{r}_t - \mathbf{r}_c)^T (\mathbf{r}_t - \mathbf{r}_c) \, d\Gamma \tag{2.15}$$

where p is the penalty term, while $\mathbf{r}_t$ and $\mathbf{r}_c$ are position vectors of the points on the overlapping boundaries of the target and contactor bodies, respectively. In the limit for infinite penalty terms, no penetration would occur, i.e.

$$\lim_{p \to \infty} U_c = 0 \tag{2.16}$$

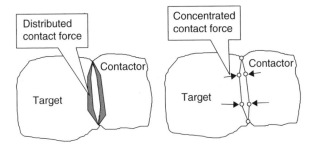

Figure 2.3a Distributed and concentrated contact force approach.

However, in practice large penalty terms are coupled with integration problems in the temporal domain, and in practical applications the penalty function method is therefore coupled with overlaps between bodies in contact.

When implemented into actual codes dealing with contact, the penalty function method in general deals with either concentrated or distributed contact force (Figure 2.3a). The concentrated contact force approach usually assumes nodal contact forces being a function of penetration of individual contactor nodes into the target, while the distributed contact force is in general evaluated from the shape and size of overlap between the contactor and target.

2.3 POTENTIAL CONTACT FORCE IN 2D

The distributed contact force is adopted for two discrete elements in contact, one of which is denoted as the contactor and the other as the target. When in contact, the contactor and target discrete elements overlap each other over area S, bounded by boundary Γ. (Figure 2.4).

It is assumed that penetration of any elemental area dA of the contactor into the target results in an infinitesimal contact force, given by

$$d\mathbf{f} = [grad\,\varphi_c(\mathrm{P}_c) - grad\,\varphi_t(\mathrm{P}_t)]d\,A \qquad (2.17)$$

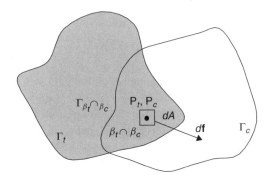

Figure 2.4 Contact force due to an infinitesimal overlap around points P_c an P_t.

where $d\mathbf{f}$ is the infinitesimal contact force due to infinitesimal overlap dA, defined by overlapping points P_c belonging to the contactor and P_t belonging to the target.

Equation (2.17) can also be written as

$$d\mathbf{f} = -d\mathbf{f}_t + d\mathbf{f}_c \tag{2.18}$$

where

$$
\begin{aligned}
d\mathbf{f}_c &= -grad\,\varphi_t(\mathbf{P}_t)dA_c, \quad dA_c = dA \\
d\mathbf{f}_t &= -grad\,\varphi_c(\mathbf{P}_c)dA_t, \quad dA_t = dA
\end{aligned}
\tag{2.19}
$$

In other words, the contact as described by (2.17) can be viewed as first the elemental area of the contactor penetrating the target, and then the elemental area of the target penetrating the contactor. Thus, for each of the discrete elements in contact, the contact force is calculated as a gradient of the corresponding potential function. The field of contact forces is therefore a conservative field for both the target penetrating contactor and the contactor penetrating target.

If point P_c of a contactor discrete element penetrates the target through any path defined by end points A and B, the total work of the contact force due to the potential function does not depend upon the path, but on the end points A and B only. If both points A and B are on the boundary of the target discrete element, a contact-contact release situation arises, i.e. point P_c comes into contact with the target at point A of the target and the contact is released through point B of the target. Preservation of the energy balance requires that the total energy of the system before and after the contact is the same, i.e. that no work is done by the contact force, which is equivalent to saying that for any points A and B on the boundary of the target discrete element,

$$\varphi_t(A) - \varphi_t(B) = 0 \tag{2.20}$$

i.e.

$$\varphi_t(A) = \varphi_t(B) \tag{2.21}$$

The same is valid for the contactor discrete element, i.e. for any points A and B from the boundary of the contactor discrete element

$$\varphi_c(A) - \varphi_c(B) = 0 \tag{2.22}$$

i.e.

$$\varphi_c(A) = \varphi_c(B) \tag{2.23}$$

Thus, provided that the potentials on the boundary of both the contactor and target discrete elements are constant, the contact force as given by (2.17) preserves the energy balance regardless of the geometry or shape of contactor and target discrete elements, the size of the penalty term or the size of penetration (overlap) when in contact.

The total contact force is obtained by integration of (2.17) over the overlapping area S

$$\mathbf{f}_c = \int_{S=\beta_t \cap \beta_c} [grad\,\varphi_c - grad\,\varphi_t]dA \tag{2.24}$$

which can also be written as an integral over the boundary of the overlapping area Γ

$$\mathbf{f}_c = \oint_{\Gamma_{\beta_t \cap \varphi_c}} n_\Gamma (\varphi_c - \varphi_t) \, d\Gamma \tag{2.25}$$

where n is the outward unit normal to the boundary of the overlapping area.

2.4 DISCRETISATION OF CONTACT FORCE IN 2D

In the combined finite-discrete element method, individual discrete elements are discretised into finite elements, and each discrete element can be represented as union of its finite elements:

$$\beta_c = \beta_{c_1} \cup \beta_{c_2} \dots \cup \beta_{c_i} \dots \cup \beta_{c_n}$$
$$\beta_t = \beta_{t_1} \cup \beta_{t_2} \dots \cup \beta_{t_j} \dots \cup \beta_{t_m} \tag{2.26}$$

where β_c and β_t are the contactor and target discrete elements, respectively, while m and n are the total number of finite elements into which the contactor and target discrete elements are discretised. In this context, the potentials φ_c and φ_t can be written as a sum of potentials associated with individual finite elements:

$$\varphi_c = \varphi_{c_1} + \varphi_{c_2} \cdots + \varphi_{c_i} \cdots + \varphi_{c_n}$$
$$\varphi_t = \varphi_{t_1} + \varphi_{t_2} \cdots + \varphi_{t_i} \cdots + \varphi_{t_m} \tag{2.27}$$

Integration over overlapping area may therefore be represented by summation over finite elements:

$$\mathbf{f}_c = \sum_{i=1}^{n} \sum_{j=1}^{m} \int_{\beta_{c_i} \cap \beta_{t_j}} [grad \, \varphi_{c_i} - grad \, \varphi_{t_j}] \, dA \tag{2.28}$$

By replacing integration over finite elements by equivalent integration over finite element boundaries (2.25), the following equation for contact force is obtained:

$$\mathbf{f}_c = \sum_{i=1}^{n} \sum_{j=1}^{m} \int_{\Gamma_{\beta_{c_i} \cap \beta_{t_j}}} \mathbf{n}_{\Gamma_{\beta_{c_i} \cap \beta_{t_j}}} (\varphi_{c_i} - \varphi_{t_j}) \, d\Gamma \tag{2.29}$$

where integration over finite element boundaries may be written as summation of integration over the edges of finite elements.

In other words, the contact force between overlapping discrete elements is calculated by summation over the edges of corresponding finite elements that overlap.

2.5 IMPLEMENTATION DETAILS FOR DISCRETISED CONTACT FORCE IN 2D

As pointed out earlier, combined finite-discrete element problems involve a large number of separate bodies that are free to move and interact with each other. Average problems may comprise a few thousand or even a few million separate bodies. In addition, each body is represented by a single discrete element, which in turn is discretised into one or more finite elements. Thus, the evaluation of contact forces at each time step may involve a large number of contacting couples of discrete elements. As each discrete element is discretised into a finite element, each contacting couple of discrete elements is in fact represented by a whole set of contacting couples of finite elements. Thus, making the total number of contacting couples of finite elements even larger.

In this context, the summation as given by (2.29) usually involves a large number of contacting couples of finite elements, and the total CPU time and overall efficiency of the contact algorithm critically depend upon implementation of part of the interaction that processes finite element to finite element contact.

It is therefore important to employ the simplest possible finite element and make that element work well for both contact and deformability. Using the simplest possible finite elements for deformability has resulted in a constant strain triangle being employed for almost any conceivable (linear and non-linear) 2D problem. Much has been published on, for instance, special techniques to avoid locking. All of these can be found in most good finite elements textbooks, and is outside the scope of this book.

The simplest possible finite element in 2D is a tri-noded triangle. The edges of a triangular finite element are straight, and its geometry is uniquely defined by the coordinates of its nodes, as shown in Figure 2.5.

As explained above, the potential φ should be constant on the boundary of a discrete element. This constraint is unconditionally met if the potential φ is defined in such a way that it is constant on the boundaries of each finite element. Thus, for a triangular finite element it is assumed that φ at point P inside the triangle is given by

$$\varphi(P) = \min\{3A_1/A, 3A_2/A, 3A_3/A\} \tag{2.30}$$

where $A_i(i = 1, 2, 3)$ is the area of corresponding sub-triangles, as shown in Figure 2.5. The potential φ equals 1 at the centre of triangle and 0 at the edges of the triangle. For any point P outside the triangle, the potential φ is zero.

According to (2.29), the problem of interaction between two triangles can be reduced to interactions between the contactor triangle and the edges of the target triangle, coupled

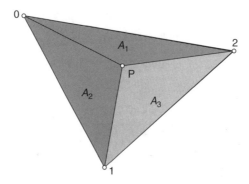

Figure 2.5   Potential φ at any point P of a triangle shaped finite element.

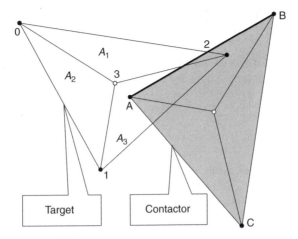

Figure 2.6 Contact of contactor and target triangles and contact of an edge of a contactor triangle with a target triangle.

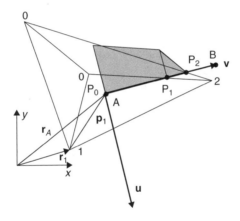

Figure 2.7 Distribution of contact force between the target triangle and an edge of contactor triangle.

with interactions between the target triangle and the edges of the contactor triangle, as shown in Figure 2.6.

Thus, in Figure 2.7 contact of edge AB of contactor triangle with target triangle is shown. To minimise the number of necessary operations and therefore CPU time to process this contact, a local coordinate system given by local axes **u** and **v** is introduced, and transformation of nodal coordinates of target triangle into local coordinate system is performed:

$$\mathbf{p}_i = ((\mathbf{r}_i - \mathbf{r}_A) \cdot \mathbf{u}, (\mathbf{r}_i - \mathbf{r}_A) \cdot \mathbf{v}) \tag{2.31}$$

Using the local nodal coordinates for the target triangle, characteristic intersection points between the edge AB and the target triangle are obtained together with the corresponding values of the potential function. The potential φ for each intersection point is calculated by interpolation between the central node (node 3, where the potential is equal to 1) and the corresponding edge node (nodes 0,1 or 2, where potential is equal to zero) for each intersection point.

The total contact force exerted by the target triangle onto the edge AB is given by the area of the diagram of potential over the edge AB, i.e.

$$\mathbf{f}_{c,AB} = \frac{1}{\mathbf{u}^2} \mathbf{u} \int_0^L p\varphi(v)dv \tag{2.32}$$

where p is the penalty term, while the term $\mathbf{u}^2$ comes from the fact that vectors $\mathbf{u}$ and $\mathbf{v}$ are not unit vectors. This is computationally convenient as, that evaluation of the integral (2.32) does not involve a square root. In addition, in between the intersection points the potential φ is given by straight lines, which reduces integration to area calculation, as shown in Figure 2.7, where shaded areas represent the potential φ. The contact forces obtained are represented by equivalent nodal forces at points A and B, together with corresponding nodal forces at nodes of target triangle, as shown in Figure 2.8.

The whole process is repeated for the remaining edges of the contactor triangle, and in this way, contact forces due to penetration of the contactor triangle into the target triangle are obtained from the potential φ_t as defined for the target triangle. To take into account the contact force due to the potential φ_c associated with the contactor triangle, contact of each edge of the target triangle to the contactor triangle is analysed, and corresponding nodal forces for both the contactor and target triangle are updated. The potential associated with the contactor triangle is defined in exactly the same way as the potential associated with the target triangle, and evaluation of the nodal forces proceeds using the same procedure.

In Figure 2.9 the combined finite-discrete element simulation aimed at demonstrating the distributed potential contact force algorithm in 2D is shown. This example involves a heap of identical rigid fragments of triangular shape. The fragments are placed close to each other in such a way that they touch with no overlap and no contact force being generated.

The initial shape of the heap looks like a simply supported beam with two smaller fragments being fixed to the ground (Figure 2.9). The heap is impacted by a rigid projectile

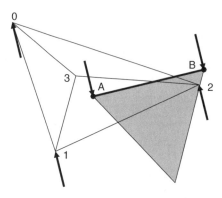

Figure 2.8   Equivalent nodal forces.

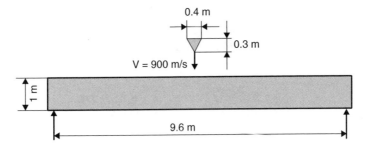

Figure 2.9   Beam-shaped heap of rigid fragments.

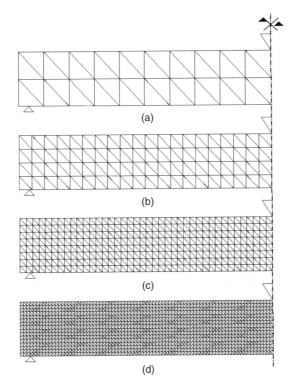

Figure 2.10   Beam-shaped heaps of fragments. Heap A−80 fragments, heap B−320 fragments, heap C−1280 fragments, heap D−5120 fragments.

moving at an initial velocity of 900 m/s. Four heaps of the same shape and size were considered (Figure 2.10). Heap A is made of a smaller number of relatively large fragments, while heaps B, C and D are comprised of a larger number of smaller fragments.

The transient motion of Heap A is given in Figure 2.11. Initially, all of the 80 heap fragments are at rest. As the projectile approaches the heap, contact between the projectile and fragments produces acceleration of the heap fragments and deceleration of the projectile. Also, interaction between the heap fragments themselves results in fragments further

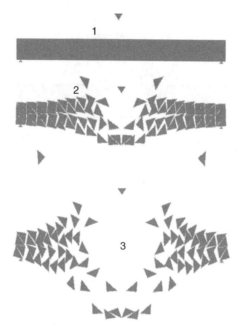

Figure 2.11   Predicted transient motion sequence for heap A.

from the point of impact acquiring velocity and moving apart from each other–thus the packing of the heap becomes looser. The actual analysis did not make use of the initial symmetry of the problem. The problem as a whole was analysed instead, and the motion of all 80 fragments, together with interaction among the fragments, was traced. Contact forces were evaluated at each time step, while the central difference explicit time marching scheme was employed to calculate the new velocity fields and new nodal coordinates. It is worth noting that after the considerable movement of the individual fragments, the symmetry of the system is still preserved.

The transient motion of heap B due to the impact of a rigid projectile is shown in Figure 2.12. Again, the initial impact induces significant disturbance close to the middle section of the heap. Through the interaction of individual rigid fragments, this disturbance is spread toward the edges of the heap. The additional constraint of small support-like triangles being fixed to the ground restricts the fragments closest to these supports. The problem consisting of 320 rigid fragments is characterised by initial symmetry, and after the impact and considerable movement of the fragments, the symmetry is preserved. This is despite the fact that no symmetry is taken into account during analysis, and the problem being analysed as a whole, i.e. by considering simultaneous transient motion of each of the 320 interacting fragments.

The motion sequence for the heap C is shown in Figure 2.13. Again, as the projectile hits the centre of the heap, the fragments close to the point of impact are accelerated. However, this time the total number of fragments is much larger (1280), and each fragment is therefore much smaller in comparison to the size of the projectile. The projectile therefore gets 'submerged' among the fragments of the heap, and the subsequent motion of the heap fragments at the initial stages reassembles a projectile going into a liquid.

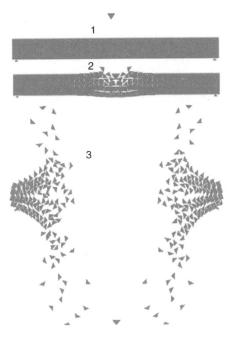

Figure 2.12 Predicted transient motion sequence for heap B.

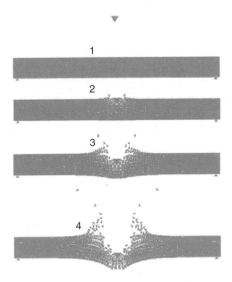

Figure 2.13 Predicted transient motion sequence for heap C.

The later stages of transient motion clearly indicate separation of individual fragments from each other and transition towards the stage where no two fragments are in contact (Figure 2.14). The reason for this is that the heap is not confined in space, and the motion of fragments is almost free except for interaction with one other.

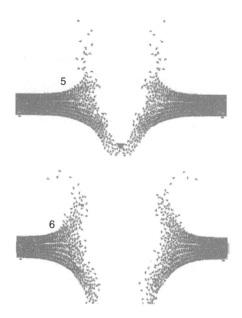

Figure 2.14   Second part of the predicted transient motion sequence for heap C.

The initial symmetry of the system is again preserved, despite the fact that no symmetry was taken into account during analysis, i.e. the problem was analysed as a whole, and motion of all 1280 fragments was traced at each time step.

Heap D consists of the smallest fragments. The size and shape of the heap remain the same as in heaps A, B and C. Thus, the total number of closely packed fragments is much larger–the heap consists of 5120 fragments, initially packed in such a way that they touch, but no two fragments overlap each other and no contact force is generated. First contact occurs after the projectile makes contact with fragments close to the centre of the heap. The projectile penetrates the heap, and fragments in the heap are initially extracted by the projectile. This results in fragments close to the point of impact moving in the opposite direction of the projectile (Figure 2.15).

This initial contact is followed by contacts between the heap fragments. Thus, the fragments move away from each other, reducing the average packing density of the heap. This motion results in the shape of the heap being gradually changed until the projectile finally goes through the heap (Figure 2.16). At this stage, the shape of the heap has changed significantly, and individual fragments have acquired significant velocity, thus the remaining contacts are released, and eventually, the projectile and fragments move at a constant velocity. This time the initial symmetry of the system is in general preserved, although localised loss of symmetry can be observed. However, it is worth noting that the total number of individual fragments is relatively large, and the local bifurcation type behaviour that is sensitive to rounding errors may occur (see Chapter 6).

The motion sequences obtained in some way converge from the motion sequence for heap A to the motion sequence for heap D, and in some sense, the motion sequence of heap A represents an approximation of the motion sequence of heap D, while the motion sequence of heap D, by similar reasoning, approximates the motion sequence of a heap consisting of an infinite number of infinitesimally small fragments. In all four heaps, the

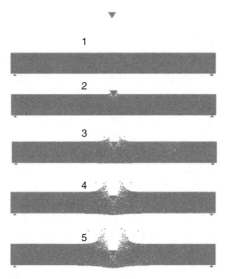

Figure 2.15   First part of the predicted motion sequence for heap D.

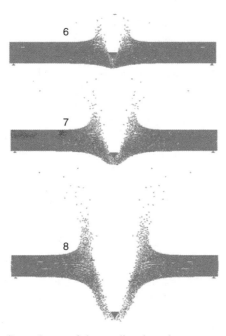

Figure 2.16   Second part of the predicted motion sequence for heap D.

system starts at a state where no two fragments are in contact, and converges toward a state where all fragments move apart from each other. Thus, the potential energy due to an overlap of individual fragments in contact is zero before the impact, and is also zero after all fragments in contact have separated from each other. This is clearly shown in Figure 2.17, where the total kinetic energy of all fragments for the four heaps is shown.

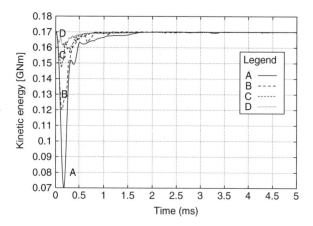

Figure 2.17   Kinetic energy of beam shaped heaps A, B, C and D as a function of time.

All four heaps start at the same initial kinetic energy, which is equal to the initial kinetic energy of the projectile. As the impact occurs, the projectile slows down, and contact between individual fragments results in some of kinetic energy of the system being taken as the potential energy, due to overlapping fragments in contact. This temporary reduction in kinetic energy is largest for heap A and smallest for heap D. In fact, in terms of kinetic energy, there is a gradual convergence from heap A to heap D, as though the kinetic energy for heap A is a crude approximation of the kinetic energy for heap D.

For all four heaps, as the transient contacts are released due to the fragments moving away from each other, the initial kinetic energy of the system is restored. Contacts have therefore resulted in some of the kinetic energy being transformed into potential energy, and finally, the transfer of potential energy into kinetic energy, due to contacts being released. Thus, the contact interaction algorithm as described in this chapter has preserved the energy for all four heaps, regardless of the number of contacts, duration of those contacts, or the size of overlap between fragments in contact, which is the result theoretically predicted. This is also demonstrated by a bar-shaped heap, shown in Figure 2.19. The heap comprises identical triangular rigid particles closely packed with each other in such a way that no contact force is generated.

All fragments of the heap move at the same initial velocity ($v = 500$ ms) towards a fixed rigid obstacle of a triangular shape, Figure 2.18.

The problem is solved for two heaps: heap A consists of 1280 identical rigid fragments of triangular shape, while heap B consists of 5120 rigid fragments (Figure 2.19).

The transient motion of heap B due to the impact is shown in Figure 2.20. The initial impact results in contact between the rigid obstacle and heap fragments. Due to the contact

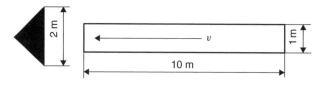

Figure 2.18   A heap of rigid fragments moving towards a rigid obstacle.

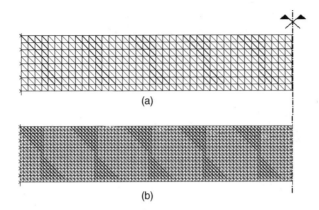

Figure 2.19   Initial configuration of bar shaped heaps A and B.

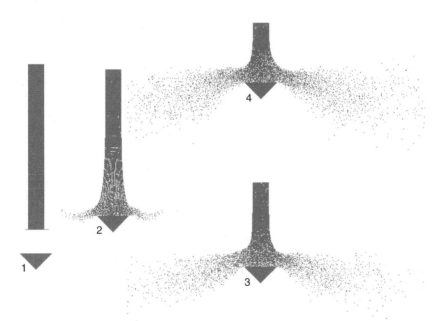

Figure 2.20   Transient motion sequence for bar shaped heap B.

forces generated, the heap fragments decelerate and move sideways around the obstacle, forming a spray of fragments behind the obstacle. This contact immediately behind the obstacle also produces contacts between the fragments further away from the obstacle. However, these contacts are not able to accelerate the fragments enough to produce a velocity field that would significantly disturb the shape of the heap. Thus, the heap disintegrates mainly due to contacts between the obstacle and heap fragments, coupled with contacts between the fragments which are close to the obstacle.

After initial contact with an obstacle, an almost steady state motion behind the obstacle is formed, in which the head of the heap disintegrates and the tail of the heap moves

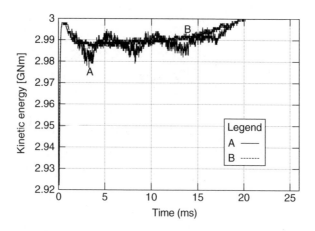

Figure 2.21   Kinetic energy of bar shaped heaps A and B.

closer to the obstacle. The final stages involve disintegration of the tail and fragments moving apart from each other, i.e. contacts between individual fragments being released. It is therefore reasonable to expect the total kinetic energy of the system to decrease after the initial impact, and to stay more or less constant until the tail starts disintegrating, and to increase again after disintegration of the tail and release of contacts. The kinetic energy for both heaps is shown in Figure 2.21. It is noticeable that the decrease of kinetic energy is larger for larger fragments (heap A). In addition, the kinetic energy for heap A oscillates indicating that no steady state is reached, while the kinetic energy for heap B stays almost constant for a relatively long time interval between 3 ms and 12 ms, indicating steady state like motion. The final kinetic energy for both heaps is equal to the initial kinetic energy, confirming that the energy balance has been preserved despite the complicated motion and interaction of a large number of separate fragments in transient motion. Again, this clearly proves the theoretical and implementation details of the potential contact force algorithm, and its superiority to the algorithms listed in the introduction to this chapter.

The advantages of discretised distributed potential contact force include:

- Distributed contact force. No artificial stress concentration due to contact is present. This has important implications in many applications, for instance, when processing the deformability of brittle materials, stress concentrations would result in fracture and non-physical behaviour of the model.

- Contact surface-driven direction of contact force as opposed to forces obtained by the pinball algorithm. This allows easy sliding between contact surfaces, and accurate representation of the physical contact conditions onto which friction, sliding, plasticity, surface roughness, wet-dry conditions, etc. can be incorporated following relatively simple rules of potential distribution over the finite element.

- Energy preservation. Without energy preservation, no proper physical behaviour can be modelled. Very often in non-potential contact algorithms, energy imbalance is coun- teracted by artificial damping, but what physical meaning does such damping have? There is no physical justification for this damping, and as the above examples have demonstrated, there is no need for it regardless of the shape of the interacting discrete elements.

- The contact force is discretised with the same algorithm and the same piece of code, regardless of the shape (concave, convex, hollow, etc.) of discrete elements. Thus, algorithmic complexities are greatly reduced.

- The contact processing is in general faster than alternative solutions. It is obvious that sphere to sphere contact is processed faster than triangle to triangle contact. However, contact between two discrete elements of general shape is processed almost as quickly as contact between two triangles. This is due to the fact that contact detection eliminates all triangular finite elements from two finite element meshes, except for the couples of finite elements that are in contact. The number of these triangles in a general shaped body is governed by the number of contact points. However, very often two such bodies (unless they are flat) touch at just one point, making the speed of interaction processing almost independent of the shape of the individual discrete elements. There is contact detection overhead, but the state of the art contact detection algorithms are so fast that contact detection as a whole consumes very little overall CPU time.

2.6 POTENTIAL CONTACT FORCE IN 3D

The easiest way of implementing the potential contact force interaction kinematics for discrete elements of general shape in 3D is by discretising the domain of each discrete element into finite elements. In this way, the potential is applied piecewise, i.e. elementwise.

To reduce implementation complexities and increase CPU efficiency, the simplest possible 3D solid finite element (i.e. four-noded tetrahedra) is chosen. The same finite element and the same discretisation is best used for deformability of discrete elements, including any non-linearity. Special procedures are often introduced to avoid phenomena such as locking. Most of them can be found in any good finite element textbook, and are outside of the scope of this book.

The potential contact force algorithm for 3D problems described in this section is based on the following assumptions:

- The domain of each discrete element is discretised into finite elements. The geometry of each element is assumed to be defined by a tetrahedron, although extension to other geometry is possible.

- Maximum allowed penetration at any point of contact is a function of the size of the finite element to which a particular point belongs. The logic behind this assumption is the fact that contact constraints at points of interface are satisfied only approximately. The same applies for field equations, for which an approximate solution is obtained by employing the finite element discretisation. The element size at each point of the domain is usually based on some kind of error estimator expressed, for instance, in terms of displacements. Relative displacements between any two points of the domain are obtained by integrating the strain field over any path connecting these two points. If this path intersects two surfaces in contact, the error in relative displacement is altered by penetration at the place of interface. According to the St. Venant principle, the closer the points to the contact surface, the greater the impact of the penetration. Thus, the greatest error is at finite elements that are closest to the surface. In this way, the maximum error introduced by the presence of penetration is a function of the maximum allowed penetration, i.e. a function of element size. Since the maximum error due to the finite element discretisation is also a function of element size, it implies that the same mesh refinements will reduce both the error due to the finite element discretisation and the error due to allowed penetration.

The main motivations behind the development of a potential contact force contact algorithm are energy and momentum balance at finite penetrations. These must be achieved by using a conventional in-core database for contact free finite element analysis.

The design of relational in-core databases for the contact free finite element analysis has reached its maturity, and similar database designs can be found in most commercial and academic codes. Most of the in-core databases are, to a large extent, normalised to enable easy manipulation of the data with minimal storage requirements. This is to a lesser extent valid for object orientated databases built mostly for object oriented codes or distributed computing with parallel post-processing and visualisation facilities. However, development in this direction is also expected to follow the logic of quick access and minimum storage requirements.

There is a number of possibilities to define the potential φ in 3D space. For the reasons explained above, it is convenient to define the potential φ in terms of the finite element discretisation employed. As explained earlier, the algorithm presented in this section uses discretisation based on tetrahedron shaped finite elements (Figure 2.22). To simplify the geometrical and computational aspects of the algorithm, the potential φ is defined on an element by element basis. First, the coordinates of the centroid of the tetrahedron are calculated:

$$\mathbf{x}_5 = \begin{bmatrix} x_5 \\ y_5 \\ z_5 \end{bmatrix} = \frac{1}{4}(\mathbf{x}_1 + \mathbf{x}_2 + \mathbf{x}_3 + \mathbf{x}_4) \qquad (2.33)$$

where

$$\mathbf{x}_1 = \begin{bmatrix} x_1 \\ y_1 \\ z_1 \end{bmatrix}, \quad \mathbf{x}_2 = \begin{bmatrix} x_2 \\ y_2 \\ z_2 \end{bmatrix}, \quad \mathbf{x}_3 = \begin{bmatrix} x_3 \\ y_3 \\ z_3 \end{bmatrix}, \quad \mathbf{x}_4 = \begin{bmatrix} x_4 \\ y_4 \\ z_4 \end{bmatrix} \qquad (2.34)$$

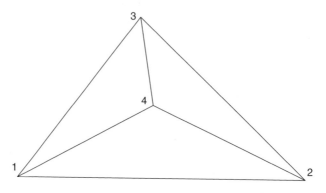

Figure 2.22 Potential definition over domain of a single tetrahedron.

are the coordinates of the four nodes of the tetrahedron finite element (Figure 2.22). Using the centroid, the tetrahedron is subdivided into four sub-tetrahedra:

$$
\begin{array}{ll}
\text{tetrahedron} & 1-2-3-5 \\
\text{tetrahedron} & 2-4-3-5 \\
\text{tetrahedron} & 3-4-1-5 \\
\text{tetrahedron} & 4-2-1-5
\end{array} \tag{2.35}
$$

For each point $\mathbf{p}$ of the sub-tetrahedron $i - j - k - l$, the potential φ is defined as

$$
\varphi(\mathbf{p}) = k \left(\frac{V_{i-j-k-p}}{4V_{i-j-k-l}} \right) \tag{2.36}
$$

where k is the penalty parameter, $V_{i-j-k-l}$ is the volume of the tetrahedron $i - j - k - l$, while $V_{i-j-k-p}$ is the volume of the sub-tetrahedron $i - j - k - p$, i.e. the tetrahedron with one of the nodes being replaced by the point $\mathbf{p}$.

2.6.1   Evaluation of contact force

Penetration of any elemental area dV of a contactor into the target results in an infinitesimal contact force, given by

$$
d\mathbf{f} = -grad\,\phi_t(\mathrm{P}_t)dV \tag{2.37}
$$

where $d\mathbf{f}$ is the infinitesimal contact force due to infinitesimal overlap dV defined by overlapping points P_c belonging to the contactor and P_t belonging to the target.

Since the target and contactor are relative attributes, to retain symmetry, a reverse situation of the target penetrating the contactor produces force:

$$
d\mathbf{f} = grad\,\phi_c(\mathrm{P}_c)dV \tag{2.38}
$$

The total infinitesimal contact force is given by

$$
d\mathbf{f} = [grad\,\phi_c(\mathrm{P}_c) - grad\,\phi_t(\mathrm{P}_t)]dV \tag{2.39}
$$

The total contact force exerted by the target onto the contactor is obtained by integration over the overlap volume:

$$
\mathbf{f} = \int_{V=\beta_t \cap \beta_c} [grad\,\varphi_c - grad\,\varphi_t]dV \tag{2.40}
$$

which can also be written as an integral over the surface S of the overlapping volume:

$$
\mathbf{f} = \int_{S_{\beta_t \cap \beta_c}} \mathbf{n}\,(\varphi_c - \varphi_t)dS \tag{2.41}
$$

where $\mathbf{n}$ is the outward unit normal to the surface of the overlapping volume.

Since individual discrete elements are discretised into finite elements, each discrete element can be represented as the union of its finite elements:

$$\beta_c = \beta_{c_1} \cup \beta_{c_2} \ldots \cup \beta_{c_i} \ldots \cup \beta_{c_n}$$
$$\beta_t = \beta_{t_1} \cup \beta_{t_2} \ldots \cup \beta_{t_j} \ldots \cup \beta_{t_m} \tag{2.42}$$

where β_c and β_t are contactor and target discrete elements, respectively, while m and n are the total number of finite elements the contactor and target discrete elements are discretised into. The potentials φ_c and φ_t are therefore a sum of potentials associated with individual finite elements:

$$\varphi_c = \varphi_{c_1} + \varphi_{c_2} \cdots + \varphi_{c_i} \cdots + \varphi_{c_n}$$
$$\varphi_t = \varphi_{t_1} + \varphi_{t_2} \cdots + \varphi_{t_i} \cdots + \varphi_{t_m} \tag{2.43}$$

Integration over the overlapping area may therefore be represented by summation over finite elements:

$$\mathbf{f} = \sum_{i=1}^{n} \sum_{j=1}^{m} \int_{\beta_{c_i} \cap \beta_{t_j}} [grad\,\varphi_{c_i} - grad\,\varphi_{t_j}]dV \tag{2.44}$$

By replacing integration over finite elements by equivalent integration over finite element boundaries (2.25), the following equation for contact force is obtained:

$$\mathbf{f}_c = \sum_{i=1}^{n} \sum_{j=1}^{m} \int_{S_{\beta_{c_i} \cap \beta_{t_j}}} \mathbf{n}\,(\varphi_{c_i} - \varphi_{t_j})dS \tag{2.45}$$

where integration over finite element boundaries may be written as summation of integration over surfaces of finite elements. In other words, the contact force between overlapping discrete elements is calculated by summation over the surfaces of corresponding finite elements that overlap.

In this context, the following solution strategies are available:

- Consider contact of discrete elements with disregard for finite element discretisation. In this case, a special data structure is needed to keep track of boundary representation.

- Consider contact of discretised domains of discrete elements with disregard to the relationship between individual discrete elements. This is the simplest path to follow. It may lead to unnecessary integration of contact forces over the inner boundaries of discrete elements, slowing down execution of the algorithm.

- Consider integration of contact forces only over boundaries (surfaces) of finite elements that are also boundaries of discrete elements. This approach requires the detection and marking of global boundaries. In return, CPU requirements are reduced.

2.6.2 Computational aspects

The numerical procedure for integration of contact forces makes use of the fact that discrete elements are discretised into tetrahedron shaped finite elements. Thus, the total contact force is the sum of contact forces between individual tetrahedrons.

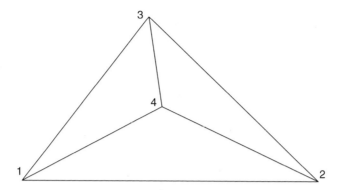

Figure 2.23 Target tetrahedron.

The numerical implementation of this integration can be summarised as follows:

1. Define the target tetrahedron (Figure 2.23) by its nodes I_1, I_2, I_3, I_4 and it centre I_5, where the position vector of the centre is given by

$$\mathbf{X}_5 = \begin{bmatrix} X_5 \\ Y_5 \\ Z_5 \end{bmatrix} = \frac{1}{4}(\mathbf{X}_1 + \mathbf{X}_2 + \mathbf{X}_3 + \mathbf{X}_4) = \frac{1}{4}\begin{bmatrix} X_1 + X_2 + X_3 + X_4 \\ Y_1 + Y_2 + Y_3 + Y_4 \\ Z_1 + Z_2 + Z_3 + Z_4 \end{bmatrix} \qquad (2.46)$$

2. Define the contactor tetrahedron by its nodes i_1, i_2, i_3, i_4 and it centre i_5, where the position vector of the centre is given by

$$\mathbf{x}_5 = \begin{bmatrix} x_5 \\ y_5 \\ z_5 \end{bmatrix} = \frac{1}{4}(\mathbf{x}_1 + \mathbf{x}_2 + \mathbf{x}_3 + \mathbf{x}_4) = \frac{1}{4}\begin{bmatrix} x_1 + x_2 + x_3 + x_4 \\ y_1 + y_2 + y_3 + y_4 \\ z_1 + z_2 + z_3 + z_4 \end{bmatrix} \qquad (2.47)$$

3. Divide the target tetrahedron into sub-tetrahedra (Figure 2.24):

$$\begin{aligned} T_1 &= (I_1, I_4, I_2, I_5) \\ T_2 &= (I_2, I_4, I_3, I_5) \\ T_3 &= (I_3, I_4, I_1, I_5) \\ T_4 &= (I_1, I_2, I_3, I_5) \end{aligned} \qquad (2.48)$$

4. Divide the contactor tetrahedron into sub-tetrahedra:

$$\begin{aligned} t_1 &= (i_1, i_4, i_2, i_5) \\ t_2 &= (i_2, i_4, i_3, i_5) \\ t_3 &= (i_3, i_4, i_1, i_5) \\ t_4 &= (i_1, i_2, i_3, i_5) \end{aligned} \qquad (2.49)$$

5. For each couple $(T_i, t_j), i = 1, 2, 3, 4; j = 1, 2, 3, 4$, perform the following operations:

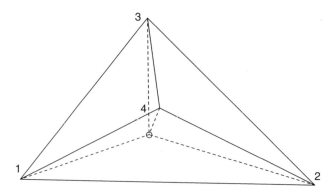

Figure 2.24   Subdivision of the target tetrahedron into sub-tetrahedrons.

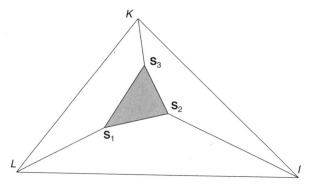

Figure 2.25   Intersection of target sub-tetrahedron (I, J, K, L) with the plane of the base of the contactor sub-tetrahedron.

(a) Find the intersection of the base of the contactor sub-tetrahedron with the target sub-tetrahedron (Figure 2.25), i.e. the convex polygon defined by points

$$S = (\mathbf{S}_1, \mathbf{S}_2, \mathbf{S}_3, \ldots, \mathbf{S}_i, \ldots, \mathbf{S}_n) \tag{2.50}$$

(b) Find the intersection of the polygon defined by these points with the base of the contactor sub-tetrahedron, as shown in Figure 2.26:

$$P = (\mathbf{B}_1, \mathbf{B}_2, \mathbf{B}_3, \ldots, \mathbf{B}_i, \ldots, \mathbf{B}_n) \tag{2.51}$$

(c) Calculate the values of potential φ of the target tetrahedron at each node of the intersection polygon P:

$$\varphi = (\varphi(\mathbf{B}_1), \varphi(\mathbf{B}_2), \varphi(\mathbf{B}_3), \ldots, \varphi(\mathbf{B}_i), \ldots, \varphi(\mathbf{B}_n)) \tag{2.52}$$

(d) Calculate the total contact force F by integrating over the intersection polygon. Integration is performed by dividing the intersection polygon into triangles, and performing summation of the contact forces over the triangles, Figure 2.27.

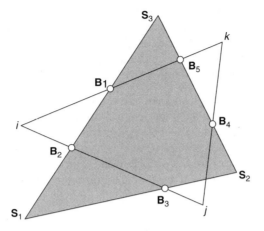

Figure 2.26 Intersection of the target sub-tetrahedron with the base of contactor sub-tetrahedron.

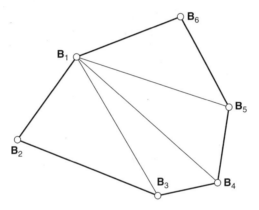

Figure 2.27 Intersection of contact force over the intersection polygon.

(e) Calculate the equivalent nodal forces on target sub-tetrahedra due to contact force F:

$$F_T = (F_I, F_J, F_K, F_L) \tag{2.53}$$

(f) Calculate the equivalent nodal forces on the contactor sub-tetrahedra due to contact force F:

$$F_C = (F_i, F_j, F_k, F_l) \tag{2.54}$$

(g) Multiply the nodal forces by the unit vector normal to the base of the contactor sub-tetrahedron:

$$\mathbf{f}_T = (F_I \mathbf{n_c}, F_J \mathbf{n_c}, F_K \mathbf{n_c}, F_L \mathbf{n_c})$$
$$\mathbf{f}_C = (F_i \mathbf{n_c}, F_j \mathbf{n_c}, F_k \mathbf{n_c}, F_l \mathbf{n_c}) \tag{2.55}$$

6. Each time contact forces are assigned to the nodes of a sub-tetrahedron, one of the nodes of the sub-tetrahedra is the centre of the target or contactor tetrahedron. Thus,

once the force on the centre of the target tetrahedron has been obtained, it is replaced by the equivalent forces at the nodes of the target tetrahedron. The same is done with the contact force assigned to the centre of the contactor tetrahedron.

It is also worth mentioning that each tetrahedron is considered twice, once as a contactor and once as a target.

There are a number of ways to numerically execute the tasks listed above. The most critical path of the algorithm is the intersection of the target sub-tetrahedron and the base of the contactor sub-tetrahedron. It is best performed in two steps. First, the intersection points of the edges of the target sub-tetrahedron with the plane containing the base of the contactor sub-tetrahedron are found. The result is a convex polygon. The intersection of this polygon with the triangular base of the contactor sub-tetrahedron is therefore a 2D problem.

2.6.3 *Physical interpretation of the penalty parameter*

As explained earlier, the maximum allowed penetration is a function of the size of finite elements at the place of contact and penalty parameter p, which can be different for different finite elements. The role of the penalty parameter is best illustrated by an example of contact between two solid finite elements, shown in Figure 2.28.

Two elements are pushed against each other by the pressure σ. Relative displacement between points A and B in the case of a small strain elasticity and a zero Poisson ratio is proportional to the pressure supplied:

$$u = \frac{\sigma h}{E} \tag{2.56}$$

In the same way, the penetration between the two solids is given by

$$d = \frac{\sigma h}{p} \tag{2.57}$$

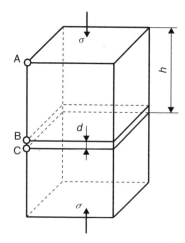

Figure 2.28   Contact between two finite elements.

To limit penetration, it is enough to select a penalty term to be proportional to the modulus of elasticity, i.e.

$$p = \alpha E \tag{2.58}$$

In this way, contribution of the allowed penetration to the displacement field is limited to

$$d = \frac{1}{\alpha} u \tag{2.59}$$

For $\alpha = 100$, for example,

$$d = \frac{1}{100} u \tag{2.60}$$

i.e. the total local error in displacements will be less than 1%.

This is another advantage of the potential contact force approach. The error in the displacements is easily controlled through setting penalty p as a function of E. In addition, the error in displacements is reduced by reducing the size of finite element h. Thus, any mesh refinements automatically reduce error introduced by contact approximation.

2.6.4 Contact damping

Any energy dissipation in contact is due to friction or plastic straining of surface asperities. Plastic straining of surface asperities can be approximated by a viscous damping model. Damping parameters for contact are defined in a similar way to the definition of penalty parameters. For the situation in Figure 2.28, the frequency of the subsystem shown can be approximated by

$$\omega = \frac{2}{h} \sqrt{\frac{p}{\rho}} \tag{2.61}$$

where ρ is the density and h is the size of the finite element, as shown in the figure. The normal contact stress due to critical viscous damping is given by

$$\sigma_c = 2\omega \dot{d} \tag{2.62}$$

while in the general case of an underdamped system,

$$\sigma_c = 2\omega \xi \dot{d} \tag{2.63}$$

where ξ is the damping ratio. If $\xi = 0$, there is no energy dissipation. If $\xi = 1$, critical damping is obtained.

After substituting the frequency from equation (2.62),

$$\sigma_c = 4\xi \frac{\sqrt{p/\rho}}{h} \dot{d} \tag{2.64}$$

This damping is due to contact only. Physical interpretation of such damping is, for instance, plastic deformation and/or breaking of surface asperities. This damping is not to be confused with damping such as energy dissipation due to deformation of the discrete

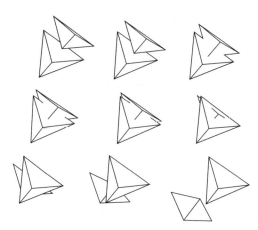

Figure 2.29   Motion sequence of the tetrahedron with initial velocity impacting a fixed tetrahedron for the penalty term $p = 1.0e + 5$ Pa.

elements. Material damping due to elastic or plastic deformation of discrete elements is naturally covered by discrete elements being discretised into finite elements.

The properties of the discretised distributed potential contact force algorithm are best demonstrated using numerical examples. In Figure 2.29, the impact of two identical tetrahedra is shown. The geometry of the tetrahedra is defined with three concurrent orthogonal edges of 2 m in length.

The material of the tetrahedra is assumed linear elastic with a modulus of elasticity $E = 100$ GPa, $\nu = 0$ and $\rho = 1500$ kg/m^3. The top tetrahedron moves with initial velocity of 550 m/s, while the bottom tetrahedron is fixed.

The problem is solved for penalty $p = 1e + 5$ Pa, $p = 7e + 5$ Pa, $p = 1e + 6$ Pa and $p = 1e + 7$ Pa. The motion sequences for penalty values of $p = 1e + 5$ Pa, $p = 7e + 5$ Pa, $p = 1e + 6$ Pa and $p = 1e + 7$ Pa are shown in Figures 2.29–2.32, respectively. The energy balance for all cases is shown in Figure 2.33. It can be observed that the energy balance is preserved irrespective of the penalty value, size of penetration or geometry of contact.

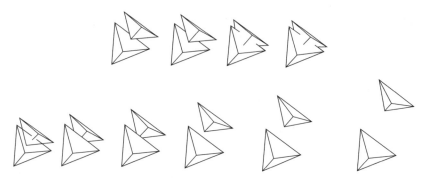

Figure 2.30   Motion sequence of a tetrahedron with initial velocity impacting on a fixed tetrahedron obtained using penalty term $p = 7.0e + 5$ Pa.

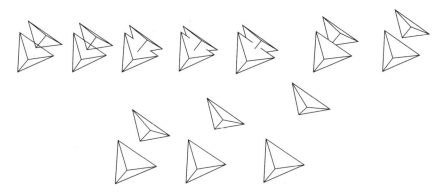

Figure 2.31 Motion sequence of a tetrahedron with initial velocity impacting on the fixed tetra-hedron–obtained using a penalty term $p = 1.0\text{e} + 6\,\text{Pa}$.

Figure 2.32 Motion sequence of a tetrahedron with initial velocity impacting on a fixed tetrahe-dron–obtained using a penalty term $p = 1.0\text{e} + 7\,\text{Pa}$.

The motion sequence shown in Figure 2.29 corresponds to a very small penalty. Thus, penetration is far from small. In fact, the tetrahedra 'go through' each other. Irrespective of this, the energy balance is preserved, as shown in Figure 2.33.

The increased value of penalty makes the moving tetrahedron bounce from the fixed tetrahedron, although only after a large overlap, as can be observed from Figure 2.30.

In Figure 2.31 the motion sequence for a slightly larger penalty term is shown. This time the penetration is still large. However, the moving tetrahedron 'bounces' from the fixed tetrahedron. Regardless of a large penetration, the energy balance is preserved, as shown in Figure 2.33.

In Figure 2.32 an even larger penalty is employed, resulting in further reduced penetration and eventual bouncing off of discrete elements from each other. Again, as Figure 2.33 shows, energy balance is preserved. It is evident that, for any value of penalty term, some of the kinetic energy is transformed into potential energy due to overlap (penetration). After the contact release, this energy is recovered and transformed back into kinetic energy. Thus, at all stages the energy balance is preserved, i.e. no energy is either 'lost'

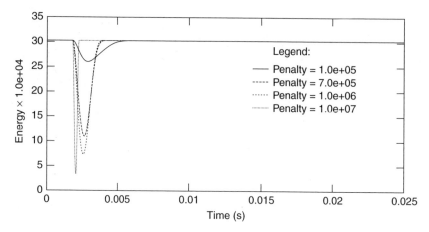

Figure 2.33 Energy balance for impact of the moving tetrahedron against a fixed tetrahedron.

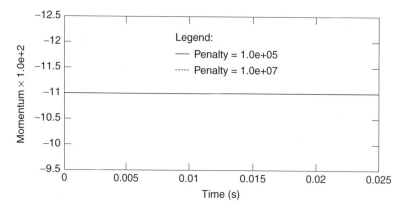

Figure 2.34 Momentum balance for impact of a tetrahedron with initial velocity against a tetrahedron at rest but free to move.

or 'created', irrespective of the size or geometry of the overlap of discrete elements in contact.

The same simulations shown above are repeated with a bottom tetrahedron being initially at rest, but free to move. The top tetrahedron moves with an initial velocity of 550 m/s, while the bottom tetrahedron is at rest, and starts moving under the impact of the top tetrahedron.

It can be observed that both energy and momentum balance are preserved, regardless of the penalty used, as shown in Figures 2.34 and 2.35. This time, both tetrahedra move, as illustrated by the motion sequence obtained using penalty $p = 1e + 7$ Pa, which is shown in Figure 2.36.

Energy balance is especially important when discrete elements are confined so that repeated contact-contact release occurs. When discrete elements are closely packed together, these contact-contact release situations cannot be avoided. If numerical

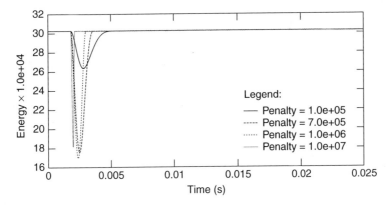

Figure 2.35   Energy balance for impact of a tetrahedron with initial velocity against a tetrahedron at rest but free to move.

Figure 2.36   Momentum balance for impact of a tetrahedron with initial velocity against a tetrahedron at rest but free to move–obtained using penalty term $p = 1.0e + 7$.

procedures employed were not to preserve the energy balance, the energy of the system would increase artificially. This energy increase is exponential, and results in the combined finite-discrete element system being 'blown up', which is another way of saying that the algorithms employed are not stable. To demonstrate that the discretised distributed potential contact force is not one of these unstable algorithms, in Figure 2.37 three tetrahedra all of the same shape and elastic properties are arranged in such a way that the two outer tetrahedra are fixed, while the inner tetrahedron is moving with an initial velocity of 500 m/s, and thus it repeatedly hits the outer tetrahedra in turn. This is therefore the case of confined contact with the middle tetrahedron oscillating between the two fixed end tetrahedra. The middle tetrahedron hits the top tetrahedron with the apex, while it hits the bottom tetrahedron with the flat base.

For small penalty terms (Figure 2.37), the end tetrahedra cannot contain the middle tetrahedron, and it simply goes through. This results in extremely large penetrations, and tests the potential contact force algorithm to the extreme. Nevertheless, as graphs in Figure 2.40 show, the energy is preserved.

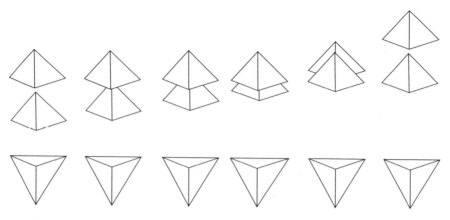

Figure 2.37   Motion sequence of a tetrahedron with initial velocity confined between two fixed tetrahedra–obtained using penalty term $p = 1.0e + 4\,\text{Pa}$, initial impact.

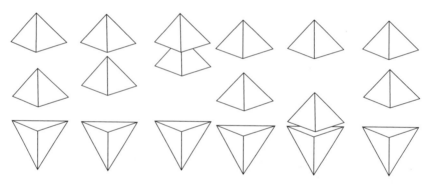

Figure 2.38   Motion sequence of a tetrahedron with initial velocity confined between two fixed tetrahedra–obtained using penalty term $p = 5.5e + 6\,\text{Pa}$, moving up and hitting top tetrahedron.

For larger penalty terms (Figures 2.38 and 2.39), the middle discrete element is contained by the end discrete elements. Thus, the middle discrete element oscillates between the end discrete elements. The penetration depends upon the size of the penalty term. However, irrespective of the size of the penalty term, the energy is preserved regardless of penalty or penetration, as shown in Figure 2.40.

As theoretically predicted, these numerical examples clearly demonstrate the most important property of the potential contact force algorithm in 3D, namely the preservation of the energy irrespective of the shape and size of discrete elements, and irrespective of the geometry, i.e. kinematics of contact. This is clearly demonstrated by energy balances shown in Figure 2.40.

It should be emphasised that the potential contact force contact algorithm has some other features apart from preservation of energy and momentum balance. For instance, due to the discretised nature of the evaluation of contacts, the algorithm is suitable for parallel or distributed computing, as discussed in Chapter 9. There is also an additional advantage from the fact that the geometry is treated in a discretised manner, which greatly

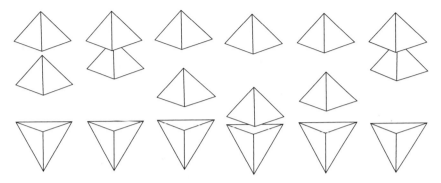

Figure 2.39 Motion sequence of a tetrahedron with initial velocity confined between two fixed tetrahedra–obtained using penalty term $p = 1.0e + 7$, moving up and bouncing back.

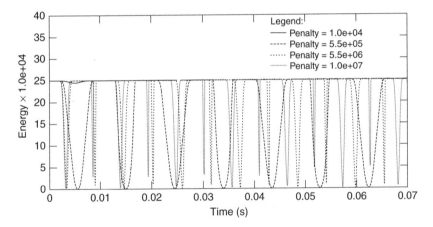

Figure 2.40 Energy balance for a repeated contact-contact release situation (tetrahedron with initial velocity confined between two fixed tetrahedra).

reduces the complexity of the algebraic expressions used to evaluate the contact forces, as discussed in Chapter 10. A further beneficial feature is a relationship between penetration and mesh size by which the penalty parameters are given physical meaning, while the effect of the penetration on the estimated domain fields can be controlled through mesh refinement, i.e. in the same way as the accuracy of the finite element approximation itself.

2.7 ALTERNATIVE IMPLEMENTATION OF THE POTENTIAL CONTACT FORCE

The finite element mesh is not the only way in which potential contact force can be defined. One of the alternative ways of defining the contact force potential is to define the potential over the domain of the discrete element irrespective of any finite element discretisation employed. Such a potential for a discrete element in 2D is shown in Figure 2.41.

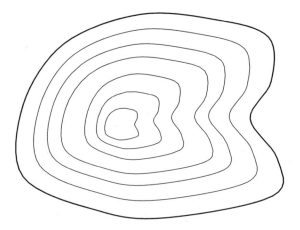

Figure 2.41   Contact force potential defined over the domain of a discrete element of a general shape.

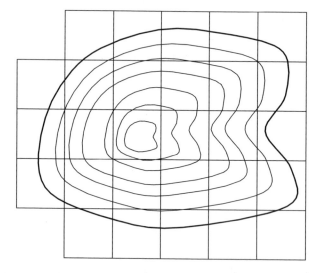

Figure 2.42   Contact force potential specified at nodes of a regular grid.

The lines connect points of constant contact force potential. The thicker line indicates the boundary of the discrete element where the potential is zero. Outside of the discrete element, the potential is negative. To represent this potential, a regular grid of points can be employed, as shown in Figure 2.42.

Four points (nodes) of the grid form a square. The potential over the domain of each of the squares is approximated using the values of the potential at the nodes (Figure 2.43).

In this way, both the boundary (shape) of the discrete element and the contact potential over a discrete element are approximated using the same regular grid. In the case of the

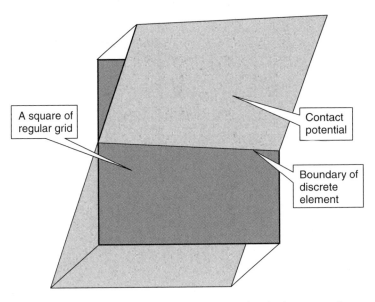

Figure 2.43   Contact force potential specified at nodes of a single square of a regular grid.

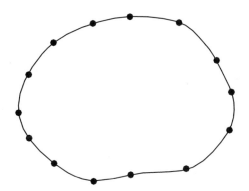

Figure 2.44   Boundary points for contact force potential defined over the domain of a discrete element of a general shape.

discrete element shown in Figure 2.42, only 35 numbers are needed to represent both the shape and contact potential.

Processing of contact can be employed in either a distributed fashion, by resolving the contact of couples of contacting squares each belonging to a different discrete element, or a simpler approach can be adopted, in which the contact force is resolved as a concentrated contact force. For this purpose, a set of boundary points on the surface (boundary) of a discrete element is specified as shown in Figure 2.44.

The contact is resolved by considering interaction of each of the boundary points with respective squares of the regular grid (Figure 2.45). For a given point **p**, the potential is

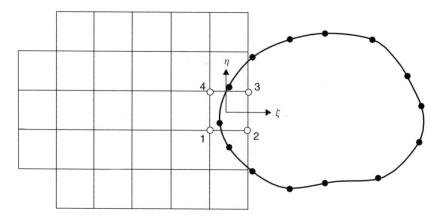

Figure 2.45   Contact force potential specified at nodes of a regular grid.

calculated using local coordinates:

$$\varphi(\xi, \eta) = N_1(\xi, \eta)\varphi_1 + N_2(\xi, \eta)\varphi_2 + N_3(\xi, \eta)\varphi_3 + N_4(\xi, \eta)\varphi_4 \tag{2.65}$$

where shape functions are given by

$$N_i(\xi, \eta) = \frac{1}{4}(1 + \xi\xi_i)(1 + \eta\eta_i); \ \ i = 1, 2, 3, 4 \tag{2.66}$$

$$\xi_1 = \xi_4 = -1; \ \ \xi_2 = \xi_3 = 1;$$

$$\eta_1 = \eta_2 = -1; \ \ \eta_3 = \eta_4 = 1.$$

3

Contact Detection

3.1 INTRODUCTION

Large-scale combined finite-discrete element simulations involve contact of a large number
of separate bodies, each body being represented by a single discrete element. It is evident
from Chapter 2 that processing of contact interaction involves the summation of contact
forces over contacting couples comprising target and contactor elements:

$$\mathbf{f} = \sum_{i=1}^{n} \sum_{j=1}^{n} \mathbf{f}_{ij} \tag{3.1}$$

Thus, processing contact interaction for all possible contacts would involve a total number of operations proportional to N^2, where N is the total number of discrete elements
comprising the problem.

This would be very CPU intensive, and would limit application of the combined finite-
discrete element method to simulations comprising a very small number (a few thousand)
discrete elements. To reduce CPU requirements of processing contact interaction, it is
necessary to eliminate couples of discrete elements that are far from each other and are
not in contact. A set of combined finite-discrete element procedures designed to detect
discrete or finite elements that are close to each other is usually called a contact detection
algorithm, or sometimes a contact search algorithm.

The contact detection algorithm must be:

- robust,
- CPU efficient,
- RAM efficient, and
- easy to implement.

The robustness of a contact detection algorithm means that it has to detect all those
couples of discrete elements that are actually in contact. In addition, it has to eliminate
most of the couples of discrete elements that are not in contact. It has to perform these
tasks irrespective of how the discrete elements are in positions relative to each other.

CPU efficiency means that the total CPU time spent on detecting all the contacting
couples must be as short as possible. The total CPU time is measured in terms of the

The Combined Finite-Discrete Element Method A. Munjiza
© 2004 John Wiley & Sons, Ltd ISBN: 0-470-84199-0

total number of discrete elements and is, for instance, said to be proportional to the total number of discrete elements, i.e.

$$T \propto N \tag{3.2}$$

or proportional to

$$T \propto N^2 \tag{3.3}$$

or

$$T \propto N \log_2 N \tag{3.4}$$

This expression means that the total CPU time necessary to perform contact detection will simply double if the size of the problem doubles, (equation (3.2)), or that doubling the size of the problem means that CPU time will increase four times, (equation (3.3)). For instance, if the size of the combined finite-discrete element problem changes from two discrete elements to 2,000,000 discrete elements, the total CPU time according to equation (3.4) would increase 20,000,000 times. Thus, if the contact detection algorithm is less CPU efficient, the total CPU time may increase much faster than the size of the problem, making very large scale problems unaffordable in terms of either available hardware performance or hardware cost and affordability.

Routine combined finite-discrete element problems may involve millions of interacting discrete elements, and in these the ideal situation is the total CPU time being proportional to the total number of discrete elements, as indicated by equation (3.2). Contact detection algorithms for which equation (3.2) applies are in general classified as linear contact detection algorithms, meaning that CPU time is a linear function of the size of the problem. Examples of linear contact detection algorithms are, for instance,

- Munjiza-NBS (No binary search) contact detection algorithm,
- Williams C-grid contact detection algorithm,
- Screening contact detection algorithm.

These algorithms have appeared relatively recently, starting with Munjiza-NBS (1995), which was the first linear contact detection algorithm employed in the combined finite-discrete element method.

All other contact detection algorithms are classified as nonlinear contact detection algorithms, meaning that CPU time increases either slower or faster than the size of the problem. Nonlinear contact detection algorithms are classified either as hypo-linear contact detection algorithms or hyper-linear contact detection algorithms. Hyper-linear contact detection algorithms have CPU time increasing faster than the size of the combined finite-discrete element problem. Examples of such algorithms are

- quadratic contact detection algorithms (equation (3.3)), and
- logarithmic contact detection algorithms (equation (3.4)).

Hypo-linear algorithms should theoretically have better performance than linear contact detection algorithms. A typical such performance would be

$$T \propto \sqrt{N} \tag{3.5}$$

or

$$T \propto N^x \tag{3.6}$$

where the exponent x is given by

$$x < 1 \tag{3.7}$$

The author is not aware that any such contact detection algorithm exists. In addition, there is no proof that such an algorithm is possible. However, before 1995 the same was true for linear contact detection algorithms, when it was not clear that such an algorithm would be feasible, until the first robust linear contact detection algorithm was designed in the form of the Munjiza-NBS contact detection algorithm.

RAM efficiency means that the total memory required for the contact detection algorithm to store all its data should be as small as possible. The total memory required for some algorithms may, for instance, depend upon the spatial distribution of discrete elements, and the total number of discrete elements.

The ideal situation is that the total RAM requirements M depend only upon the total number of discrete elements, and are proportional to the total number of discrete elements

$$M \propto N \tag{3.8}$$

The optimal contact detection algorithm is in general dependent upon the problem to be solved. The properties of different contact detection algorithms make them suited for different types of problems, such as dense packing and loose packing, as shown in Figure 3.1. In addition, one algorithm may be better suited for quasi-static problems, where the relative motion of individual discrete elements is restricted, (Figure 3.2), while the altogether different algorithm is suited for dynamic problems where individual discrete elements move significantly, as shown in Figure 3.3.

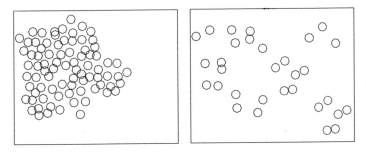

Figure 3.1 Packing density: (left) dense packing; (right) loose packing.

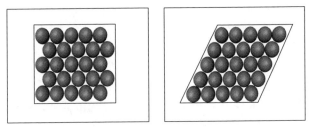

Figure 3.2 Typical quasi-static problem.

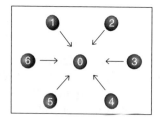 

Figure 3.3 Typical dynamic problem.

For instance, there is a whole class of contact algorithms routinely employed with finite element static systems involving contact. Most of these algorithms can handle only problems where the relative motion of contacting boundaries is restricted. Contact detection is performed only once. Thus, the CPU performance of such contact detection algorithms is not very important.

Real problems where contact detection is a big issue are dynamic problems, comprising large numbers of discrete elements that are free to move significantly. Contact detection itself can take a considerable proportion of the total CPU time required to analyse such problems (over 60% in some cases).

In this context, when developing the contact detection algorithm, it is important to:

- Minimise the CPU requirements, i.e. total detection time T defined as the total CPU time needed to detect all couples close to each other.

- Minimise the total memory (RAM) requirements M, expressed in terms of the total memory size as a function of the total number of discrete elements N and packing density.

- Maximise performance expressed in terms of rate of M and T change with change in packing density.

In this chapter, a number of contact detection algorithms currently available are described in detail. However, the list of contact detection algorithms covered in this chapter is far from complete, although it does represent different types of contact detection algorithms that have been or are used to speed up the CPU intensive combined finite-discrete simulations. Some of the algorithms described are suited for dense packs, while others perform better with loose packs. Dense packs of discrete elements are characterised by:

- discrete elements being close to each other, and
- most of the physical space being occupied by discrete elements.

Loose packs of discrete elements are characterised by:

- discrete elements being far from each other, and
- most of the physical space is not occupied by discrete elements, i.e. is empty.

Towards the end of this chapter, the Munjiza-NBS algorithm, suited for both dense and loose packs or systems where the density of the pack changes with time, is also presented in detail, together with the recently developed Williams-C-grid algorithm. For the sake of clarity, all algorithms are described in 2-dimensional space, while at the end of the chapter extensions to 3-dimensional (and in some cases extensions to n-dimensional) spaces are given.

3.2 DIRECT CHECKING CONTACT DETECTION ALGORITHM

The direct checking contact detection algorithm is the simplest contact detection algorithm possible. The algorithm is usually implemented in two steps:

- a bounding object for each discrete element is defined, and
- a simple intersection check for bounding objects is made, and if bounding objects of any two discrete elements are found to be intersecting each other, it is assumed that the two discrete elements are in contact.

For bodies of similar size, a few different implementations are possible, depending on the bounding object selected. Two types of implementation are explained here, namely the circular bounding object and the rectangular bounding object.

3.2.1 Circular bounding box

It is assumed that all discrete elements are circles of constant diameter d, chosen in such a way that no point of any discrete element is outside the circle. Thus, the diameter of the bounding circle is defined by the largest discrete element present (Figure 3.4).

A loop over all bounding circles is performed, and each bounding circle is checked against all others for intersection:

> **Loop over discrete elements** (i=1; i<N; i++)
> {   **Loop over remaining discrete elements** (j=i+1; j<N; j++)
> {   **contact check**
> }
> }

Contact check is a simple operation:

$$(x_i - x_j)^2 + (y_i - y_j)^2 < d^2 \tag{3.9}$$

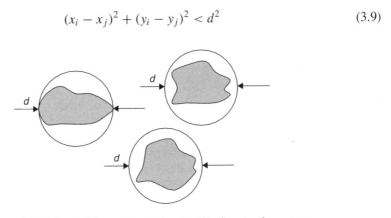

Figure 3.4   Identical bounding circles for 2D discrete elements.

where x and y denote the current coordinates of the centres of the bounding circles. Each check involves two float point multiplications. Thus, the total number of float point multiplications necessary to detect all the contacts is given as

$$n_N = \frac{N^2 - N}{2} \propto N^2 \tag{3.10}$$

The implementation of this algorithm takes a few minutes of a programmer's time. However, CPU times for solving large-scale combined finite-discrete element problems involving even thousands of bodies are large. Consequently, systems comprising millions of discrete elements are not feasible on present day computers due to the fact that contact detection is performed millions of times, and the CPU times required can be estimated from the total number of float-point operation performed, being given by

$$F \propto 1{,}000{,}000 N^2 = 1{,}000{,}000 \cdot 1{,}000{,}000^2 = 10^{18} \tag{3.11}$$

Such CPU demanding jobs for large enough systems would probably run for centuries on the most expensive machines. Although this algorithm is very often implemented in the finite element analysis of contact problems, it is never a part of any serious combined finite-discrete element code.

3.2.2 Square bounding object

It is assumed that all discrete elements are squares of constant edge a, chosen in such a way that no point of any discrete element is outside the bounding square. Thus, the edge of the bounding square is defined by the largest discrete element present, (Figure 3.5).

A loop over all bounding squares is performed, and each bounding square is checked against all others for intersection:

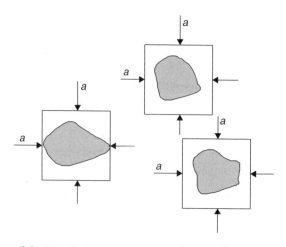

Figure 3.5 Identical bounding squares for 2D discrete elements.

Loop over discrete elements *(i=1; i<N;i++)*
{   Loop over remaining discrete elements *(j=i+1; j<N;j++)*
{   contact check
}
}

Contact check is a simple operation, given as follows:

$$\left(x_i + \frac{a}{2}\right) < \left(x_j - \frac{a}{2}\right) \text{ for } x_i \leq x_j \tag{3.12}$$

$$\left(x_j + \frac{a}{2}\right) < \left(x_i - \frac{a}{2}\right) \text{ for } x_i > x_j$$

$$\left(y_i + \frac{a}{2}\right) < \left(y_j - \frac{a}{2}\right) \text{ for } y_i \leq y_j$$

$$\left(y_j + \frac{a}{2}\right) < \left(y_i - \frac{a}{2}\right) \text{ for } y_i > y_j$$

where x and y denote the current coordinates of the centres of the bounding squares. Contact check involves no float point multiplication, and is therefore usually a cheaper operation than in the case of the bounding box being a circle. However, the total number of operations is still given as

$$n_N \propto N^2 \tag{3.13}$$

Again, the implementation of this algorithm takes minutes of a programmer's time. However, CPU times for solving problems involving as little as a few thousand discrete elements are extremely large.

3.2.3   Complex bounding box

It is possible to introduce a more complex bounding box than a square or circle, for instance, a rectangle of constant edges a and b chosen in such a way that no point of any discrete element is outside the bounding rectangle (Figure 3.6). Contact check is again a simple operation as follows:

$$\left(x_i + \frac{a}{2}\right) < \left(x_j - \frac{a}{2}\right) \text{ for } x_i \leq x_j \tag{3.14}$$

$$\left(x_j + \frac{a}{2}\right) < \left(x_i - \frac{a}{2}\right) \text{ for } x_i > x_j$$

$$\left(y_i + \frac{b}{2}\right) < \left(y_j - \frac{b}{2}\right) \text{ for } y_i \leq y_j$$

$$\left(y_j + \frac{b}{2}\right) < \left(y_i - \frac{b}{2}\right) \text{ for } y_i > y_j$$

It is in principle possible to select any shape of the bounding box. In addition, different shapes and sizes of the bounding boxes for each discrete element could be selected. These improvements could in some cases result in CPU times being reduced by 10%, 20% or even 500%. However, the major problem would still remain–irrespective of the bounding

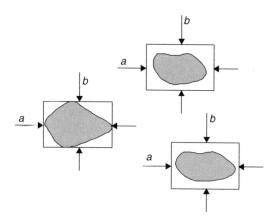

Figure 3.6   Identical bounding rectangles for 2D discrete elements.

box, the total CPU time is proportional to N^2. Thus, if the total number of discrete elements is increased by 100 times, the total CPU time increases by 10,000 times. This is a major problem with the direct checking approach, and the main reason why direct checking based contact detection algorithms are almost never used in FEM/DEM codes.

3.3   FORMULATION OF CONTACT DETECTION PROBLEM FOR BODIES OF SIMILAR SIZE IN 2D

As explained in the previous section, various bounding objects can be defined for each discrete element, and then contact detection and checking for contact performed in terms of these bounding objects.

Use of bounding boxes greatly simplifies the contact detection problem, and makes contact detection algorithms based on bounding boxes more robust and more general. For the sake of simplicity and clarity of explanation, in the rest of this chapter, the simplified contact detection problem limited to two-dimensional space is defined as follows:

> *A set of discrete elements is given in a 2D space. Every discrete element is bounded by a circle of the same diameter d. All discrete elements are contained inside the space bounded by a square of the edge s (Figure 3.7). The task is to detect all discrete elements that are close to each other so that the bounding circles either touch or overlap.*

The contact detection algorithms considered in the following sections will assume a simplified contact detection problem in 2D, unless otherwise stated. All the algorithms will be explained in 2D for the sake of clarity. Towards the end of this chapter, an extension to 3D will be given.

As demonstrated by the direct checking approach from the previous section, changing the shape or size of the bounding box, or using different shapes and sizes of bounding boxes, is a relatively simple task, which does not change the essence or the key procedures of any contact detection algorithm. Some contact detection algorithms are very complex, and are not easy to explain. Different shapes and sizes of bounding boxes would complicate the situation even further, making the explanation difficult to follow.

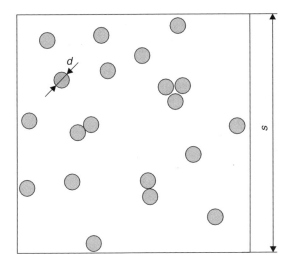

Figure 3.7   Simplified definition of the contact detection problem in 2D.

3.4 BINARY TREE BASED CONTACT DETECTION ALGORITHM FOR DISCRETE ELEMENTS OF SIMILAR SIZE

In this section, a simplified contact detection problem is assumed with all discrete elements being bounded by a circle of constant diameter d. For contact detection purposes, each discrete element is therefore uniquely identified by the centre of its bounding circle.

Binary tree based search is a space based search, where space is subdivided into hierarchical cells. The size of the smallest cell is made large enough so that the largest discrete element can fit into a cell, i.e. the size of the smallest single cell is equal to the diameter of the bounding circle d.

For 2D problems, the space is assumed to coincide with the square large enough to contain all the discrete elements. At level 1 this space is subdivided into two cells of a rectangular shape. At level 2 each rectangle is further subdivided into two squares. This process is continued until the smallest cell is reached, as shown in Figure 3.8.

Thus, for instance, the particular discrete element shown in Figure 3.8 is found to belong to the successive cells as shown. In other words, each discrete element is mapped onto b cells, where b is the number of successive subdivisions (levels). In the case shown in Figure 3.8, there are five levels of space subdivision, thus

$$b = 5 \tag{3.15}$$

The mapping of discrete elements onto cells is best represented by a binary tree, as shown in Figure 3.9. At each level, the left-hand cell and lower cell are represented by the left-hand node, while the upper cell and right-hand cell are represented by the right-hand node.

A discrete element shown in Figure 3.8 is mapped onto the nodes of the binary tree, as shown in Figure 3.9. Level one cells are represented by level one nodes, level two cells by level two nodes, etc.

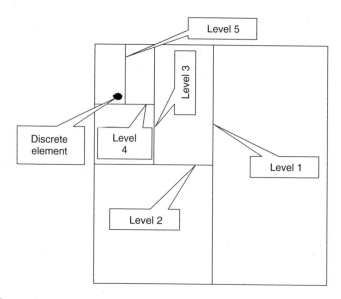

Figure 3.8   Successive space subdivisions until the cell containing a discrete element is located.

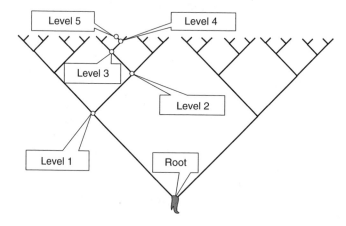

Figure 3.9   Successive space subdivisions represented by a binary tree.

Contact detection consists of three steps:

- In the first step, all the discrete elements are mapped onto the cells.
- In the second step, this mapping is represented by the binary tree (i.e. binary tree is built).
- In the third step, contact detection itself is performed.

It is worth mentioning that only cells with discrete elements mapped into them are represented by the binary tree. Thus, the size of the binary tree depends upon the number of discrete elements present.

Initially, only the root of the binary tree is necessary. As the discrete elements are mapped onto the cells, the binary tree is being built node by node until all discrete elements are 'loaded' onto the tree.

Contact detection for each discrete element consists in identifying which cells a particular discrete element is mapped onto, and thus identifying the smallest cell together with its neighbouring cells. Contact detection itself is then done in a way similar to the direct checking approach, except that only discrete elements in the neighbouring cells are considered.

The algorithm for creating the binary tree can be summarised as follows:

> **Loop over all discrete elements** $(i=1; i\leq N; i++)$
> **{   Loop over cell levels** $(j=1; j\leq b; j++)$
> **{   check in which of the two cells the centre of the discrete element
>   is located
>   if the corresponding node of the binary tree does not exist,
>   create it
>   if** $(j=b)$ **add discrete element** i **onto the connected list
>   of discrete elements that are mapped onto the current cell**
>
> **}**
>
> **}**

It can be seen that the creation of the binary tree involves b loops over cell levels for each discrete element. Inside each loop, the following operations are performed:

- *Step 1:* calculation of the boundary between cells. This involves operation of the type (see Figure 3.10):

$$x_{mid} = \frac{x_{min} + x_{max}}{2} \tag{3.16}$$

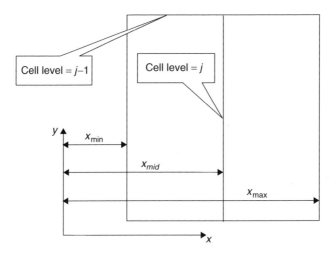

Figure 3.10   Calculation of boundary between cells; parent cell (level $j-1$) is divided into two child cells (level j).

- *Step 2:* check which cell the current discrete element belongs to:

$$\text{left–hand cell if } x_i > x_{mid}$$
$$\text{right–hand cell otherwise} \tag{3.17}$$

- *Step 3:* place the current discrete element onto a list of discrete elements if the current level equals b, i.e.

$$\text{if } j = b,$$
$$\text{place current discrete element (element } i) \text{ onto the list} \tag{3.18}$$

The easiest way to achieve Step 3 is to use singly connected lists. For example, for the discrete elements shown in Figure 3.11, a singly connected list for all cells at level 4 contain four discrete elements, as shown in Figure 3.12.

This list is easiest achieved by using an array of integer numbers of size N, where N is the total number of discrete elements comprising the system. The array is called **E** (short for element). The head of the list is remembered with the node of the binary tree, as shown in Figure 3.12.

All successive discrete elements are shown using array **E**, as shown in Figure 3.13. All the numbers of array **E** are initially assigned as -1, meaning the end of the list, i.e.

$$\mathbf{E}[i] = -1; \quad i = 1, 2, 3, \ldots, N \tag{3.19}$$

while all the heads are initially assigned -1, meaning an empty list.

Placing a new discrete element i onto the list is done by setting

$$\mathbf{E}[i] = HEAD$$
$$HEAD = i \tag{3.20}$$

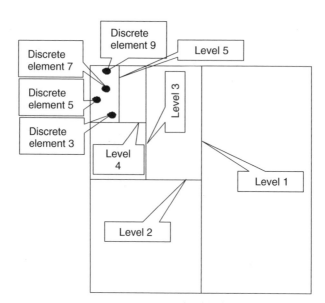

Figure 3.11   Four discrete elements in a single cell.

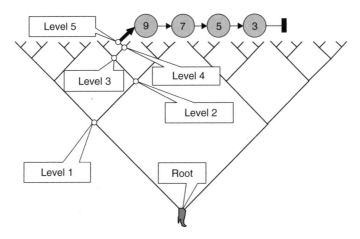

Figure 3.12   Singly connected list showing discrete elements mapped onto the cell at level 5.

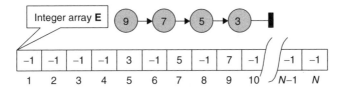

Figure 3.13   Singly connected list.

Thus, for instance, if discrete element number 1 is to be added onto the list, one simply sets

$$\mathbf{E}[1] = HEAD; \quad i.e. \ 9$$
$$HEAD = 1$$

(3.21)

Thus, the new list becomes as shown in Figure 3.14.

It is worth mentioning that HEADs are stored at the leaf node level of the binary tree, i.e. they are an integral part of the binary tree. Each leaf node has one HEAD, represented by one integer number, which is the first discrete element in the list. It is also worth noting that in C and/or C++ implementations, very often pointers are used instead of integer numbers. These usually make the code run faster, save memory space, but may also lead to implementation errors (bugs) difficult to detect.

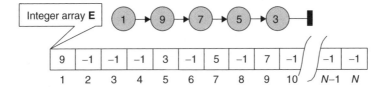

Figure 3.14   The singly connected list after addition of discrete element 1.

The total number of operations per discrete element necessary to make a binary tree is approximately given by

$$n_1 = (7 \text{ additions} + 1 \text{ division}) \cdot b \tag{3.22}$$

The total number of levels b necessary is a function of the total number of smallest cells. The lower limit for this number is given by

$$b = \log_2 N \tag{3.23}$$

The upper limit is obviously a function of the distribution of discrete elements in space. In the cases when discrete elements are sparsely distributed in space this can be very large. Thus, a lower limit of the total number of operations for building binary tree is given by

$$n_N = n_1 N = (7 \text{ additions} + 1 \text{ division})b$$
$$> (7 \text{ additions} + 1 \text{ division})(N \log_2 N) \tag{3.24}$$

Once the binary tree has been built, contact detection itself can be performed. It is done in a similar way to how the binary tree has been built. A loop over discrete elements is performed, and hierarchical cells onto which discrete element is mapped are identified. Thus, corresponding nodes of the binary tree are followed until the leaf level node corresponding to the smallest cell is reached. Direct check for contact against discrete elements mapped onto the neighbouring cells is finally performed:

> **Loop over all discrete elements** $(i=1; \; i{\leq}N; i++)$
> **{   Loop over cell levels**$(j=1; \; j{\leq}b; j++)$
> **{   check which of the two cells the centre of discrete element is in**
>   **if**$(j{=}b)$ **collect neighbouring cells at level** b **and check for contact**
>   **against all discrete elements in those cells**
>
> **}**
>
> **}**

Building a binary tree does not have to be separated from detecting contact. Two approaches are available:

- The first approach is to completely delete the binary tree after detection is done and then to rebuild it again. To delete a binary tree, it is enough to set the root to -1, denoting an empty binary tree, which is an extremely cheap operation in terms of CPU. In addition, the singly connected lists of discrete elements need to be deleted, which is done by simply setting all elements of the array $\mathbf{E}$ to -1

$$\mathbf{E}[i] = -1; \quad i = 1, 2, 3, \ldots, N \tag{3.25}$$

  meaning the end of lists, i.e. empty lists.

- The second approach is to remember the position of the centre of each discrete element at the time it was put onto the binary tree. Thus, each discrete element would be removed from the tree before putting it back onto the tree. This would be done only if the discrete element has moved significantly. This approach enables rebuilding of the binary tree, and is used, for instance, when most of the discrete elements do not move. In such a case, it is easier to simply remove those discrete elements that have moved and place them onto the tree again. The contact check is necessary only for these discrete elements. This way the contact detection in these particular cases can be speeded up.

3.5 DIRECT MAPPING ALGORITHM FOR DISCRETE ELEMENTS OF SIMILAR SIZE

It is not always necessary to use a binary tree to get an efficient spatial based search. Space subdivision into cells and subsequent mapping of discrete elements onto cells lends itself well to being used with sorting algorithms. The basic idea is relatively simple. Instead of dividing the space into a hierarchy of cells, as in the case of the binary tree, the space is divided into cells all of the same size, as shown in Figure 3.15. The size of individual cells is chosen so that the largest discrete element contained by the system can be fitted into a single cell. Thus, the size of the cells is equal to the diameter of the bounding circle, which for a simplified contact detection problem is the same for all discrete elements (Figure 3.16).

Contact detection is performed in two steps:

Step 1: Map discrete elements onto cell: discrete elements are mapped onto cells according to the current position of the centre of the individual discrete element. Thus, each discrete element is mapped to one and only one cell. This in essence involves only integerisation

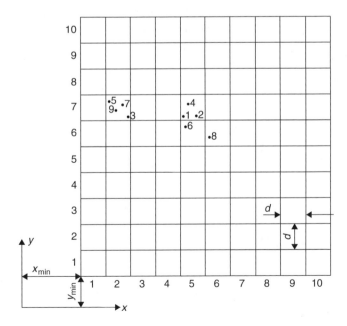

Figure 3.15 Space divided into identical cells large enough to contain the largest discrete element comprising the system. Discrete element centres are marked with dots.

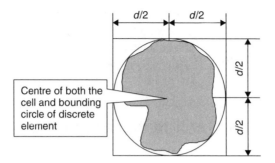

Figure 3.16   Cells are large enough to contain a whole discrete element.

of the current coordinates of the centre of each discrete element:

$$^{int}x_i = 1 + Int \left(\frac{x_i - x_{min}}{d} \right) \qquad (3.26)$$

$$^{int}y_i = 1 + Int \left(\frac{y_i - y_{min}}{d} \right)$$

Step 2: Find discrete elements that may be in contact: it is worth noting that every single discrete element can fit into a single cell in such a way that if the centre of a particular discrete element coincides with the centre of the cell, no point of discrete element is outside of the cell, as shown in Figure 3.16. Two discrete elements mapped onto cells that share either nodes or edges (neighbouring cells) can be in contact. In addition, two discrete elements mapped onto cells that share neither nodes nor edges (non-neighbouring cells) cannot be in contact, as shown in Figure 3.17.

No discrete elements mapped onto the central cell in Figure 3.17 can be in contact to discrete elements outside either the central cell or neighbouring cells. Thus, detection of

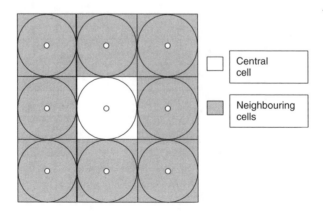

Figure 3.17   Cells neighbouring the central cell are marked. All other cells are non-neighbouring cells.

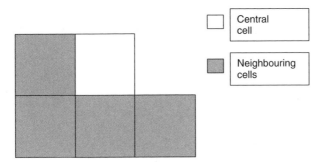

Figure 3.18 Contact check mask for discrete elements mapped to the central cell.

contact for a particular discrete element from the central cell is reduced to a direct check with discrete elements from either central or neighbouring cells.

In fact, it is enough to check against the discrete elements from cells shown in Figure 3.18. This is because all cells containing a discrete element are in turn considered to be central cells.

Contact detection described above involves only integerisation of current coordinates of the centre for each discrete element. In addition, the total number of algebraic operations per discrete element does not depend upon the total number of discrete elements comprising the system.

The total number of algebraic operations for mapping of all discrete elements onto the cells is given by

$$n_1 = (2 \text{ additions} + 1 \text{ division})N \tag{3.27}$$

Thus, the total number of algebraic operations for detecting all contacts is proportional to the total number of discrete elements, i.e.

$$n_n \propto N \tag{3.28}$$

The total CPU time is also proportional to the total number of discrete elements. In other words, if the total number of discrete elements is increased by tenfold, the total CPU time will also increase by tenfold. Note that this was not the case with binary tree based search. This algorithm is therefore much more efficient in its use of CPU time than the binary tree based search. However, there is a problem to be solved. The algorithm is based on the mapping of discrete elements onto identical cells. This mapping somehow has to be represented in the memory of a computer. Depending on the representation of this mapping, a whole family of contact detection algorithms can be devised. These include the screening contact detection algorithm, sorting contact detection algorithm, Munjiza-NBS contact detection algorithm and the Williams-C-Grid contact detection algorithm, all described in the following sections.

3.6 SCREENING CONTACT DETECTION ALGORITHM FOR DISCRETE ELEMENTS OF SIMILAR SIZE

The simplest way of representing the mapping of discrete elements onto identical cells is to use one integer number per cell. The total number of integer numbers is equal to the total number of cells. All these integer numbers are best represented by a two-dimensional array of integer numbers (called a screening array):

$$\mathbf{C} = \begin{bmatrix} c_{1,1} & c_{1,2} & c_{1,3} & \cdots & c_{1,j} & \cdots & c_{1,n_{cel}} \\ c_{2,1} & & & & & & \\ c_{3,1} & & & & & & \\ \cdots & & & & & & \\ c_{i,1} & & & & i,j & & \\ \cdots & & & & & & \\ c_{n_{cel},1} & & & & & & c_{n_{cel},n_{cel}} \end{bmatrix} \qquad (3.29)$$

Mapping of discrete elements onto the cells is represented by associating a singly connected list of discrete elements with each cell. The first discrete element in the list (head of the list) is represented by array $\mathbf{C}$ (short for cell), while the rest of the list is represented by the integer array called $\mathbf{E}$ (short for element).

As each discrete element is mapped to one and only one cell, $\mathbf{E}$ is a one-dimensional array of size N, where N is the total number of discrete elements comprising the discrete element system, i.e.

$$\mathbf{E} = \begin{bmatrix} e_1 & e_2 & e_3 & \cdots & e_i & \cdots & e_N \end{bmatrix} \qquad (3.30)$$

For the discrete element system shown in Figure 3.19, the representation of mapping of discrete elements onto cells is achieved by setting array $\mathbf{C}$ to point to the heads of singly connected lists for each of the cells, i.e.

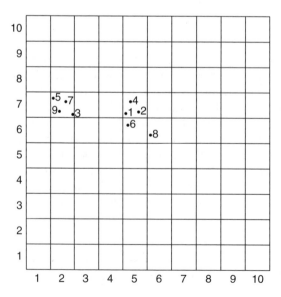

Figure 3.19 Mapping of discrete elements onto the cells for the screening contact detection algorithm.

$$C = \begin{bmatrix} -1 & -1 & -1 & -1 & -1 & -1 & -1 & -1 & -1 & -1 \\ -1 & -1 & -1 & -1 & -1 & -1 & 9 & -1 & -1 & -1 \\ -1 & -1 & -1 & -1 & -1 & -1 & -1 & -1 & -1 & -1 \\ -1 & -1 & -1 & -1 & -1 & -1 & -1 & -1 & -1 & -1 \\ -1 & -1 & -1 & -1 & -1 & 6 & 4 & -1 & -1 & -1 \\ -1 & -1 & -1 & -1 & -1 & 8 & -1 & -1 & -1 & -1 \\ -1 & -1 & -1 & -1 & -1 & -1 & -1 & -1 & -1 & -1 \\ -1 & -1 & -1 & -1 & -1 & -1 & -1 & -1 & -1 & -1 \\ -1 & -1 & -1 & -1 & -1 & -1 & -1 & -1 & -1 & -1 \\ -1 & -1 & -1 & -1 & -1 & -1 & -1 & -1 & -1 & -1 \end{bmatrix} \qquad (3.31)$$

The rest of the lists are represented by array **E**, indicating the next discrete element in the list, as shown in Figure 3.20. It is worth noting that most of the cells have no discrete elements mapped onto them. These cells are termed *empty cells*. The empty cells are represented by setting the corresponding number of array **C** to -1. The number -1 is also used to indicate the end of a singly connected list. Thus, the end of each singly connected non-empty list is indicated by setting the corresponding number of array **E** to -1. Array **E** shown in Figure 3.20 represents four singly connected lists. At the same time, array **C** represents 100 singly connected lists, out of which 96 are empty.

It can be seen that most of the numbers of array **C** represent empty lists. Thus, for space sparsely populated with discrete elements, array **C** may be very large. This is the reason for calling the algorithm the screen array algorithm. Array **C** 'screens' discrete elements into correct cells.

However, in situations where discrete elements are closely packed together, such as packing problems, most of the cells are not empty. In such cases, this algorithm makes sense. Otherwise, the RAM space needed to store array **C** may not always be available, or alternatively, the size of the problem (total number of discrete elements) may be limited by the available RAM space if expensive operations such as memory paging are to be avoided (see Chapter 9). Array **C** is a two-dimensional array. In 3D space it is a three-dimensional array. Accessing elements of such an array requires multiplication. Thus, in

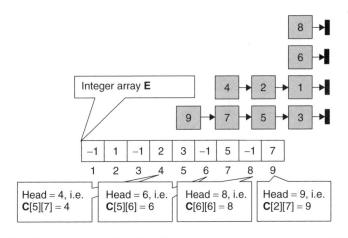

Figure 3.20 Singly connected lists for discrete elements system shown in Figure 3.19.

implementations using computer languages such as C or C++, which allow pointers, it is better to use dynamic, purpose-built arrays. More on this can be found in the section dealing with computational aspects of the combined finite-discrete element method in Chapter 9.

To achieve the maximum CPU efficiency of the algorithm, it is necessary to minimise the number of operations when mapping discrete elements onto the cells. Thus the mapping is done as follows:

Step 0: Initialisation: this is done by setting all numbers of arrays **C** and **E** to -1 (empty lists and/or ends of lists). This step is done only once, i.e. when the computer program is started:

Loop over all rows of cells$(i=1;\ i\leq n_{cel};\ i++)$
{   **Loop over all cells in a row**$(j=1;j\leq n_{cel};\ j++)$
   {   set

$$C[i][j] = -1 \tag{3.32}$$

   }
}
Loop over all discrete elements$(k=1;\ k\leq N;\ k++)$
{   set

$$E[k] = -1 \tag{3.33}$$

}

Step 1: Mapping of discrete elements onto cells: this operation is done each time discrete elements move, for instance at each time step. It involves building of a singly connected list for each cell. This is achieved by looping over all discrete elements, integerising its coordinates and placing the discrete element onto the corresponding singly connected list:

Loop over all discrete elements $(k=1;\ k\leq N;k++)$
{   **Integerise current coordinates using equation (26) and set**

$$i = {}^{int}x_k = 1 + Int\left(\frac{x_k - x_{min}}{d}\right) \tag{3.34}$$

$$j = {}^{int}y_k = 1 + Int\left(\frac{y_k - y_{min}}{d}\right)$$

   Place the discrete element onto the
   corresponding singly connected list by

setting $E[k] = C[i][j]$

and afterwards $C[i][j] = k$
$$\tag{3.35}$$

}

Further mapping of discrete elements onto the cells after the discrete elements have moved does not require initialisation. Instead, discrete elements are simply removed from the singly connected lists, and all singly connected lists are thus made empty before the mapping process (step 1) starts again.

Removing of discrete elements from the singly connected lists is done by looping over all discrete elements, integerising its coordinates and setting the corresponding number of array **C** and also the corresponding number of array **E** to -1, indicating empty lists:

Loop over discrete elements $(k=1; k \leq N; k++)$
{   Integerise current coordinates using equation (26) and set

$$i = {}^{int}x_k = 1 + Int\left(\frac{x_k - x_{min}}{d}\right)$$

$$j = {}^{int}y_k = 1 + Int\left(\frac{y_k - y_{min}}{d}\right)$$

(3.36)

Empty the corresponding singly connected list by

setting $\mathbf{E}[k] = -1$

and $\mathbf{C}[i][j] = -1$

(3.37)

}

Once the discrete elements have been mapped onto the cells, detection of contacts is performed. This involves looping over all discrete elements, and finding out which cell a particular discrete element is mapped onto. Once the particular cell has been identified, direct check against all discrete elements from the central and neighbouring cells (as shown in Figure 3.18) is performed:

Loop over discrete elements $(k=1; k \leq N; k++)$
{   Integerise current coordinates using equation (26) and set

$$i = {}^{int}x_k = 1 + Int\left(\frac{x_k - x_{min}}{d}\right)$$

(3.38)

$$j = {}^{int}y_k = 1 + Int\left(\frac{y_k - y_{min}}{d}\right)$$

if C*[i][j]*≤N

(3.39)

{ C*[i][j]*=C*[i][j]*+N
loop over all discrete elements from C*[i][j]*/**list**
{ loop over all discrete elements from neighbouring
cells, i.e. lists C*[i-1][j-1]*,C*[i][j-1]*,
C*[i+1][j-1]*,C*[i-1][j]*,C*[i][j]*
{ direct check for contact between discrete elements
} } }
}

Note that no loop over cells is used. Loops over discrete elements are used instead. In this way, the total number of algebraic operations involved is made independent of the total number of cells, i.e. independent of the size of array **C**.

In summary, the screening contact detection algorithm has the following characteristics:

- It has CPU time proportional to the total number of discrete elements.
- It is relatively simple to implement.
- Extension to 3D is straightforward.
- It can be very demanding in terms of RAM space, and is therefore suitable only for simulations involving very densely packed discrete elements in either static or transient dynamic problems.

3.7 SORTING CONTACT DETECTION ALGORITHM FOR DISCRETE ELEMENTS OF A SIMILAR SIZE

The screening algorithm works well for discrete element systems with large spatial density of discrete elements. However, if discrete elements are spaced at relatively large distances from each other, the size of the screening array may be very large, and may result in either large systems not fitting into the available RAM space or CPU overheads (virtual memory, paging, etc.) being too large. Thus, there is a need to either reduce the size of the screening array or eliminate it all together.

One way of eliminating the screening array is by using sorting arrays instead of a screening array. One array for each coordinate axis is used together with one array for a discrete element identifier. Thus, for cells in the x-direction, the $\mathbf{X}$ array is used, while for cells in the y-direction, a $\mathbf{Y}$ array is used. Both $\mathbf{X}$ and $\mathbf{Y}$ arrays are of size N, where N is the total number of discrete elements comprising the discrete element system. A third array $\mathbf{D}$ indicates the discrete element number, and is also of size N. Extension to 3D would require an additional $\mathbf{Z}$ array.

The contact detection is done in three steps:

Step 1: Mapping of discrete elements onto cells: discrete elements are mapped onto cells through the integerisation of coordinates:

$$i = {}^{int}x = 1 + Int\left(\frac{x - x_{min}}{d}\right) \tag{3.40}$$

$$j = {}^{int}y = 1 + Int\left(\frac{y - y_{min}}{d}\right)$$

where i and j indicate a cell to which the discrete element of the current centre coordinates x and y is mapped, while operation Int simply indicates truncation of a real number.

For the discrete element system shown in Figure 3.21, mapping of discrete elements onto cells is represented by arrays $\mathbf{X}$, $\mathbf{Y}$ and $\mathbf{D}$, as shown in Figure 3.22. Array $\mathbf{D}$ indicates the discrete element, while array $\mathbf{X}$ indicates the integerised coordinate x and array $\mathbf{Y}$ indicates the integerised coordinate y of the centre of the bounding box of the discrete element, i.e. arrays $\mathbf{X}$ and $\mathbf{Y}$ indicate the cell onto which a particular element is mapped. For instance, the discrete element 1 is mapped onto cell (4,7), while discrete element 3 is mapped onto cell (4,6) and discrete element 7 is mapped onto cell (3,7). Thus,

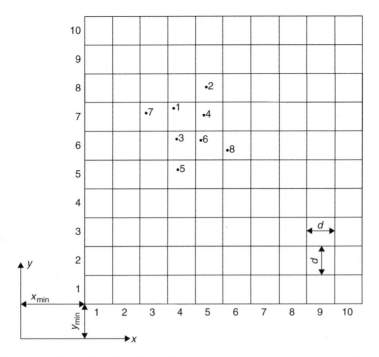

Figure 3.21 Space divided into identical cells large enough to contain the largest discrete element comprising the system. Centres of discrete elements are marked with dots; the thicker line marks the bounding box of physical space.

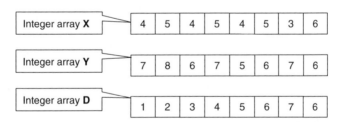

Figure 3.22 mapping of discrete elements onto cells for discrete element system shown in Figure 3.21.

array **D** contains the 'name' of the discrete element, and arrays **X** and **Y** contain the corresponding cell.

Step 2: Sorting of arrays representing mapping: after discrete elements have been mapped onto cells, arrays **X**, **Y** and **D** are obtained. Array **X** contains the integerised current x coordinates of all the discrete elements; array **Y** contains the integerised current y coordinates of all the discrete elements; and array **D** contains the discrete element numbers. It is worth noting that array **D** contains sequence $1, 2, 3, \ldots, N$, where N is the total number of discrete elements comprising the combined finite-discrete element system. Arrays **X** and **Y** contain a more or less random series of integer numbers, as shown in Figure 3.22.

For array **D** to contain a meaningful representation of spatial distribution of discrete elements, all three arrays are sorted relative to the integerised x and y coordinates, i.e. relative to the content of arrays **X** and **Y**.

The first sorting relative to the contents of array X is performed as shown in Figure 3.23.

In the case where the content of array **X** is the same, sorting relative to the content of array **Y** is performed as shown in Figure 3.25.

After the sorting operation is completed, no information is lost. Array **D** still contains the discrete element 'names' (i.e. numbers), while arrays **X** and **Y** contain the cell onto which a particular discrete element is mapped. For instance, after the sorting operation, it is still evident from arrays **D, X** and **Y** that discrete element 1 is mapped onto cell (4,7), while the discrete element 3 is mapped onto cell (4,6) and discrete element 7 is mapped onto cell (3,7) (Figure 3.24).

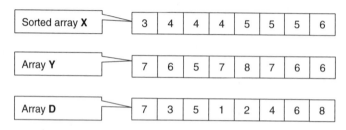

Figure 3.23   Arrays sorted relative to content of array **X**.

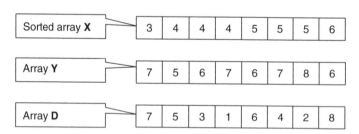

Figure 3.24   Sorting arrays relative to the content of array **Y**.

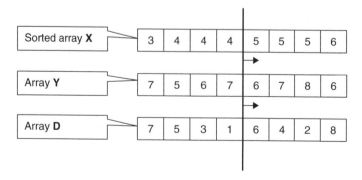

Figure 3.25   First stage of the binary search for cell (5,7).

The following formula is used in sorting arrays:

$$\text{discrete element } i \text{ is before discrete element } j \text{ if} \qquad (3.41)$$

$$^{int}x_i < {}^{int}x_j$$

or if

$$^{int}x_i = {}^{int}x_j \text{ and } {}^{int}y_i < {}^{int}y_j$$

Step 3: Binary search of sorted arrays for contact: after the sorting process, array **D** acquires a spatial dimension. Discrete elements closer to the beginning of array **D** are positioned closer to the first column of cells, while the discrete elements closer to the end of array **D** are positioned closer to the last column of cells. Finding any particular cell (i, j) involves searching the arrays.

To detect contacts for particular discrete elements, it is enough to search for cells neighbouring the cell to which a particular element is mapped. Once the neighbouring cells are identified, a contact check is performed between the particular discrete element and all the discrete elements from neighbouring cells and the central cell.

Detection of contacts is done by looping over discrete elements:

Loop over discrete elements $(k=1; k\leq N; k++)$
{   Integerise current coordinates of the discrete element
and identify the cell onto which it is mapped (central cell)

$$i = {}^{int}x_k = 1 + Int\left(\frac{x_k - x_{min}}{d}\right) \qquad (3.42)$$

$$j = {}^{int}y_k = 1 + Int\left(\frac{y_k - y_{min}}{d}\right)$$

Use binary search to search the sorting arrays to find
the central and neighbouring cells
check for contact against discrete elements in these cells
}

Once the central cell (i, j) for a particular discrete element k is identified, the neighbouring cells are simply: $(i - 1, j - 1), (i, j - 1), (i + 1, j - 1)$ and $(i - 1, j)$. All discrete elements mapped onto these cells have to be found by searching arrays **X, Y** and **D**. This can be done using the binary search.

The first stage of the search process for a particular cells is shown in Figure 3.25. The arrays are searched for cell (5,7). Thus, in Figure 3.25, arrays **X, Y** and **D** are divided into two arrays of equal (or near equal) size. The part that does not contain the cell that is being searched for is rejected.

The remaining parts of arrays **X, Y** and **D** (non-shaded area in Figure 3.26) are further divided into two equal parts, as shown in Figure 3.26. Again, the part that does not contain cell (5,7) is rejected.

Further subdivision (shown in Figure 3.27) finally identifies the cell that is being searched for, i.e. cell (5,7). It is evident from array **D** that this cell contains the discrete element identified as discrete element number 4.

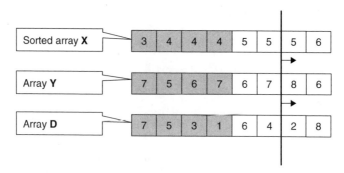

Figure 3.26   Second stage of the binary search for cell (5,7).

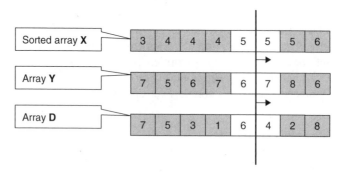

Figure 3.27   Third (final) stage of the binary search for cell (5,7).

Once a particular cell and discrete element are identified, a direct check for contact can be performed.

A search for a particular cell in general requires b steps, where

$$b = \log_2 N \tag{3.43}$$

and N is the total number of discrete elements comprising the system, i.e. the size of arrays **X, Y** and **D**.

In the example given, N was 8, thus the search for cell (5,7) was completed in three steps:

$$b = \log_2 N = \log_2 8 = 3 \tag{3.44}$$

The sorting contact detection algorithm comprises three steps. The first step (mapping of discrete elements onto the cells) involves a constant number of operations per discrete element. The second step (sorting of integer arrays) involves the total number of operations per discrete element proportional to

$$\log_2 N \tag{3.45}$$

The third step (searching for particular cells for each discrete element) again involves the total number of operations per discrete element proportional

$$\log_2 N \tag{3.46}$$

Thus, the total number of operations necessary to perform all three steps (i.e. to detect all the contacts for a single discrete element) is proportional to $\log_2 N$.

The total CPU time for detecting all the contacts is therefore given by

$$T \propto N \log_2 N \tag{3.47}$$

The total RAM requirements of the sorting contact detection algorithm are proportional to the total number of discrete elements N, and do not depend upon how the discrete elements are distributed (packed) in space. In fact, the total RAM requirements are equal to $3N$ integer numbers. In 3D space these increase to $4N$ integer numbers, which is usually 16 bytes per discrete element.

The CPU performance of the sorting algorithm is not a function of spatial distribution of the discrete elements. Arrays **X**, **Y** and **D** are the same regardless of whether discrete elements are close to each other or far away. For instance, if the only change to the discrete element system is the doubling of the distance between discrete elements (i.e. doubling the size of the edge of the space bounding box), the search operations together with arrays **X**, **Y** and **D** would not change at all.

However, to achieve the above described CPU and RAM performance, it is necessary to develop an efficient procedure for sorting arrays **X**, **Y** and **D**. There are a number of ways to sort these arrays. The procedure described below does not require any additional RAM space, and achieves sorting in CPU time proportional to

$$T_{sort} \propto N \log_2 \tag{3.48}$$

The procedure is explained using the example given in Figure 3.28.

The first step in sorting is to identify the maximum and minimum number in the array, as shown in Figure 3.29. This operation requires a CPU time proportional to the total number of discrete elements N.

The next step is to calculate the middle point between the identified maximum and minimum number:

$$x_{mid} = \frac{x_{min} + x_{max}}{2} = \frac{1 + 16}{2} = 8 \tag{3.49}$$

Now the array is parsed from both sides simultaneously. Each number is compared to the middle point:

$$x_{mid} = 8 \tag{3.50}$$

| 3 | 16 | 7 | 11 | 2 | 14 | 5 | 9 | 4 | 8 | 12 | 6 | 13 | 15 | 1 | 10 |

Figure 3.28   A hypothetical array to be sorted.

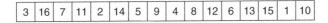

Figure 3.29   Identifying the maximum and minimum number.

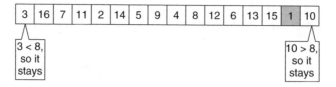

Figure 3.30   Sorting concept.

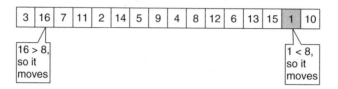

Figure 3.31   Number 16 and 1 need to swap places.

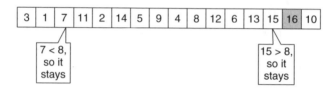

Figure 3.32   No swapping needed.

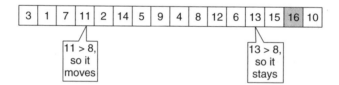

Figure 3.33   No pair to swap exists yet.

The first number is supposed to be smaller than 8, while the second is supposed to be greater than 8, as shown in Figure 3.30.

If this is not the case, as shown in Figure 3.31, the two numbers are swapped.

The process is repeated until a new pair of numbers that require swapping is found (Figures 3.32, 3.33).

The swapping sequence is continued as shown in Figures 3.35–3.37, finally resulting in the original array being split into two sub-arrays, as shown in Figure 3.38. The first array has all numbers smaller than the medium number (i.e. smaller than 8), and the second array contains all numbers grater than 8.

The total number of operations performed so far is proportional to the size of the array, i.e. to the total number of discrete elements N. Now the array is split into two arrays, as shown in Figure 3.38. Array $X_{1,A}$ contains all numbers smaller than 8, while array $X_{1,B}$ contains all numbers grater or equal to 8.

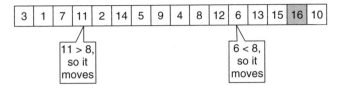

Figure 3.34   A new pair that needs to be swapped is found (6 and 11 need to swap places).

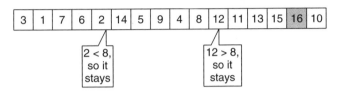

Figure 3.35   No swapping needed.

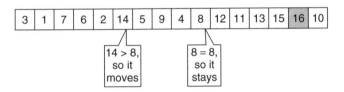

Figure 3.36   Swapping needed, but not possible yet.

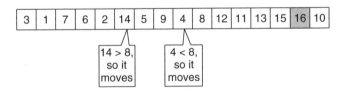

Figure 3.37   The new pair to be swapped is identified.

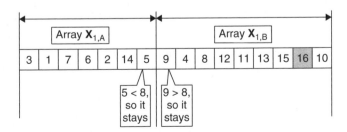

Figure 3.38   The swapping sequence has resulted in two sub-arrays.

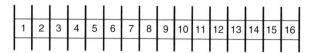

Figure 3.39   The final stage of the sorting process.

The process of identifying the largest and smallest number followed by the swapping procedure as described above is now repeated for each of the sub-arrays in turn, until each of them is split into two arrays.

This process of splitting arrays into two array is recursively performed until the size of the individual arrays is reduced to a single number (Figure 3.39). At this point array **X** is sorted.

It is evident that the total number of successive subdivisions is equal to

$$b = \log_2 N \tag{3.51}$$

During each subdivision, the total number of operations is proportional to N. Thus, the total number of algebraic operations necessary to sort an array of size N is proportional to $N \log_2 N$, and the total CPU time necessary to sort the array is therefore given by

$$T_s = N \log_2 N \tag{3.52}$$

In summary, the sorting contact detection algorithm has the following properties:

- The total CPU time is not dependent on the spatial distribution of discrete elements. Thus, the algorithm has the same performance for both dense and loose packs.

- The total RAM requirements are not a function of the spatial distribution of discrete elements.

- The total CPU time is proportional to

$$T \propto N \log_2 N \tag{3.53}$$

- The total RAM space needed is given by

$$M = 3N; \text{ in 2D}$$

$$M = 4N; \text{ in 3D, i.e.} \tag{3.54}$$

$$M \propto N$$

3.8   MUNJIZA-NBS CONTACT DETECTION ALGORITHM IN 2D

The screening contact detection algorithm is very efficient in terms of CPU. However, memory requirements may be very significant, and may emerge as a limiting factor in determining the size of the problem that can be solved on a particular computer. The sorting algorithm is very efficient in terms of the size of arrays, i.e. in terms of RAM requirements. However, it is not as efficient in terms of the CPU time required to solve a particular problem, because the CPU time is greater than the CPU time required by either the screening or binary tree based contact detection algorithms. In addition, the CPU time is not a linear function of the total number of discrete elements.

The aim of this section is to present an algorithm that is as efficient as the screening algorithm in terms of CPU, and as efficient as the sorting algorithm in terms of RAM requirements. The algorithm was developed in 1995 by Munjiza, and is called the Munjiza No Binary Search (i.e. Munjiza-NBS) contact detection algorithm.

The Munjiza-NBS contact detection algorithm assumes a simplified contact detection problem in 2D, i.e. a system comprising N identical discrete elements occupying a finite space of rectangular shape (Figure 3.40), although extensions to non-identical discrete elements are possible. The task is to find all the discrete element couples that are close to each other in a sense that the distance between their closest points is less than or equal to zero, in other words, that they overlap or touch.

3.8.1 Space decomposition

The NBS contact detection algorithm is based on space decomposition. The space is subdivided into identical square cells of size d (Figure 3.40).

For the sake of clarity, each discrete element is assigned an integer identification number $1, 2, 3, \ldots, N-1, N$. In a similar way, each cell is assigned an identification couple of integer numbers (i_x, i_y), where $i_x = 1, 2, 3 \ldots n_x$ and $i_y = 1, 2, 3 \ldots n_y$, where n_x and n_y are the total number of cells in the x and y directions, respectively:

$$n_x = \frac{x_{\max} - x_{\min}}{d} \tag{3.55}$$

and

$$n_y = \frac{y_{\max} - y_{\min}}{d} \tag{3.56}$$

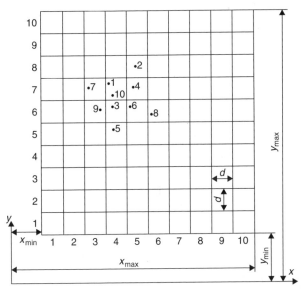

Figure 3.40 Space divided into identical cells large enough to contain the largest discrete element comprising the system. Centres of discrete elements are marked with dots. Thicker line marks the bounding box of physical space.

3.8.2 Mapping of discrete elements onto cells

Mapping from the set of discrete elements

$$E = \{1, 2, 3, 4, 5, \ldots, N\} \tag{3.57}$$

to the set of cells

$$C = \left\{ \begin{array}{cccccc} (1, 1), & (1, 2), & (1, 3), & \ldots & (1, n_y), \\ (2, 1), & (2, 2), & (2, 3), & \ldots & (2, n_y), \\ (3, 1), & (3, 2), & (3, 3), & (3, 4), & (1, n_y), \\ \ldots & \ldots & \ldots & \ldots & \ldots \\ (n_x, 1), & (n_x, 2), & (n_x, 3), & \ldots & (n_x, n_y) \end{array} \right\} \tag{3.58}$$

is introduced and defined in such a way that each discrete element is assigned to one and only one cell. This is done through each particular discrete element with coordinates (x, y) being assigned to the cell (i_x, i_y), where

$$i_x = Int \left(\frac{x - x_{\min}}{d} \right) \tag{3.59}$$

$$i_y = Int \left(\frac{y - y_{\min}}{d} \right)$$

In essence, i_x and i_y are integerised relative coordinates of the centre of the bounding circle for each discrete element, and are thus referred to as integerised coordinates. An example is given in Figure 3.40, where for instance discrete element 1 is assigned to cell (4,7), while discrete element 2 is assigned to cell (5,8), and discrete element 3 is assigned to cell (4,6).

3.8.3 Mapping of discrete elements onto rows and columns of cells

In addition to the mapping of discrete elements onto cells, mapping of discrete elements onto columns and rows of cells is also introduced.

A discrete element is said to be mapped to a particular row of cells if it is mapped to any cell from that row. For instance, discrete element 1 is mapped to row 7 of cells, discrete element 2 is mapped to row 8, and discrete element 3 is mapped to row 6 of cells (Figure 3.40).

In a similar way, a discrete element is said to be mapped to a particular column of cells if it is mapped to any cell from that column. For instance, discrete element 1 is mapped to the column 4 of cells, discrete element 2 is mapped to column 5 and discrete element 3 is mapped to column 4 of cells (Figure 3.40).

3.8.4 Representation of mapping

In the previous sections, the binary tree, screening arrays and sorting arrays have all been used to represent different types of mapping between discrete elements and cells. In the case of the Munjiza-NBS algorithm, linked lists are used instead. This is what makes the Munjiza-NBS algorithm very efficient both in terms of RAM and CPU requirements. Linked lists reduce memory requirements significantly.

To further reduce CPU time, singly connected lists are used. The representation of mapping is performed in two stages:

STAGE 1: Mapping of all discrete elements onto the rows of cells (*y*-direction) is performed, and a singly connected list of discrete elements for each row i_y (where $i_y = 1, 2, 3 \ldots n_y$) is formed (Figure 3.41). There exist n_y of these lists in total. Some of the lists are empty, others contain only one discrete element, etc. These lists are referred to as the *y*-lists, and each *y*-list is identified by the row of cells it is formed for, and is thus referred to as the y_{i_y} list. For instance, the y_{10} list is empty, the y_7 list contains discrete elements 7, 4 and 1, etc.

It is worth noting that the discrete elements in each of the lists follow each other in a descending numerical order. For instance, the first discrete element on the list y_6 is discrete element 9, while the last discrete element on this list is discrete element 3.

This is because the discrete elements are mapped using a loop over discrete elements in ascending numerical order, i.e.

Loop over all discrete elements $(i{=}1;\ i{\le}N;i{+}{+})$
  {   **calculate integerised coordinate**

$$i_y = Int \left(\frac{y - y_{min}}{d} \right) \tag{3.60}$$

    and place the discrete element onto the y_{i_y} list

}

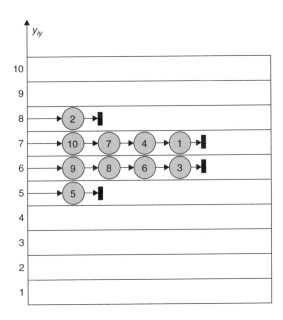

Figure 3.41   Singly connected lists for rows of cells.

Thus, the list for row $i_y = 6$ (i.e. the list y_6) in Figure 3.41 was assembled by placing discrete element 3 onto the list. It is then 'pushed' by discrete element 6, which in turn is 'pushed' by discrete element 8, which in turn is 'pushed' by discrete element 9, which was the last discrete element added to the list.

Digital representation of all y-lists (empty and non-empty) is achieved through two integer arrays. The arrays are common for all the y-lists, and contain enough information to read all the discrete elements from each of the lists.

The first array **B** contains the number of the last discrete element mapped to each row, i.e. the head of the singly connected list for each row of cells. The array **B** is a 1D array of size n_y, where n_y is the total number of cells in the y direction, i.e. the total number of rows of cells.

The second array **Y** is 1D array of size N, where N is the total number of discrete elements. For each discrete element, the array **Y** contains the next discrete element in the singly connected list.

For both arrays a negative number is used as termination of a singly connected list. So if there are no elements in a particular row, a negative number is assigned to the corresponding element of the array **B**.

Both arrays are shown in Figure 3.42, where the list y_6 is highlighted. It is evident that the array **B** contains a 'pointer' to a singly connected list of discrete elements for each i-row (where $i_y = 0, 1, 2, 3, \ldots, n_y - 1$) of cells. In a similar way, the array **Y** points to the next element in the list, terminating with a negative number.

For the example shown in Figure 3.40, row 1 of cells has no discrete elements mapped to it, so a singly connected list of discrete elements mapped to this row (the list y_0) is empty, which is achieved by setting **B**[1] $= -1$.

In a similar way, the last discrete element mapped onto row 6 of cells (the list y_6) is discrete element number 9, thus **B**[6] $= 9$. The next discrete element on this list is 8, thus **Y**[9] $= 8$; the next discrete element is 6, thus **Y**[8] $= 6$; the next discrete element is 3, thus **Y**[6] $= 3$; the last discrete element is 3, thus **Y**[3] $= -1$, (Figure 3.40).

All y_{iy} lists are considered as 'new lists' at this point. Later on, some of these lists will be marked as 'old lists'.

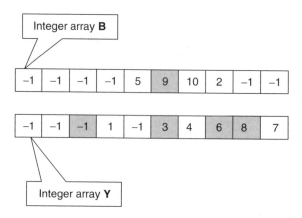

Figure 3.42   All singly connected lists for rows of cells represented by two arrays.

STAGE 2: Mapping of discrete elements to individual cells: For each row of cells, discrete elements are now ready to be mapped onto the individual cells. For each row of cells, the discrete elements mapped onto that row are all listed in singly connected lists, as explained in Stage 1. The head of the list is represented by the corresponding number $\mathbf{B}[i_y]$ for the y_{iy} list (the list containing discrete elements mapped onto the row i_y).

It would be very inefficient to consider all of the lists. Only non-empty lists are thus considered. This is achieved by looping over all discrete elements, instead of looping over rows of cells:

Loop over all discrete elements $(i=1; i \leq N; i++)$
{   calculate integerised coordinate

$$i_y = Int\left(\frac{y - y_{\min}}{d}\right) \tag{3.61}$$

if the y_{iy} list is marked as "new":
{   mark it as "old"
map discrete elements from the corresponding y-list
onto the cells and use these for detection of contacts

}

}

As a so-called 'new' y-list is detected, it is first marked as a so-called 'old' list, so that no other discrete element will be considering this row of cells again.

Each discrete element from the current y_{iy}-list is mapped onto the cells by performing a loop over all discrete elements from the y_{iy} list:

Loop over all discrete elements from the list y_{iy}
$(j=\mathbf{B}[i_y]$ *and* $j=\mathbf{Y}[j]$ *until the end of the list is reached, i.e. until j=-1)*
{   calculate integerised coordinate for the discrete element j

$$i_x = Int\left(\frac{x - x_{\min}}{d}\right) \tag{3.62}$$

add the discrete element j onto $x_{ix,iy}$ list,
i.e. onto the list of discrete elements mapped onto the cell *(ix,iy)*
consider all $x_{ix,iy}$ list as "new list"

}

A particular $x_{ix,iy}$ list contains all discrete elements with integerised coordinates i_x and i_y. In addition, all singly connected lists $x_{ix,iy}$ (where $i_x = 1, 2, 3, \ldots n_x$) contain all discrete elements from the y_{iy}-list, and are represented by two arrays of integer numbers.

The first array is a 1D array $\mathbf{A}$ of size n_x, where n_x is the total number of cells in the x-direction, i.e. the total number of columns of cells. The second array is a 1D array $\mathbf{X}$ of size N, where N is the total number of discrete elements comprising the combined finite-discrete element system.

Singly connected lists $x_{ix,iy}$ (where $i_x = 0, 1, 2 \ldots n_x$), for example from Figure 3.42, are shown in Figure 3.43. Numerical representation of those lists by arrays of integer numbers is shown in Figure 3.44, where the list $x_{4,7}$ is highlighted.

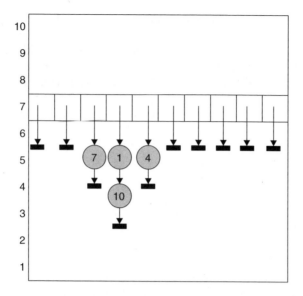

Figure 3.43   All x-lists of discrete elements for row 7 of cells.

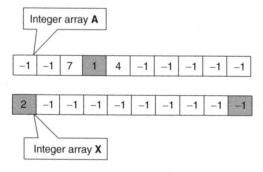

Figure 3.44   Representation of all singly connected lists for row 7 of cells using two integer arrays.

Discrete elements are mapped by looping over all the discrete elements in ascending order.

List y_7 (Figure 3.42) contains all discrete elements in row 2 of cells, i.e. discrete elements 10, 7, 4 and 1. No discrete element from list y_2 has an integerised coordinate i_x equal to 1 or 2, thus singly connected lists $x_{1,7}$ and $x_{2,7}$ are empty, i.e. $A[1] = -1$ and $A[2] = -1$. Only discrete element 4 has an integerised coordinate i_x equal to 5, thus $A[5] = 4$ and $X[4] = -1$. Only discrete element 7 has an integerised coordinate i_x equal to 3, thus $A[3] = 7, X[7] = -1$. Discrete elements 1 and 10 have integerised coordinates i_x equal to 4, thus $A[4] = 1, X[1] = 10, X[10] = -1$.

It is important to note how lists y_{iy} and x_{ix} are assembled:

- First a loop over all discrete elements is performed, and inside the loop a particular discrete element j is added to the corresponding y_{iy} list, depending on its integerised coordinate i_y.

- Secondly, a loop over all discrete elements from a particular y_{iy} list is performed, and inside this loop a particular discrete element is added to the corresponding $x_{ix,iy}$ list, depending on its integerised coordinate i_x.

To assemble new lists, discrete elements are removed from the old lists in a similar fashion—a loop over all discrete elements is performed, and for each discrete element, $B[i_y] = -1$ is set, and in a similar way, a loop over all discrete elements from a particular y_{iy} list is performed, and for each discrete element, $A[i_x] = -1$ is set.

Thus, no loop over cells is involved in assembling the lists, which leads to the conclusion that the total CPU time needed to perform all the operations described so far in this section is proportional to the total number of discrete elements N, and is neither a function of the total number of cells in the x-direction n_x, nor a function of the total number of cells in the y-direction n_y. In other words, it is not a function of packing density ρ.

Detection of contact. Detection of contact is accomplished by checking all the discrete elements mapped to a particular cell against all discrete elements in neighbouring cells. For instance, discrete elements mapped to the cell (i_x, i_y), shown in Figure 3.45, are checked for contact against all discrete elements mapped to cells (i_x, i_y), $(i_x - 1, i_y)$, $(i_x - 1, i_y - 1)$, $(i_x, i_y - 1)$ and $(i_x + 1, i_y - 1)$. This is equivalent to checking all discrete elements from list $x_{ix,iy}$ against all discrete elements from lists $x_{ix,iy}, x_{ix-1,iy}, x_{ix-1,iy-1}, x_{ix,iy-1}$ and $x_{ix+1,iy-1}$.

In this way, discrete elements mapped to any non-empty cell are checked against all discrete elements mapped to neighbouring cells (only discrete elements from neighbouring cells can touch each other).

Thus, it is necessary at any given time to have singly connected lists $x_{ix,iy}$ only for two neighbouring rows of cells i_y and $i_y - 1$, i.e. for discrete elements from lists y_{iy} and y_{iy-1}. Thus, two parallel arrays A are needed, and this is accomplished through a 2D array $A[2][n_x]$ of size $2n_x$ integer numbers. The array $A[0]$ points to all singly connected lists $x_{ix,iy}$ (where i_y is fixed and $i_x = 0, 1, 2, \ldots, n_x$). The 1D array $A[1]$ points to all singly connected lists comprising discrete elements mapped onto row $i_y - 1$ of cells, i.e. all lists $x_{ix,iy-1}$ (where $i_y - 1$ is fixed and $i_x = 0, 1, 2, \ldots, n_x$).

Detection of contact is performed only for cells that have one or more discrete elements mapped to them, i.e. for cells with a non-empty $x_{ix,iy}$ list of discrete elements. This is accomplished by employing a loop over discrete elements from list y_{iy} to find a cell

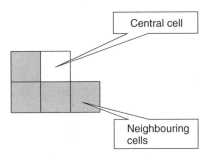

Figure 3.45   Contact checking mask.

(i_x, i_y) with one or more discrete elements assigned to it, i.e. a 'new' $x_{ix,iy}$ list. Discrete elements mapped to each such cell are checked for contact against all discrete elements mapped to neighbouring cells, as shown in Figure 3.45. For instance, for the list y_7 from Figure 3.41, the loop over all discrete elements from the list will include discrete elements 10, 7, 4 and 1, resulting in 'new' lists $x_{3,7}, x_{4,7}$ and $x_{5,7}$ being found. Thus, all discrete elements from lists $x_{3,7}, x_{4,7}$ and $x_{5,7}$ are checked for contact against discrete elements from lists of discrete elements mapped to neighbouring cells.

It is worth noting that no loop over cells is performed for any operation described so far, which leads to the conclusion that the total CPU time needed to perform detection of contact described is independent of either the total number of cells in the x-direction n_x, or the total number of cells in the y-direction n_y.

Implementation. The implementation of the NBS algorithm in the form of pseudo-code can be summarised as follows:

1. Loop over all discrete elements$(k=0;k\leq N;k++)$
{   calculate integerised y coordinates of its centre
place the discrete element
onto a list for the corresponding row of cells (y-list)
}
consider all y-lists to be "new" lists
2. Loop over discrete elements$(k=0;k\leq N;)$
{   calculate integerised y coordinates of its centre
if discrete element belongs to a new y-list
{   mark the y-list as an old list and call it central y-list
{   3. loop over all discrete elements from central y-list
{   integerise x coordinate of the discrete element and place
it onto the corresponding x-list
}
4. loop over all discrete elements from neighbouring (y-)-list
{   integerise x coordinate of the discrete element and place
discrete element onto the corresponding x-list
}
5. loop over all discrete elements from the central y-list
{   if the discrete element belongs to a new list (x,y) list
{   mark the list (x,y) as old and call it central (x,y)-list
check for contact between discrete elements from
the central list (x,y) and lists
$(x,y),(x-1,y),(x-1,y-1),(x,y-1),(x+1,y-1)$
}
}
}
6. loop over all discrete element from the central y-list
{   remove the corresponding x-list
i.e. set the head of the list to zero
}
7. loop over all discrete element from list (y-1)
{   remove the corresponding x-list,

            i.e. set head of the list to zero

       }

    }

  }

8. loop over all discrete elements
  {   **remove the corresponding** y**-lists i.e. set the head of the list to zero**
  }

CPU and RAM requirements. Detailed analysis of the Munjiza-NBS contact detection algorithm as explained in the previous section leads to the conclusion that the total detection time is proportional to the total number of discrete elements present:

$$T \propto N \qquad (3.63)$$

In other words, the total CPU time needed to detect all the contacts is given by

$$T \propto cN \qquad (3.64)$$

where c is a constant independent of either packing density or number of discrete elements.

    This can be proved by analysing each step of the algorithm presented in the previous section.

    To make all singly connected lists y_{iy} (where $i_y = 1, 2, 3, \ldots, n_y$), a loop over all elements is involved. Inside the loop the current discrete element is added to the corresponding singly connected list, as specified by its integerised coordinate i_y.

    To make all singly connected lists $x_{ix,iy}$ (where $i_x = 1, 2, 3, \ldots, n_x$), for a particular value of i_y, a loop over discrete elements from the corresponding y_{iy} list is employed, and inside the loop the current discrete element is added to the corresponding singly connected list $x_{ix,iy}$, as specified by its integerised coordinate i_x.

    To find a 'new list' y_{iy}, a loop over all discrete elements is performed, while to find a 'new list' $x_{ix,iy}$, a loop over discrete elements from the y_{iy} list is performed. It is worth noting that a loop over all y_{iy} lists is equivalent to a single loop over all discrete elements, because each y_{iy} list contains only discrete elements mapped onto one row of cells, while all y_{iy} lists together contain all discrete element comprising the combined finite-discrete element problem.

    At no place is a loop over cells employed; loops over discrete elements are employed instead. Those consist of:

- Three loops over all discrete elements, i.e. loops 1, 2 and 8.

- Three loops over discrete elements from central y lists, i.e. loops 3, 5 and 6 (as explained above, all these loops are in total equivalent to three loops over all discrete elements).

- Two loops over discrete elements from neighbouring $y - 1$ list, i.e. loops 4 and 7 (as explained above, all these loops are in total equivalent to two loops over all discrete elements).

In total, an equivalent of eight loops over all discrete elements is performed. Operations performed inside the loops do not depend upon either the number of cells in the x-direction n_x, or the number of cells in the y-direction n_y. This leads to the conclusion that the total CPU time to detect all contacts is not dependent on the number of cells. In other words, it is not dependent on the size of the finite space within which discrete elements are distributed. Operations performed inside the loops do not depend upon the number of discrete elements N either. The consequence of this is that the total CPU time taken to detect all the contact is proportional to the total number of discrete elements.

RAM requirements in terms of required RAM space M related to contact detection are easily calculated as the total space occupied by arrays $\mathbf{A}[2][n_x]$, $\mathbf{B}[n_y]$, $\mathbf{X}[N]$ and $\mathbf{Y}[N]$. Thus

$$M = n_y + 2n_x + 2N \text{ integer numbers} \tag{3.65}$$

In FORTRAN-based implementation integer arrays are used, as explained above, resulting in total memory requirements of approximately 20 bytes per discrete element. In C and C++ implementations, very often arrays of pointers are used in place of integer arrays. For instance, a particular discrete element may not be identified by its number but by its address. Arrays $\mathbf{X}$ and $\mathbf{Y}$ therefore contain the address of the next discrete element in the list, and are accordingly the array of addresses (pointers). Arrays $\mathbf{A}$ and $\mathbf{B}$ contain the address of the first discrete element in the list, and are consequently arrays of pointers. RAM space can be saved by using pointers. In addition, C codes run faster when pointers are used.

Numerical examination of CPU performance. The CPU and RAM requirements for the NBS algorithm can be demonstrated by numerical examples.

Example I consists of N circular discrete elements of diameter D spaced at distance D in the x-direction and at a distance 2D in the y-direction (Figure 3.46).

Contact detection is solved 10 times for the problem, and each time all contacting couples are detected. The cumulative CPU times for all 10 contact detections as a function of the total number of discrete elements comprising the problem are shown in Figure 3.47.

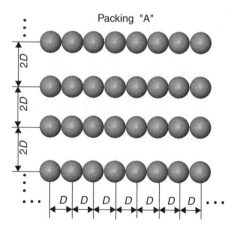

Figure 3.46   Example I: packing A (NBS Contact Detection Algorithm for bodies of similar size, A. Munjiza and K.R.F. Andrews, *International Journal for Numerical Methods in Engineering*, 43/1). (Reproduced by permission of John Wiley & Sons, Ltd.).

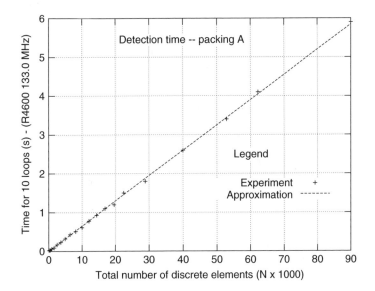

Figure 3.47 Example I: CPU time as a function of the total number of discrete elements (A. Munjiza and K.R.F. Andrews, *International Journal for Numerical Methods in Engineering*, 43/1). Reproduced by permission John Wiley & Sons, Ltd).

The results shown are obtained by changing N from $N = 100$ to $N = 90,000$. The results accurately fit a linear relation, which confirms that the total detection time is indeed proportional to the total number of discrete elements.

Example II consists of $N = 10,000$ circular discrete elements of diameter D spaced at distance $2D$ in the y-direction and at variable distance S (spacing) in the x-direction (packing 'B'), (Figure 3.48).

The packing density ρ changes with spacing S as

$$\rho \propto \frac{1}{S} \tag{3.66}$$

i.e. the packing density is reduced by increasing spacing S. In Figure 3.49, the total CPU time for 10 repeated detections of all contacts is shown as a function of spacing.

The results obtained in this numerical experiment confirm that the total detection time does not depend upon packing density (which is the same as predicted in Chapter 6).

Example III consists of $N = 10,000$ circular discrete elements of diameter D spaced at distance S in the x-direction and at distance S in the y-direction. (Figure 3.50).

The packing density ρ changes with spacing S as

$$\rho \propto \frac{1}{S^2} \tag{3.67}$$

Contact detection for various values of S is repeated 10 times, and the total CPU time for all 10 detections of all contacts is recorded as shown in Figure 3.51.

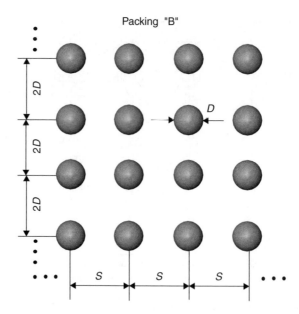

Figure 3.48 Example II, packing B (A. Munjiza and K.R.F. Andrews, *International Journal for Numerical Methods in Engineering*, 43/1). (Reproduced by permission John Wiley & Sons, Ltd).

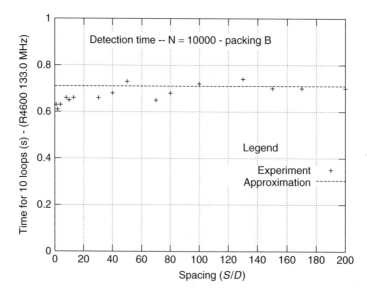

Figure 3.49 Example II: total CPU time as function of packing density for pack B (A. Munjiza and K.R.F. Andrews, *International Journal for Numerical Methods in Engineering*, 43/1). (Reproduced by permission John Wiley & Sons, Ltd).

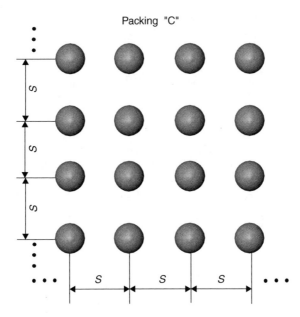

Figure 3.50   Example III, packing C (A. Munjiza and K.R.F. Andrews, *International Journal for Numerical Methods in Engineering*, 43/1). (Reproduced by permission of John Wiley & Sons, Ltd).

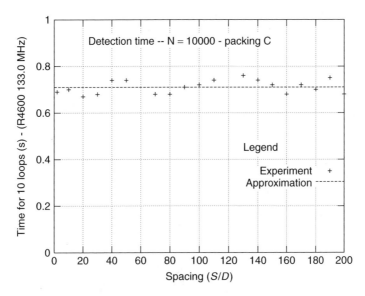

Figure 3.51   Example III: total CPU time as a function of packing density for pack C (A. Munjiza and K.R.F. Andrews, *International Journal for Numerical Methods in Engineering*, 43/1). (Reproduced by permission of John Wiley & Sons, Ltd).

The results obtained show that the total detection time is not dependent on packing density, which changes from

$$1 \text{ to } \frac{1}{200^2} \text{ i.e. from 1 to } \frac{1}{40{,}000} \tag{3.68}$$

Example IV consists of $N = 10{,}000$ circular discrete elements of diameter D spaced at distance $2D$ in the y-direction and at distance D and S in the x-direction (Figure 3.52). For each given value of S, contact detection is performed 10 times, and the total CPU time for all 10 detections of all contacts is recorded using profiled debugging version of the code. The results are shown in Figure 3.53.

The numerical results shown above clearly demonstrate that the total detection time is constant for a large range of packing densities. This is the case for all packs considered, as theoretically predicted.

In summary:

- Memory requirements of the Munjiza-NBS algorithm are insignificant, and do not change significantly with considerable changes in packing density.

- Total detection time for the Munjiza-NBS algorithm does not depend upon packing density, a result confirmed by both theoretical investigations and numerical experiments.

- The relationship between the total detection time and the total number of discrete elements comprising the problem is linear. This result is confirmed by both theoretical investigations and numerical experiments.

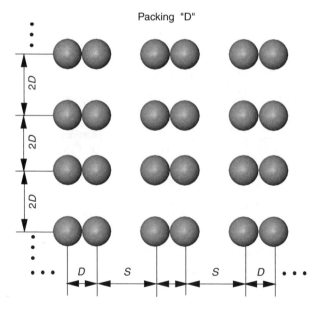

Figure 3.52 Example IV, packing D (A. Munjiza and K.R.F. Andrews, *International Journal for Numerical Methods in Engineering*, 43/1). (Reproduced by permission of John Wiley & Sons, Ltd).

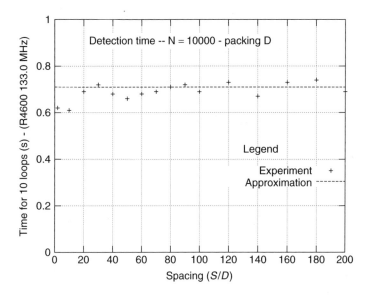

Figure 3.53 Example IV: total CPU time as a function of packing density for pack D (A. Munjiza and K.R.F. Andrews, *International Journal for Numerical Methods in Engineering*, 43/1). (Reproduced by permission of John Wiley & Sons, Ltd).

- In terms of CPU requirements, the Munjiza-NBS contact detection algorithm has a better performance than either binary search based contact detection algorithms or sorting contact detection algorithms.

- The Munjiza-NBS contact detection algorithm uses less RAM space than the binary search based contact detection algorithm.

- The Munjiza-NBS contact detection algorithm uses slightly more RAM space than the sorting contact detection algorithm.

- Both the Munjiza-NBS contact detection algorithm and sorting contact detection algorithm have RAM requirements proportional to the total number of discrete elements.

3.9 SELECTION OF CONTACT DETECTION ALGORITHM

A whole range of contact detection algorithms is available for large scale combined finite-discrete element simulations. The most efficient algorithm in terms of CPU time required to detect all contacts is the screening array based contact detection algorithm. The problem with this algorithm is that RAM requirements are most often prohibitive.

The most efficient algorithm in terms of RAM requirements is the sorting contact detection algorithm. The problem with this algorithm is that the CPU requirements are not a linear function of the total number of discrete elements. This algorithm belongs to the hyper-linear category of contact detection algorithms, which means that for very large scale problems, CPU times can be prohibitive

The Munjiza-NBS contact detection algorithm is slightly less efficient than the sorting algorithm in terms of RAM requirements. However, this algorithm, while being gentle in terms of RAM requirements, is superior in terms of CPU requirements.

The binary tree based contact detection algorithms were among the first contact detection algorithms used in the combined finite-discrete element method. Now, the binary tree based contact detection algorithms are mostly of historical value for most of the large scale combined finite-discrete element simulations. These algorithms have been superseded by the new generation of algorithms, which are superior in terms of both RAM and CPU requirements. It is also worth noting that the implementation of the binary tree based contact detection algorithms is much more complex than that of most of the new generation of algorithms.

3.10   GENERALISATION OF CONTACT DETECTION ALGORITHMS TO 3D SPACE

All the contact detection algorithms listed in this chapter have conveniently been explained using 2D space. In this section, an extension to 3D space is given for each of the algorithms.

3.10.1   Direct checking contact detection algorithm

Extension of the direct checking contact detection algorithm into 3D space is straightforward. The bounding box becomes a 3D object (say sphere), while the check for contact becomes

$$(x_i - x_j)^2 + (y_i - y_j)^2 + (z_i - z_j)^2 < d^2 \tag{3.69}$$

where x, y and z denote the centre coordinates of bounding spheres, while d is the diameter of the spheres.

3.10.2   Binary tree search

Extension of the binary tree search based contact detection algorithm to 3D space involves the addition of successive subdivisions of the space in the z-direction, as shown in Figure 3.54.

The space is first divided into two halves through division in the x-direction (level 1). Subsequent subdivision in the y-direction (level 2) is followed by subdivision in the z-direction (level 3). This in turn is followed by another subdivision in the x-direction (level 4), which is followed by subdivision in the y-direction (level 5) and subdivision in the z-direction (level 6), followed by subdivision in the x-direction (level 7), etc. until the smallest cell is reached.

3.10.3   Screening contact detection algorithm

Extension of the screening contact detection algorithm into 3D space involves subdivision of the 3D space into identical cube shaped cells, as shown in Figure 3.55.

Mile End Library
Queen Mary, University of London
Christmas Vacation 15th Dec - 7th Jan

Extended Vacation Loans
Ordinary Loans borrowed or renewed
from Saturday 17th November
will be due back on Friday 11th January

One Week Loans borrowed or renewed
from Saturday 8th December
ll be due back on Wednesday 9th Januar

Borrowed Items 08/01/2013 12:21
XXXXXX3244

tem Title	Due Date
The physics of medical imag	09/01/2013
The combined finite-discre	09/01/2013
Fluid flow : a first course in	15/01/2013

Amount Outstanding : £3.00

Indicates items borrowed today
PLEASE NOTE
f you still have overdue books on loan
ou may have more fines to pay

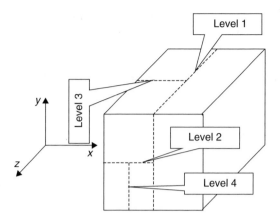

Figure 3.54   Successive space subdivisions in 3D space.

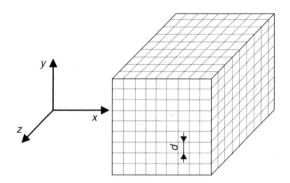

Figure 3.55   3D space divided into identical cells large enough to contain the largest discrete element comprising the systems.

Mapping of discrete elements onto the cells is achieved through integerisation of x, y and z coordinates of the centre of each discrete element:

$$^{int}x_i = 1 + Int\left(\frac{x_i - x_{min}}{d}\right) \qquad (3.70)$$

$$^{int}y_i = 1 + Int\left(\frac{y_i - y_{min}}{d}\right)$$

$$^{int}z_i = 1 + Int\left(\frac{z_i - z_{min}}{d}\right)$$

The mapping of discrete elements onto the cells is represented by a 3D integer array:

$$\mathbf{C}[n_{cel}][n_{cel}][n_{cel}] \qquad (3.71)$$

The problems with memory requirements associated with the array $\mathbf{C}$ are even more extensive than in 2D space.

Extension of the screening contact detection algorithm into 4D space or multi-dimensional space is almost impossible due to RAM requirement.

3.10.4 Direct mapping contact detection algorithm

Extension of the direct mapping contact detection algorithm into 3D space involves sub-division of the 3D space into identical cube shaped cells (Figure 3.55).

Mapping of discrete elements onto the cells is achieved through integerisation of x, y and z coordinates of the centre of each discrete element:

$$^{int}x_i = 1 + Int\left(\frac{x_i - x_{min}}{d}\right) \tag{3.72}$$

$$^{int}y_i = 1 + Int\left(\frac{y_i - y_{min}}{d}\right)$$

$$^{int}z_i = 1 + Int\left(\frac{z_i - z_{min}}{d}\right)$$

In 2D the screening array was replaced by three sorting arrays. In 3D four arrays are used; one array for each coordinate axis and one array for the discrete element number (identifier). Thus for cells in the x-direction, the **X** array is used, for cells in the y-direction the **Y** array is used, while for cells in the z-direction the **Z** array is used. Arrays **X**, **Y** and **Z** are of size N, where N is the total number of discrete elements comprising the discrete element system. The third array, **D**, indicates discrete element number, and is also of size N.

After mapping of discrete elements onto the cells, sorting of arrays **X** and **Y** and **Z** is performed to give spatial meaning to array **D**.

Contact detection is performed through binary search of arrays **X, Y** and **Z** for a specific cell.

This algorithm is linear in terms of RAM in both 2D and 3D space. For the same number of discrete elements, the CPU time for 3D contact detection is on average 30% larger than the CPU time for 2D contact detection. This is due to the processing of the **Z** array. The algorithm remains in 3D hyper-linear space in terms of CPU requirements.

Extensions to 4D and multidimensional spaces are possible, and these extensions preserve the major CPU and RAM properties of 2D and 3D algorithms.

3.11 GENERALISATION OF MUNJIZA-NBS CONTACT DETECTION ALGORITHM TO MULTIDIMENSIONAL SPACE

The contact mask for the NBS contact detection algorithm in 3D space is shown in Figure 3.56.

There are 26 neighbouring cells in total. However, because each cell with at least one discrete element mapped onto it is considered in turn as the central cell, all neighbouring cells for a given central cell need not be considered. It is enough to consider cells as shown in Figure 3.56.

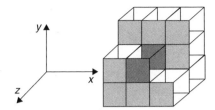

Figure 3.56 The contact mask in 3D space.

The contact mask shown includes the following cells:

$$(i_x - 1, i_y - 1, i_z - 1); \quad (i_x, i_y - 1, i_z - 1); \quad (i_x + 1, i_y - 1, i_z - 1); \quad (3.73)$$
$$(i_x - 1, i_y, i_z - 1); \quad (i_x, i_y, i_z - 1); \quad (i_x + 1, i_y, i_z - 1);$$
$$(i_x - 1, i_y + 1, i_z - 1); \quad (i_x, i_y + 1, i_z - 1); \quad (i_x + 1, i_y + 1, i_z - 1);$$
$$(i_x - 1, i_y - 1, i_z); \quad (i_x, i_y - 1, i_z); \quad (i_x + 1, i_y - 1, i_z);$$
$$(i_x - 1, i_y, i_z);$$

There are a total of 13 cells for each central cell. These cells are distributed in two layers of cells (z-direction). In the lower layer (layer $i_z - 1$), cells are distributed in three rows (rows $i_y - 1$, i_y, and $i_y + 1$). In layer i_z, cells are distributed over two rows (rows $i_y - 1$ and i_y).

Implementation of the NBS contact detection algorithm in 3D requires that all discrete elements be mapped onto layers of cells (z-direction). This mapping is performed using integerised coordinates of the centre of the bounding box for each discrete element:

$$i_x = Int \left(\frac{x - x_{min}}{d} \right) \tag{3.74}$$

$$i_y = Int \left(\frac{y - y_{min}}{d} \right)$$

$$i_z = Int \left(\frac{z - z_{min}}{d} \right)$$

Mapping of all discrete elements onto the layers of cells is represented using singly connected lists (Figure 3.57).

One list corresponds to each layer of cells, i.e. there are n_z singly connected lists in total, where n_z is the total number of cells in the z-direction, which is the same as the total number of layers of cells.

All of these lists can be represented by two arrays:

- One-dimensional array **C** of size n_z integer numbers (**C**$[n_z]$) represents the head of each singly connected list. The head of each list is identical to the first discrete element in the list.

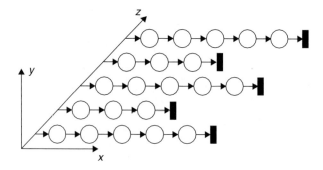

Figure 3.57   Singly connected lists for layers of cells represented by two integer arrays.

- One-dimensional array **Z** of size N integer numbers ($\mathbf{Z}[N]$), where N is the total number of discrete elements, represents for each discrete element the next discrete element that is in the same singly connected list (Figure 3.58).

Once all the lists for the z-layers of cells have been formed, they are considered to be 'new lists'. To locate a non-empty list, a loop over all discrete elements is performed:

Loop over all discrete elements ($i=1$; $i \leq N$; $i++$)
{   calculate integerised coordinate

$$i_z = Int\left(\frac{z - z_{\min}}{d}\right)$$ (3.75)

if the z_{iy} list is marked as "new":
{   mark it as "old" and call it central z-list
}
}

and a particular 'new list' found is marked as an 'old list'.

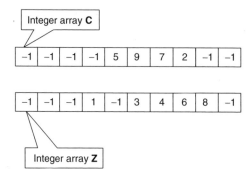

Figure 3.58   Layers of cells with all discrete elements in a particular layer i_z being represented by a singly connected z-list z_{iz}.

It is evident from the contact mask that all discrete elements immediately below the current layer ought to be located as well. All these discrete elements belong to the list z_{iz-1}.

Thus, both discrete elements from the list z_{iz-1} and discrete elements from the list z_{iz} are mapped onto rows of cells (y-direction):

Loop over all discrete elements from the list z_{iz}
    $(k=C[i_z]$ and $k=Z[k]$ *until the end of the list is reached, i.e. until k=-1)*
    {   **calculate integerised coordinate for the discrete element** *j*

$$i_y = Int\left(\frac{y - y_{\min}}{d}\right) \tag{3.76}$$

    add the discrete element *k* **onto** $y_{iy,iz}$ **list,**
    i.e. onto the list of discrete elements mapped onto
    the row of cells (i_y, i_z)
    consider the $y_{iy,iz}$ **list as "new list"**

}

A particular $y_{iy,iz}$ list contains all discrete elements with integerised coordinates i_y and i_z.

In a similar way, all discrete elements from the list z_{iz} are mapped onto rows of cells (*y*-direction):

Loop over all discrete elements from the list z_{iz-1}
    $(k=C[i_z-1]$ and $k=Z[k$ *until the end of the list is reached, i.e. until k=-1)*
    {   **calculate integerised coordinate for the discrete element** *j*

$$i_y = Int\left(\frac{y - y_{\min}}{d}\right) \tag{3.77}$$

    add the discrete element *k* **onto** $y_{iy,iz-1}$ **list,**
    i.e. onto the list of discrete elements mapped onto
    the row of cells $(i_y, i_z\text{-}1)$

}

There is a total of n_y y-lists in each layer. The heads of all of these lists are represented by a two-dimensional array $\mathbf{B}[2][n_y]$. Array $\mathbf{B}[0]$ represents the heads of all y-lists for layer i_z and array $\mathbf{B}[1]$ represents the heads of all y-lists for layer $i_z - 1$. For a particular discrete element, the discrete element that is the next discrete element in the same list is represented by the array $\mathbf{Y}[N]$, where N is the total number of discrete elements. Array $\mathbf{B}$ comprises $2n_y$ integer numbers, while array $\mathbf{Y}$ comprises N integer numbers.

It is worth mentioning that assembling all lists of layers of discrete elements (z-lists) is equivalent to one loop over all discrete elements. In a similar way, assembling all lists of rows of discrete elements for all the layers is equivalent to two loops over all discrete elements. This is because no discrete element can be mapped onto two layers at the same time.

Once the lists of rows of discrete elements are assembled, for all non-empty y-lists of rows of discrete elements, lists of discrete elements for individual cells are assembled.

Finding nonempty y-lists involves looping over all discrete elements from the z_{iz} list. Thus, no loop over cells is involved. For a given non-empty $y_{iy,iz}$ list only x-lists for rows

$$
\begin{aligned}
&(i_y - 1, i_z); \quad (i_y, i_z); \\
&(i_y - 1, i_z - 1); \quad (i_y, i_z - 1); \quad (i_y + 1, i_z - 1);
\end{aligned}
\tag{3.78}
$$

need to be assembled. It is important to separate elements from row $(i_{y,iz})$ from elements from the other rows. Thus, two sets of singly connected x-lists are assembled — one set of lists for discrete elements from row $(i_{y,iz})$, and one set of lists for discrete elements from rows

$$
\begin{aligned}
&(i_y - 1, i_z); \\
&(i_y - 1, i_z - 1); \quad (i_y, i_z - 1); \quad (i_y + 1, i_z - 1);
\end{aligned}
\tag{3.79}
$$

All these lists can be represented by two arrays:

- The two-dimensional array $\mathbf{A}[2][n_x]$ containing the head of each of the lists, i.e. the first discrete elements in a particular list.

- The one-dimensional array $\mathbf{X}[N]$; for each discrete element this array contains the next discrete element in the same x-list.

To further aid understanding of the above described procedures, a visual representation of the NBS contact detection algorithm in 3D is given in Figures 3.59–3.63. First, discrete elements are mapped onto layers of cells. (Figure 3.59) and a nonempty layer is identified and called the central layer (Figure 3.60).

The non-empty z-layer i_z together with the layer immediately below it (layer i_z-1) is divided into y-rows of cells (Figure 3.61).

By looping over discrete elements from layer i_z, a non-empty row (i_y, i_z) is identified and called the central row (Figure 3.62).

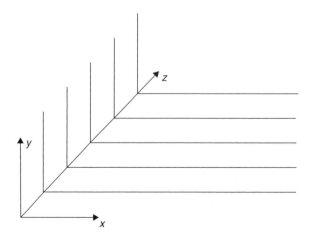

Figure 3.59 Layers of cells.

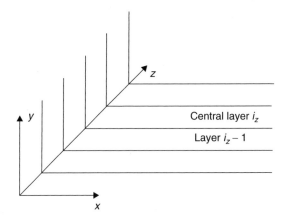

Figure 3.60 A non-empty layer i_z is detected, with all discrete elements mapped onto it being represented by a singly connected z-list z_{i_z}.

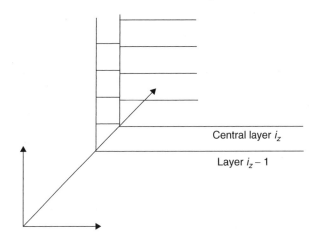

Figure 3.61 Layers i_z and $i_z - 1$ are divided into rows of cells.

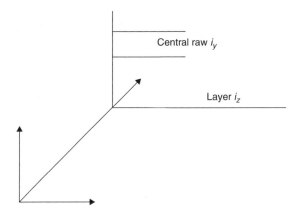

Figure 3.62 A non-empty row of cells is detected (i_y, i_z).

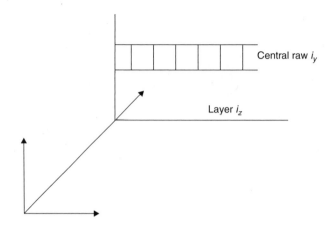

Figure 3.63   The non-empty row of cells (i_y, i_z) is divided into individual cells.

The central row (i_y, i_z) is divided into individual cells (Figure 3.63). For each of the cells, a singly connected list containing all discrete elements mapped onto that cell is assembled.

By looping over all discrete elements in row (i_y, i_z), a nonempty cell (i_x, i_y, i_z) is identified (Figure 3.64). Contact interaction with discrete elements from neighbouring cells according to the contact mask is performed. Each non-empty cell in the row (i_y, i_z) in turn becomes the central cell. In a similar way, each non-empty row from layer i_z in turn becomes the central row. In the same way, each non-empty layer in turn becomes the central layer. Only non-empty layers, non-empty rows and non-empty cells are visited. The pseudo code for NBS contact detection algorithm in 3D is as follows:

1. Loop over discrete elements *(k=0;k≤N;)*
{   calculate integerised z coordinates of the centre and
place the discrete element onto the corresponding z-list

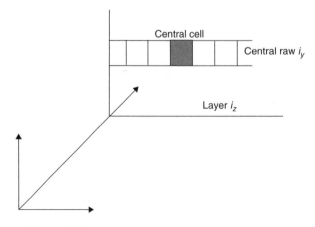

Figure 3.64   A non-empty cell (i_x, i_y, i_z) is detected.

}
consider all z**-lists to be "new" lists**
2. Loop over discrete elements *(k=0;k≤N;)*
{   calculate integerised z coordinates of the centre
     if discrete element belongs to a new z **list**
     {   mark the list as an old list and call it the central z list
          {   **3. loop over all discrete elements on the central** z **list**
               {   integerise y coordinate of the discrete element and place
                    it onto the corresponding list in y direction
               }
               4. loop over all discrete elements from list neighbouring
                    to central z**-list according to the contact mask**
               {   integerise y coordinate of the discrete element and place
                    it onto the corresponding list in y direction
               }
               consider all y**-lists to be "new"** y**-lists**
               5. loop over all discrete elements from the central z**-list**
               {   **if the discrete element belongs to a new list** *(y,z)*
                    {   mark the list *(y,z)* as old and call it central y-list
                         6. loop over all discrete elements from central *(y,z)*-**list**
                         {   integerise x coordinate of the discrete element
                              and place it onto the corresponding x-list
                         }
                         7. loop over all discrete elements from neighbouring
                              (y,z)-**lists according to the contact mask**
                         {   integerise x coordinate of the discrete element
                              and place it onto the corresponding x-list
                         }
                         consider all x-**lists to be "new"** x-**lists**
                         8. loop over all discrete elements from central *(y,z)*-**list**
                         {   **if the discrete element belongs to a "new"** x-**list**
                              {   mark the list *(x,y,z)* as old and call it central x-list
                                   check for contact using contact mask
                              }
                         }
                         9. loop over all discrete elements from central *(y,z)*-**list**
                         {   remove corresponding x-list
                         }
                         10. loop over all discrete elements from
                              neighbouring *(y,z)*-**lists according to contact mask**
                         {   remove corresponding x-list
                         }
                    }
               }
               11. loop over all discrete elements from the central z-**list**
               {   remove corresponding y-lists
               }

**12. loop over all discrete elements from the list neighbouring
to the central z-list according to contact mask**
{   **remove corresponding y-lists**
}

}

}

13. Loop over discrete elements *($k=0;k\leq N;$)*
{   **remove corresponding z-list**
}

RAM requirements of the NBS algorithm in 3D are given by

$$M = 3N + n_z + 2n_x + 2n_y \tag{3.80}$$

which represents a negligible increase in comparison to the RAM requirements of the NBS contact detection algorithm in 2D. In a similar way, the CPU requirements are greater in 2D than in 3D for the same number of discrete elements. Both RAM and CPU requirements increase linearly with the increase in the number of discrete elements.

In theory, similar extensions of NBS to 4-dimensional and multi-dimensional spaces are relatively easy to implement. By generalising the NBS contact detection algorithm to multi-dimensional space, the contact mask changes and the arrays to store the singly connected lists change accordingly. At the top is always a set of singly connected lists containing all discrete elements. All of these lists are represented by two arrays $\mathbf{H}[n_{cel}]$ and $\mathbf{D}_n$. These lists are obtained by discretising the space along the nth dimension. A loop over each such hyper-layer is performed, a non-empty hyper-layer is detected and discrete elements are put into singly connected lists representing hyper-rows of cells. The process is continued until the last level representing individual hyper-cells is reached, when contact interaction is processed using the contact mask. Both the number of nested loops and number of arrays needed to represent all the lists increase with an increase in the dimension of space. Thus, the Munjiza-NBS algorithm in multi-dimensional space requires more RAM space and more CPU time to process the same number of discrete elements. However, the most important property of RAM and CPU linearity remains regardless of the dimension of the space.

3.12 SHAPE AND SIZE GENERALISATION–WILLIAMS C-GRID ALGORITHM

There are two approaches to shape generation. The first is to employ a discretised contact solution similar to the discretised contact solution employed for contact interaction. The basic idea is very simple; each discrete element is discretised into finite elements or grid squares, and contact detection is performed on the finite element level or grid square level. The contact detection algorithm therefore detects the finite elements or grid squares that are likely to be in contact, regardless of what discrete element they may belong to. The shape and size of finite elements is governed by the accuracy of deformability analysis, which makes finite elements less elongated and more uniform in size than the discrete elements. In this way, complexities associated both with the shape and size variation are

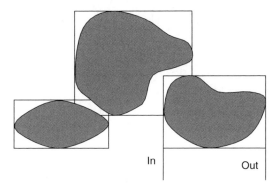

Figure 3.65   Taking into account the rectangular bounding box when placing discrete elements onto the lists or taking them out of the lists.

minimised, and the Munjiza-NBS contact detection algorithm in its original form can be employed.

The second approach is to modify the Munjiza-NBS contact detection algorithm to take into account variation in the size of discrete elements. It could be argued that similar modifications would be required for all the other contact detection algorithms listed in this chapter if theoretical performance is to be achieved. One particular modification of the NBS contact detection algorithm that preserves the theoretical performance of the original NBS contact detection algorithm was first proposed by a MIT group headed by J.R. Williams, and is termed the Williams-C-grid contact detection algorithm. The C-grid algorithm preserves the most important property of RAM and CPU linearity, regardless of the size of discrete elements.

The Williams-C-grid contact detection algorithm is in essence the same as the Munjiza-NBS contact detection algorithm, except that when assembling the singly connected lists, discrete elements are taken into the lists according to the maximum and minimum x, y and z coordinates of the rectangular bounding box (Figure 3.65).

Similar extensions can be added to the other contact detection algorithms described in this chapter. For instance, by using the rectangular bounding box, the sorting contact detection algorithm can be implemented without using the decomposition of space into cells, and sorting can be done directly using actual minimum and maximum coordinates of the bounding box. It is beyond the scope of this book to go into the details of all possible variations of the contact detection algorithms described in this chapter. However, it is assumed that a particular implementation of any of the above described contact detection algorithms should take into account the specific details of the problem for which it is designed.

4

Deformability of Discrete Elements

4.1 DEFORMATION

Discrete elements were originally introduced to model problems and processes that continuum-based models cannot model correctly. A large class of such problems and processes are generally termed 'problems of discontinua'. However, a large class of problems of discontinua involves individual bodies (discrete elements) that can deform, fail, fracture and even fragment. Such discrete elements are termed 'deformable discrete elements'.

Each discrete element represents a single deformable body, which at any instance of time occupies a region of space (Figure 4.1). Special meaning is given to some of these regions. For instance, one such region $\mathcal{B}$ is termed the *initial* or *reference configuration*, while

$$\mathbf{p} \in B \tag{4.1}$$

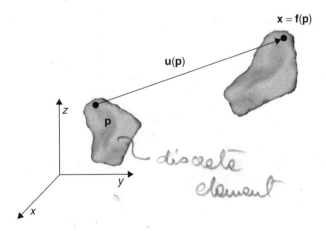

Figure 4.1 Deformation of a discrete element.

The Combined Finite-Discrete Element Method   A. Munjiza
© 2004 John Wiley & Sons, Ltd   ISBN: 0-470-84199-0

are called material points, and bounded subregions of the body are called parts. Body deforms via mapping

$$x = f(p) \qquad (4.2)$$

where **f** is one to one smooth mapping which maps $\mathcal{B}$ onto a closed region $\mathcal{E}$, and which satisfies

$$\det \nabla f(p) > 0 \qquad (4.3)$$

for any material point **p**. This condition simply states that no part with nonzero volume can map into zero volume space, i.e. parts of the body occupy space before and after deformation. The volume of such space may defer, but is always greater than zero.

4.2   DEFORMATION GRADIENT

Deformation can also be written as

$$x = f(p) = p + u(p) \qquad (4.4)$$

where **u(p)** is called displacement. Mapping

$$F(p) = \nabla f(p) = I + \nabla u \qquad (4.5)$$

describes change in deformation in the vicinity of each material point, and is referred to as the deformation gradient.

4.2.1   Frames of reference

To describe the deformation of a particular discrete element in the vicinity of the material point **p**, four reference frames are used in this chapter (Figure 4.2):

- *Global frame:* this reference frame is the frame defined by a triad of orthogonal unit vectors that coincide with the axes of Cartesian coordinate system

$$(\mathbf{i}, \mathbf{j}, \mathbf{k}) \qquad (4.6)$$

- *Local frame:* this reference frame is the frame associated with the initial position of a particular discrete element. It is therefore fixed in space and does not move with the discrete element. Very often this frame is made to coincide with the major axes of inertia of a particular discrete element. This frame is defined by a triad of orthogonal unit vectors

$$(\bar{\mathbf{i}}, \bar{\mathbf{j}}, \bar{\mathbf{k}}) \qquad (4.7)$$

- *Deformed local frame:* this frame is fixed to the material point **p** of the discrete element and moves with that point. The base vectors of this frame also follow the deformation in the vicinity of the point **p**. The origin of this frame thus coincides at all times with the point $\mathbf{x} = \mathbf{f}(\mathbf{p})$, while the direction of triad vectors and magnitude of this vectors

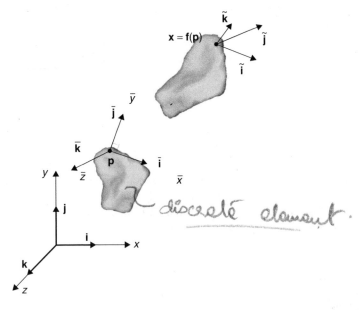

Figure 4.2   Frames of reference.

changes with deformation of the discrete element. The orientation of the initial frame is defined by a triad of generally non-unit vectors that are, in general, not parallel to the respective axes of the Cartesian coordinate system. In addition, these vectors are in general not orthogonal to each other. Initially, this frame coincides with the local frame. As the body deforms, this frame begins to differ from the local frame. As this frame is obtained through deformation of the local frame, it is in further text referred to as a *deformed local frame.* A deformed local frame is defined by a triad of non-orthogonal non-unit vectors:

$$(\tilde{\mathbf{i}}, \tilde{\mathbf{j}}, \tilde{\mathbf{k}}) \tag{4.8}$$

- *Initial frame:* in the same way as the local frame, this reference frame is linked to the material point and is associated with the initial position of a particular discrete element. It is therefore fixed in space, and does not move with the discrete element. Very often this frame is made to coincide with, say, the edges of a finite element on a particular discrete element. Thus, base vectors of this frame are in general not orthogonal to each other. The magnitude of the base vectors is, for instance, equal to the length of the corresponding edges of finite elements, i.e. the base vectors are not unit vectors. The initial frame is defined by a triad of non-orthogonal non-unit vectors:

$$(\hat{\mathbf{i}}, \hat{\mathbf{j}}, \hat{\mathbf{k}}) \tag{4.9}$$

- *Deformed initial frame:* this frame is fixed to the material point **p** of the discrete element, and moves with that point. The base vectors of this frame also follow the deformation in the vicinity of the point **p**. The origin of this frame thus coincides at all times with the point $\mathbf{x} = \mathbf{f}(\mathbf{p})$, while the direction of triad vectors and magnitude of this vector changes with the deformation of the discrete element. The orientation of the initial frame is defined by a triad of non-unit vectors that are in general not parallel to the respective axes of the Cartesian coordinate system. In addition, these vectors are in general not orthogonal to each other. Initially, this frame coincides with the initial frame. As the body deforms, this frame begins to differ from the initial frame. As this frame is obtained through deformation of the initial frame, it is in further text referred to as a *deformed initial frame*. A deformed initial frame is defined by a triad of non-orthogonal non-unit vectors:

$$(\check{\mathbf{i}}, \check{\mathbf{j}}, \check{\mathbf{k}}) \tag{4.10}$$

The global, local and initial frames are inertial frames, while the deformed local frame and deformed initial frame are non-inertial frames.

The relationship between a local and deformed local frame can be obtained using the deformation gradient. In the global frame, the deformation gradient tensor can be written using the following matrix:

$$\mathbf{F} = \nabla \mathbf{f} = \nabla \begin{bmatrix} x + u(x, y, z) \\ y + v(x, y, z) \\ z + w(x, y, z) \end{bmatrix} = \begin{bmatrix} 1 + \dfrac{\partial u}{\partial x} & \dfrac{\partial u}{\partial y} & \dfrac{\partial u}{\partial z} \\[2mm] \dfrac{\partial v}{\partial x} & 1 + \dfrac{\partial v}{\partial y} & \dfrac{\partial v}{\partial z} \\[2mm] \dfrac{\partial w}{\partial x} & \dfrac{\partial w}{\partial y} & 1 + \dfrac{\partial w}{\partial z} \end{bmatrix} \tag{4.11}$$

The same deformation gradient tensor can be written in the local frame using the following matrix:

$$\mathbf{F} = \nabla \mathbf{f} = \nabla \begin{bmatrix} \bar{x} + \bar{u}(\bar{x}, \bar{y}, \bar{z}) \\ \bar{y} + \bar{v}(\bar{x}, \bar{y}, \bar{z}) \\ \bar{z} + \bar{w}(\bar{x}, \bar{y}, \bar{z}) \end{bmatrix} = \begin{bmatrix} 1 + \dfrac{\partial \bar{u}}{\partial \bar{x}} & \dfrac{\partial \bar{u}}{\partial \bar{y}} & \dfrac{\partial \bar{u}}{\partial \bar{z}} \\[2mm] \dfrac{\partial \bar{v}}{\partial \bar{x}} & 1 + \dfrac{\partial \bar{v}}{\partial \bar{y}} & \dfrac{\partial \bar{v}}{\partial \bar{z}} \\[2mm] \dfrac{\partial \bar{w}}{\partial \bar{x}} & \dfrac{\partial \bar{w}}{\partial \bar{y}} & 1 + \dfrac{\partial \bar{w}}{\partial \bar{z}} \end{bmatrix} \tag{4.12}$$

Triad vectors of the local frame are parallel to the axes of the corresponding Cartesian coordinate system, and can be expressed in a matrix form:

$$\bar{\mathbf{i}} = 1\bar{\mathbf{i}} + 0\bar{\mathbf{j}} + 0\bar{\mathbf{k}} = \begin{bmatrix} 1 \\ 0 \\ 0 \end{bmatrix} \tag{4.13}$$

$$\bar{\mathbf{j}} = 0\bar{\mathbf{i}} + 1\bar{\mathbf{j}} + 0\bar{\mathbf{k}} = \begin{bmatrix} 0 \\ 1 \\ 0 \end{bmatrix} \tag{4.14}$$

$$\bar{\mathbf{k}} = 0\bar{\mathbf{i}} + 0\bar{\mathbf{j}} + 1\bar{\mathbf{k}} = \begin{bmatrix} 0 \\ 0 \\ 1 \end{bmatrix} \tag{4.15}$$

Since the triad vectors of the deformed frame are fixed to the material points in the vicinity of material point **p**, it follows that the magnitude and orientation of this vectors changes with deformation gradient in the vicinity of the point **p**, and is given by

$$
\tilde{\mathbf{i}} = \mathbf{F}\bar{\mathbf{i}} =
\begin{bmatrix}
1 + \dfrac{\partial \bar{u}}{\partial \bar{x}} & \dfrac{\partial \bar{u}}{\partial \bar{y}} & \dfrac{\partial \bar{u}}{\partial \bar{z}} \\[2mm]
\dfrac{\partial \bar{v}}{\partial \bar{x}} & 1 + \dfrac{\partial \bar{v}}{\partial \bar{y}} & \dfrac{\partial \bar{v}}{\partial \bar{z}} \\[2mm]
\dfrac{\partial \bar{w}}{\partial \bar{x}} & \dfrac{\partial \bar{w}}{\partial \bar{y}} & 1 + \dfrac{\partial \bar{w}}{\partial \bar{z}}
\end{bmatrix}
\begin{bmatrix} 1 \\ 0 \\ 0 \end{bmatrix}
=
\begin{bmatrix}
1 + \dfrac{\partial \bar{u}}{\partial \bar{x}} \\[2mm]
\dfrac{\partial \bar{v}}{\partial \bar{x}} \\[2mm]
\dfrac{\partial \bar{w}}{\partial \bar{x}}
\end{bmatrix}
\tag{4.16}
$$

$$
= \left(1 + \frac{\partial \bar{u}}{\partial \bar{x}}\right)\bar{\mathbf{i}} + \frac{\partial \bar{v}}{\partial \bar{x}}\bar{\mathbf{j}} + \frac{\partial \bar{w}}{\partial \bar{x}}\bar{\mathbf{k}}
$$

In the same way,

$$
\tilde{\mathbf{j}} = \mathbf{F}\bar{\mathbf{j}} =
\begin{bmatrix}
1 + \dfrac{\partial \bar{u}}{\partial \bar{x}} & \dfrac{\partial \bar{u}}{\partial \bar{y}} & \dfrac{\partial \bar{u}}{\partial \bar{z}} \\[2mm]
\dfrac{\partial \bar{v}}{\partial \bar{x}} & 1 + \dfrac{\partial \bar{v}}{\partial \bar{y}} & \dfrac{\partial \bar{v}}{\partial \bar{z}} \\[2mm]
\dfrac{\partial \bar{w}}{\partial \bar{x}} & \dfrac{\partial \bar{w}}{\partial \bar{y}} & 1 + \dfrac{\partial \bar{w}}{\partial \bar{z}}
\end{bmatrix}
\begin{bmatrix} 0 \\ 1 \\ 0 \end{bmatrix}
=
\begin{bmatrix}
\dfrac{\partial \bar{u}}{\partial \bar{y}} \\[2mm]
\dfrac{\partial \bar{v}}{\partial \bar{y}} \\[2mm]
\dfrac{\partial \bar{w}}{\partial \bar{y}}
\end{bmatrix}
\tag{4.17}
$$

$$
= \frac{\partial \bar{u}}{\partial \bar{y}}\bar{\mathbf{i}} + \left(1 + \frac{\partial \bar{v}}{\partial \bar{y}}\right)\bar{\mathbf{j}} + \frac{\partial \bar{w}}{\partial \bar{y}}\bar{\mathbf{k}}
$$

and

$$
\tilde{\mathbf{k}} = \mathbf{F}\bar{\mathbf{k}} =
\begin{bmatrix}
1 + \dfrac{\partial \bar{u}}{\partial \bar{x}} & \dfrac{\partial \bar{u}}{\partial \bar{y}} & \dfrac{\partial \bar{u}}{\partial \bar{z}} \\[2mm]
\dfrac{\partial \bar{v}}{\partial \bar{x}} & 1 + \dfrac{\partial \bar{v}}{\partial \bar{y}} & \dfrac{\partial \bar{v}}{\partial \bar{z}} \\[2mm]
\dfrac{\partial \bar{w}}{\partial \bar{x}} & \dfrac{\partial \bar{w}}{\partial \bar{y}} & 1 + \dfrac{\partial \bar{w}}{\partial \bar{z}}
\end{bmatrix}
\begin{bmatrix} 0 \\ 0 \\ 1 \end{bmatrix}
=
\begin{bmatrix}
\dfrac{\partial \bar{u}}{\partial \bar{z}} \\[2mm]
\dfrac{\partial \bar{v}}{\partial \bar{z}} \\[2mm]
\dfrac{\partial \bar{w}}{\partial \bar{z}}
\end{bmatrix}
\tag{4.18}
$$

$$
= \frac{\partial \bar{u}}{\partial \bar{z}}\bar{\mathbf{i}} + \frac{\partial \bar{v}}{\partial \bar{z}}\bar{\mathbf{j}} + \left(1 + \frac{\partial \bar{w}}{\partial \bar{z}}\right)\bar{\mathbf{k}}
$$

In other words, through the deformation process, vectors of the local triad

$$
(\bar{\mathbf{i}}, \bar{\mathbf{j}}, \bar{\mathbf{k}})
\tag{4.19}
$$

are mapped onto the vectors of the deformed local triad:

$$
(\tilde{\mathbf{i}}, \tilde{\mathbf{j}}, \tilde{\mathbf{k}})
\tag{4.20}
$$

with

$$\tilde{\mathbf{i}} = \begin{bmatrix} 1 + \dfrac{\partial \overline{u}}{\partial \overline{x}} \\[6pt] \dfrac{\partial v}{\partial \overline{x}} \\[6pt] \dfrac{\partial w}{\partial \overline{x}} \end{bmatrix} ; \quad \tilde{\mathbf{j}} = \begin{bmatrix} \dfrac{\partial \overline{u}}{\partial \overline{y}} \\[6pt] 1 + \dfrac{\partial \overline{v}}{\partial \overline{y}} \\[6pt] \dfrac{\partial \overline{w}}{\partial \overline{y}} \end{bmatrix} ; \quad \tilde{\mathbf{k}} = \begin{bmatrix} \dfrac{\partial \overline{u}}{\partial \overline{z}} \\[6pt] \dfrac{\partial \overline{v}}{\partial \overline{z}} \\[6pt] 1 + \dfrac{\partial \overline{w}}{\partial \overline{z}} \end{bmatrix} \tag{4.21}$$

The physical meaning of the deformation gradient can be explained by taking an infinitesimal material element in the vicinity of point **p**, as shown in Figure 4.3. It is assumed that this material element coincides with a cube of edge of unit length. By choosing a very small unit for the length, the edge of the cube is made infinitesimally small.

With such an assumption, the local triad at point **p** is given by:

$$\overline{\mathbf{i}} = \begin{bmatrix} 1 \\ 0 \\ 0 \end{bmatrix} ; \quad \overline{\mathbf{j}} = \begin{bmatrix} 0 \\ 1 \\ 0 \end{bmatrix} ; \quad \overline{\mathbf{k}} = \begin{bmatrix} 0 \\ 0 \\ 1 \end{bmatrix} \tag{4.22}$$

Because the unit for length is conveniently chosen to be infinitesimally small, the base vectors of the local triad coincide with the edges of the material element. As the material in the vicinity of point **p** deforms, these base vectors are mapped through deformation into corresponding base vectors of the deformed local triad:

$$\tilde{\mathbf{i}} = \left(1 + \frac{\partial \overline{u}}{\partial \overline{x}} \right) \overline{\mathbf{i}} + \frac{\partial \overline{v}}{\partial \overline{x}} \overline{\mathbf{j}} + \frac{\partial \overline{w}}{\partial \overline{x}} \overline{\mathbf{k}} \tag{4.23}$$

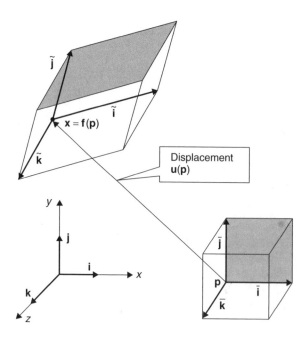

Figure 4.3   Physical meaning of deformation gradient.

$$\tilde{\mathbf{j}} = \frac{\partial \overline{u}}{\partial \overline{y}} \mathbf{i} + \left(1 + \frac{\partial \overline{v}}{\partial \overline{y}}\right) \mathbf{j} + \frac{\partial \overline{w}}{\partial \overline{y}} \mathbf{k}$$

$$\tilde{\mathbf{k}} = \frac{\partial \overline{u}}{\partial \overline{z}} \mathbf{i} + \frac{\partial \overline{v}}{\partial \overline{z}} \mathbf{j} + \left(1 + \frac{\partial \overline{w}}{\partial \overline{z}}\right) \mathbf{k}$$

As can be seen from Figure 4.3, these vectors are in general non-orthogonal to each other. In addition, these vectors are not unit vectors. Thus, a cube shaped material element of unit volume changes both its volume and it original cubic shape.

Special types of deformation include the deformation with constant displacement and deformation with constant deformation gradient.

The deformation with constant displacement

$$\mathbf{u}(\mathbf{p}) = const \tag{4.24}$$

is referred to as translation (Figure 4.4). As can be seen from the figure, the initial material element is identical in shape, size and orientation to the deformed initial volume, except that it is translated. Translation therefore does not produce any straining of the material.

The deformation with constant deformation gradient

$$\mathbf{F}(\mathbf{p}) = const \tag{4.25}$$

is referred to as homogeneous. It can be expressed as

$$\mathbf{f}(\mathbf{p}) = \mathbf{f}(\mathbf{q}) + \mathbf{F}(\mathbf{p} - \mathbf{q}) \tag{4.26}$$

Two important examples of homogeneous deformation are stretch from $\mathbf{q}$ and rotation about $\mathbf{q}$. Stretch from $\mathbf{q}$ can be written as follows:

$$\mathbf{f}(\mathbf{p}) = \mathbf{q} + \mathbf{U}(\mathbf{p} - \mathbf{q}) \tag{4.27}$$

where $\mathbf{U}$ is a symmetric and positive definite tensor. Spectral decomposition of $\mathbf{U}$ in the form

$$\mathbf{Ue} = s\mathbf{e} \tag{4.28}$$

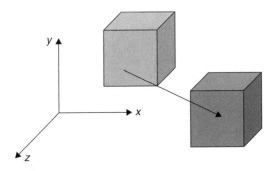

Figure 4.4   Translation.

(where s is a scalar) yields three real eigenvalues (principal values) corresponding to three mutually orthogonal directions (principal directions), such that

$$\mathbf{U}\mathbf{e}_1 = s_1\mathbf{e}_1$$
$$\mathbf{U}\mathbf{e}_2 = s_2\mathbf{e}_2 \qquad (4.29)$$
$$\mathbf{U}\mathbf{e}_3 = s_3\mathbf{e}_3$$

The matrix of components of tensor $\mathbf{U}$ relative to the basis

$$(\mathbf{e}_1, \mathbf{e}_2, \mathbf{e}_3) \qquad (4.30)$$

is therefore diagonal:

$$\mathbf{U} = \begin{bmatrix} s_1 & 0 & 0 \\ 0 & s_2 & 0 \\ 0 & 0 & s_3 \end{bmatrix} \qquad (4.31)$$

The physical meaning of stretch is illustrated in Figure 4.5. The material element is elongated or shrunk in three orthogonal directions. The stretch is generally different in each direction, and is in essence similar to scaling in computer graphics.

The rotation about $\mathbf{q}$ is given by

$$\mathbf{f}(\mathbf{p}) = \mathbf{q} + \mathbf{R}(\mathbf{p} - \mathbf{q}) \qquad (4.32)$$

where $\mathbf{R}$ is a proper orthogonal tensor, i.e.

$$\mathbf{R}^{-1} = \mathbf{R}^T \quad \text{and} \quad \det(\mathbf{R}) = 1 \qquad (4.33)$$

and for any vector $\mathbf{a}$

$$\mathbf{R}^{-1}(\mathbf{R}\mathbf{a}) = \mathbf{R}^{-1}\mathbf{R}\mathbf{a} = \mathbf{R}^T\mathbf{R}\mathbf{a} = \mathbf{I}\mathbf{a} \qquad (4.34)$$

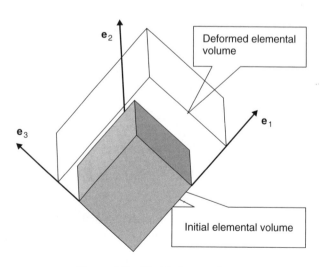

Figure 4.5   Stretch from point $\mathbf{q}$.

Any rotation in three-dimensional space is characterised by the axis $\boldsymbol{\psi}$ of rotation and the angle of rotation ψ:

$$\psi = |\boldsymbol{\psi}| \tag{4.35}$$

For any vector $\mathbf{a}$, the rotated configuration using this alternative way of calculating rotation is given as follows:

$$\mathbf{Ra} = \frac{1}{\psi^2}(\boldsymbol{\psi}\mathbf{a})\boldsymbol{\psi} + \left[\mathbf{a} - \frac{1}{\psi^2}(\boldsymbol{\psi}\mathbf{a})\boldsymbol{\psi}\right]\cos(\psi) + \tag{4.36}$$

$$\frac{1}{\psi}(\boldsymbol{\psi} \times \mathbf{a})\sin(\psi)$$

The matrix of rotation tensor with respect to the orthonormal basis

$$\left(\mathbf{e}_1, \mathbf{e}_2, \frac{\boldsymbol{\psi}}{|\boldsymbol{\psi}|}\right) \tag{4.37}$$

(where $\mathbf{e}_1$ and $\mathbf{e}_2$ are arbitrary unit vectors orthogonal to each other and orthogonal to vector $\boldsymbol{\psi}$) is given by

$$\mathbf{R} = \begin{bmatrix} \cos\psi & -\sin\psi & 0 \\ \sin\psi & \cos\psi & 0 \\ 0 & 0 & 1 \end{bmatrix} \tag{4.38}$$

Physical interpretation of rotation is given in Figure 4.6. Material element does not change its shape or size. It does not translate in space either. The material element rotates about axis $\boldsymbol{\psi}$ instead

4.2.2 Transformation matrices

Base vectors of a deformed local frame can be expressed using base vectors of the local frame:

$$\tilde{\mathbf{i}} = \tilde{i}_{\bar{x}}\mathbf{i} + \tilde{i}_{\bar{y}}\mathbf{j} + \tilde{i}_{\bar{z}}\mathbf{k} \tag{4.39}$$

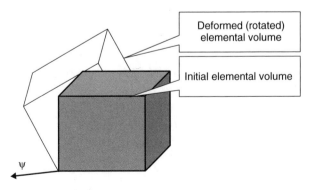

Figure 4.6   Rotation about point **q**.

$$\tilde{\mathbf{j}} = \tilde{j}_x\mathbf{i} + \tilde{j}_y\mathbf{j} + \tilde{j}_z\mathbf{k}$$

$$\tilde{\mathbf{k}} = \tilde{k}_x\mathbf{i} + \tilde{k}_y\mathbf{j} + \tilde{k}_z\mathbf{k}$$

In a similar way, base vectors of the local frame are expressed using base vectors of the deformed local frame:

$$\mathbf{i} = \bar{i}_x\tilde{\mathbf{i}} + \bar{i}_y\tilde{\mathbf{j}} + \bar{i}_z\tilde{\mathbf{k}} \tag{4.40}$$

$$\mathbf{j} = \bar{j}_x\tilde{\mathbf{i}} + \bar{j}_y\tilde{\mathbf{j}} + \bar{j}_z\tilde{\mathbf{k}}$$

$$\mathbf{k} = \bar{k}_x\tilde{\mathbf{i}} + \bar{k}_y\tilde{\mathbf{j}} + \bar{k}_z\tilde{\mathbf{k}}$$

Any particular vector $\mathbf{a}$ can be expressed using either frame:

$$\mathbf{a} = a_{\bar{x}}\mathbf{i} + a_{\bar{y}}\mathbf{j} + a_{\bar{z}}\mathbf{k} \tag{4.41}$$

$$= a_{\bar{x}}(\bar{i}_x\tilde{\mathbf{i}} + \bar{i}_y\tilde{\mathbf{j}} + \bar{i}_z\tilde{\mathbf{k}})$$

$$+ a_{\bar{y}}(\bar{j}_x\tilde{\mathbf{i}} + \bar{j}_y\tilde{\mathbf{j}} + \bar{j}_z\tilde{\mathbf{k}})$$

$$+ a_{\bar{z}}(\bar{k}_x\tilde{\mathbf{i}} + \bar{k}_y\tilde{\mathbf{j}} + \bar{k}_z\tilde{\mathbf{k}})$$

As explained above, base vectors of the deformed local frame are obtained from the base vectors of the local frame, thus

$$\mathbf{a} = a_{\tilde{x}}\tilde{\mathbf{i}} + a_{\tilde{y}}\tilde{\mathbf{j}} + a_{\tilde{z}}\tilde{\mathbf{k}} \tag{4.42}$$

$$= a_{\bar{x}}\left[\left(1 + \frac{\partial\bar{u}}{\partial\bar{x}}\right)\bar{\mathbf{i}} + \frac{\partial\bar{v}}{\partial\bar{x}}\bar{\mathbf{j}} + \frac{\partial\bar{w}}{\partial\bar{x}}\bar{\mathbf{k}}\right]$$

$$+ a_{\bar{y}}\left[\frac{\partial\bar{u}}{\partial\bar{y}}\bar{\mathbf{i}} + \left(1 + \frac{\partial\bar{v}}{\partial\bar{y}}\right)\bar{\mathbf{j}} + \frac{\partial\bar{w}}{\partial\bar{y}}\bar{\mathbf{k}}\right]$$

$$+ a_{\bar{z}}\left[\frac{\partial\bar{u}}{\partial\bar{z}}\bar{\mathbf{i}} + \frac{\partial\bar{v}}{\partial\bar{z}}\bar{\mathbf{j}} + \left(1 + \frac{\partial\bar{w}}{\partial\bar{z}}\right)\bar{\mathbf{k}}\right]$$

$$= a_{\bar{x}}\bar{\mathbf{i}} + a_{\bar{y}}\bar{\mathbf{j}} + a_{\bar{z}}\bar{\mathbf{k}}$$

The components of vector $\mathbf{a}$ in the local frame are therefore calculated from the vector components in the deformed local frame as follows:

$$\begin{bmatrix} a_{\bar{x}} \\ a_{\bar{y}} \\ a_{\bar{z}} \end{bmatrix} = \begin{bmatrix} \tilde{i}_x & \tilde{j}_x & \tilde{k}_x \\ \tilde{i}_y & \tilde{j}_y & \tilde{k}_y \\ \tilde{i}_z & \tilde{j}_z & \tilde{k}_z \end{bmatrix} \begin{bmatrix} a_{\bar{x}} \\ a_{\bar{y}} \\ a_{\bar{z}} \end{bmatrix} \tag{4.43}$$

where transformation matrix is given by

$$
\begin{bmatrix} \tilde{i}_{\bar{x}} & \tilde{j}_{\bar{x}} & \tilde{k}_{\bar{x}} \\ \tilde{i}_{\bar{y}} & \tilde{j}_{\bar{y}} & \tilde{k}_{\bar{y}} \\ \tilde{i}_{\bar{z}} & \tilde{j}_{\bar{z}} & \tilde{k}_{\bar{z}} \end{bmatrix} = \begin{bmatrix} 1 + \dfrac{\partial \overline{u}}{\partial \overline{x}} & \dfrac{\partial \overline{u}}{\partial \overline{y}} & \dfrac{\partial \overline{u}}{\partial \overline{z}} \\[2mm] \dfrac{\partial \overline{v}}{\partial \overline{x}} & 1 + \dfrac{\partial \overline{v}}{\partial \overline{y}} & \dfrac{\partial \overline{v}}{\partial \overline{z}} \\[2mm] \dfrac{\partial \overline{w}}{\partial \overline{x}} & \dfrac{\partial \overline{w}}{\partial \overline{y}} & 1 + \dfrac{\partial \overline{w}}{\partial \overline{z}} \end{bmatrix} \tag{4.44}
$$

By analogy, the components of vector **a** in the local deformed frame are calculated from the vector components in the local frame using the following transformation:

$$
\begin{bmatrix} a_{\tilde{x}} \\ a_{\tilde{y}} \\ a_{\tilde{z}} \end{bmatrix} = \begin{bmatrix} \tilde{i}_{\tilde{x}} & \tilde{j}_{\tilde{x}} & \tilde{k}_{\tilde{x}} \\ \tilde{i}_{\tilde{y}} & \tilde{j}_{\tilde{y}} & \tilde{k}_{\tilde{y}} \\ \tilde{i}_{\tilde{z}} & \tilde{j}_{\tilde{z}} & \tilde{k}_{\tilde{z}} \end{bmatrix} \begin{bmatrix} a_{\overline{x}} \\ a_{\overline{y}} \\ a_{\overline{z}} \end{bmatrix} \tag{4.45}
$$

where the transformation matrix is given by

$$
\begin{bmatrix} \tilde{i}_{\bar{x}} & \tilde{j}_{\bar{x}} & \tilde{k}_{\bar{x}} \\ \tilde{i}_{\bar{y}} & \tilde{j}_{\bar{y}} & \tilde{k}_{\bar{y}} \\ \tilde{i}_{\bar{z}} & \tilde{j}_{\bar{z}} & \tilde{k}_{\bar{z}} \end{bmatrix} = \begin{bmatrix} \tilde{i}_{\bar{x}} & \tilde{j}_{\bar{x}} & \tilde{k}_{\bar{x}} \\ \tilde{i}_{\bar{y}} & \tilde{j}_{\bar{y}} & \tilde{k}_{\bar{y}} \\ \tilde{i}_{\bar{z}} & \tilde{j}_{\bar{z}} & \tilde{k}_{\bar{z}} \end{bmatrix}^{-1} = \begin{bmatrix} 1 + \dfrac{\partial \overline{u}}{\partial \overline{x}} & \dfrac{\partial \overline{u}}{\partial \overline{y}} & \dfrac{\partial \overline{u}}{\partial \overline{z}} \\[2mm] \dfrac{\partial \overline{v}}{\partial \overline{x}} & 1 + \dfrac{\partial \overline{v}}{\partial \overline{y}} & \dfrac{\partial \overline{v}}{\partial \overline{z}} \\[2mm] \dfrac{\partial \overline{w}}{\partial \overline{x}} & \dfrac{\partial \overline{w}}{\partial \overline{y}} & 1 + \dfrac{\partial \overline{w}}{\partial \overline{z}} \end{bmatrix}^{-1} \tag{4.46}
$$

4.3 HOMOGENEOUS DEFORMATION

Homogeneous deformation can be expressed as a composition of rotation **g** and stretch **s**:

$$
\mathbf{f}(\mathbf{p}) = \mathbf{g} \circ \mathbf{s}_1 = \mathbf{s}_2 \circ \mathbf{g} \tag{4.47}
$$

The deformation gradient for homogeneous deformation is therefore given by

$$
\mathbf{F} = \mathbf{RU} = \mathbf{VR} \tag{4.48}
$$

where

$$
\mathbf{R} = \nabla \mathbf{g}
$$
$$
\mathbf{U} = \nabla \mathbf{s}_1 \tag{4.49}
$$
$$
\mathbf{V} = \nabla \mathbf{s}_2
$$

It is worth mentioning that by definition of homogeneous deformation tensors **F**, **R** and **U** are constant tensors, i.e. they do not change from point to point (over the spatial domain).

Tensor $\mathbf{U}$ is called the right stretch tensor. Tensor $\mathbf{V}$ is called the left stretch tensor. Tensors $\mathbf{U}$ and $\mathbf{V}$ are symmetric and positive definite tensors with

$$\det \mathbf{U} = \det \mathbf{V} = |\det \mathbf{F}| > 0 \qquad (4.50)$$

representing the ratio between the volume of the deformed material element and the initial material element. Both right and left stretch tensors can be decomposed into a succession of three extensions in three mutually orthogonal directions:

$$\mathbf{U} = \sum_{i=1}^{3} \lambda_i \bar{\mathbf{e}}_i \otimes \bar{\mathbf{e}}_i \qquad (4.51)$$

$$= \mathbf{U}_1 \mathbf{U}_2 \mathbf{U}_3$$

$$\mathbf{U}_i = \mathbf{I} + (s_i - 1)\bar{\mathbf{e}}_i \otimes \bar{\mathbf{e}}_i$$

$$\mathbf{V} = \sum_{i=1}^{3} \lambda_i \tilde{\mathbf{e}}_i \otimes \tilde{\mathbf{e}}_i \qquad (4.52)$$

$$= \mathbf{V}_1 \mathbf{V}_2 \mathbf{V}_3$$

$$\mathbf{V}_i = \mathbf{I} + (s_i - 1)\tilde{\mathbf{e}}_i \otimes \tilde{\mathbf{e}}_i$$

where scalars s_1, s_2 and s_3 represent principal stretches. Principal stretches are in essence elongation in the principal directions, i.e. the ratio between the deformed length and initial length. Principal stretches are the same for both right and left stretch tensor.

The right stretch tensor $\mathbf{U}$ therefore represents successive stretching of the material element in three mutually orthogonal directions. This stretching is applied before any rotation. In contrast, left stretch tensor $\mathbf{V}$ represents successive stretching of the material element in three mutually orthogonal directions applied after rotation. Thus, the principal directions of left stretch tensor $\mathbf{V}$ are obtained by simply rotating the principal directions associated with the right stretch tensor $\mathbf{U}$:

$$\tilde{\mathbf{e}}_1 = \mathbf{R}\bar{\mathbf{e}}_1$$

$$\tilde{\mathbf{e}}_2 = \mathbf{R}\bar{\mathbf{e}}_2 \qquad (4.53)$$

$$\tilde{\mathbf{e}}_3 = \mathbf{R}\bar{\mathbf{e}}_3$$

4.4 STRAIN

Using stretch tensors $\mathbf{U}$ and $\mathbf{V}$, different strain tensors can be defined. For instance,

$$\mathbf{C} = \mathbf{F}^T \mathbf{F} = (\mathbf{R}\mathbf{U})^T (\mathbf{R}\mathbf{U}) = \mathbf{U}^T \mathbf{R}^T \mathbf{R}\mathbf{U} = \mathbf{U}^T \mathbf{U}$$

$$\mathbf{B} = \mathbf{F}\mathbf{F}^T = (\mathbf{V}\mathbf{R})(\mathbf{V}\mathbf{R})^T = \mathbf{V}\mathbf{R}\mathbf{R}^T \mathbf{V}^T = \mathbf{V}\mathbf{V}^T = \mathbf{V}^2 \qquad (4.54)$$

are the right and left Cauchy–Green strain tensor, respectively. In the case of the right Cauchy–Green strain tensor $\mathbf{C}$, rotation occurs after stretch. Thus, the left Cauchy–Green strain tensor is best represented using a local frame, i.e. using configuration before any rotation has taken place

$$(\bar{\mathbf{i}}, \bar{\mathbf{j}}, \bar{\mathbf{k}}) \tag{4.55}$$

In the case of the left Cauchy–Green strain tensor **B**, stretch occurs after rotation. Thus, the left Cauchy–Green strain tensor is best represented using a deformed local frame, i.e. using configuration after the rotation has taken place:

$$(\tilde{\mathbf{i}}, \tilde{\mathbf{j}}, \tilde{\mathbf{k}}) \tag{4.56}$$

Other strain tensors (also called strain measures) can be derived using stretch tensors, for instance strain measure in the form

$$e = \begin{cases} \dfrac{s^m - 1}{m} & \text{for } m \neq 0 \\ \ln(s) & \text{for } m = 0 \end{cases} \tag{4.57}$$

(where s is a stretch). Depending on the parameter m, the following strain tensors are obtained:

- For $m = 2$ a Green–St. Venant strain tensor is obtained:

$$\overline{\mathbf{E}}_2 = \tfrac{1}{2}(\mathbf{U}^2 - \mathbf{I}) = \tfrac{1}{2}(\mathbf{C} - \mathbf{I}) \tag{4.58}$$

- For $m = -2$ a Almanasi–Hmel strain tensor is obtained:

$$\tilde{\mathbf{E}}_{-2} = \tfrac{1}{2}(\mathbf{I} - \mathbf{V}^{-2}) = \tfrac{1}{2}(\mathbf{I} - \mathbf{B}^{-1}) \tag{4.59}$$

- For $m = 0$ a logarithmic strain tensor is obtained:

$$\overline{\mathbf{E}}_0 = \ln \mathbf{U} = \tfrac{1}{2} \ln \mathbf{C} \tag{4.60}$$

- For $m = 1$ a Biot strain tensor is obtained:

$$\overline{\mathbf{E}}_1 = \mathbf{U} - \mathbf{I} \tag{4.61}$$

4.5 STRESS

4.5.1 Cauchy stress tensor

Cauchy's theorem makes it possible for integral relations of momentum balance to be replaced by partial differential equations. The necessary and sufficient condition for the momentum balance law to be satisfied is the existence of a spatial tensor field **T** (also called Cauchy stress) such that:

- for a vector **m**, the surface traction force is given by

$$\mathbf{s}(\mathbf{m}) = \mathbf{T}\mathbf{m} \tag{4.62}$$

- $\mathbf{T}$ is symmetric and positive, i.e. for any vector $\mathbf{a}$

$$\mathbf{a}\cdot\mathbf{T}\mathbf{a} > 0 \text{ unless } \mathbf{a} = \mathbf{0} \tag{4.63}$$

- $\mathbf{T}$ satisfies the equation of motion

$$div\,\mathbf{T} + \mathbf{b} = \rho\dot{\mathbf{v}} \tag{4.64}$$

where $\mathbf{s}$ is the traction force corresponding to the surface of deformed configuration, $\mathbf{b}$ is the body force per unit volume of the deformed configuration, ρ is the density measured per unit volume of the deformed configuration, and $\mathbf{m}$ is the normal to the boundary of deformed configuration.

The Cauchy stress tensor in essence represents a linear mapping where a given outward surface normal $\mathbf{m}$ is mapped onto a total surface traction force $\mathbf{s}$. The surface normal $\mathbf{m}$ is of magnitude equal to the surface area it represents. Thus, for instance, if the magnitude of $\mathbf{m}$ is doubled, the total surface traction is doubled. This is easily understood, for doubling the normal $\mathbf{m}$ is equivalent to doubling the surface area.

The matrix of Cauchy stress tensor in the global frame

$$(\mathbf{i}, \mathbf{j}, \mathbf{k}) \tag{4.65}$$

is given by

$$\mathbf{T} = \begin{bmatrix} t_{xx} & t_{xy} & t_{xz} \\ t_{yx} & t_{yy} & t_{yz} \\ t_{zx} & t_{zy} & t_{zz} \end{bmatrix} \tag{4.66}$$

where the first index indicates the direction of the stress component (direction of traction force) and the second index denotes the corresponding surface normal. Thus, t_{xy} is the traction force in the x-direction on the surface 'in the y-direction', i.e.

$$\text{force } t_{xy}\,\mathbf{i} \quad \text{on the surface } 1\mathbf{j} \tag{4.67}$$

Cauchy stress refers to the force per unit area of the deformed configuration. Components of Cauchy stress tensor are shown in Figure 4.7.

For any given surface defined by surface normal $\mathbf{m}$ (Figure 4.8), the surface traction in global orthonormal frame

$$(\mathbf{i}, \mathbf{j}, \mathbf{k}) \tag{4.68}$$

is obtained by simply multiplying the matrix of tensor $\mathbf{T}$ with the matrix of vector $\mathbf{m}$, i.e.

$$\mathbf{s} = s_x\mathbf{i} + s_y\mathbf{j} + s_z\mathbf{k} = \begin{bmatrix} s_x \\ s_y \\ s_z \end{bmatrix} = \begin{bmatrix} t_{xx} & t_{xy} & t_{xz} \\ t_{yx} & t_{yy} & t_{yz} \\ t_{zx} & t_{zy} & t_{zz} \end{bmatrix} \begin{bmatrix} m_x \\ m_y \\ m_z \end{bmatrix} \tag{4.69}$$

From this expression, it is obvious that the Cauchy stress is a linear mapping from the space of normals (surfaces) into a space of forces, where each normal is mapped onto a corresponding traction force. This is logical, because in mathematical terms a tensor is

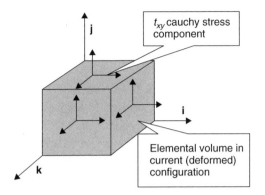

Figure 4.7   Cauchy stress tensor components in deformed configuration. Note that the material element is taken in the directions of the global base vectors.

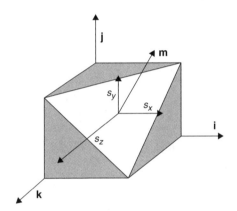

Figure 4.8   Surface traction force.

defined as the linear mapping from one vector space into another vector space. In simple terms, for a given normal it returns traction force.

4.5.2   First Piola-Kirchhoff stress tensor

Cauchy stress is defined for the deformed body, i.e. stress is defined as traction force per unit area of the deformed body. Sometimes it is easier to deal with the initial configuration than with the deformed configuration. Thus, stress called 'Piola-Kirchhoff stress' is introduced. The first Piola-Kirchhoff stress is defined by the expression

$$\mathbf{S}_1 = (\det \mathbf{F})\mathbf{T}\mathbf{F}^{-T} \tag{4.70}$$

To understand the above definition, it is necessary to investigate each element in the above formula. First, the term $(\det \mathbf{F})$ represents the ratio of volume of deformed material element and volume of undeformed material element (initial configuration). Thus, if a material element in the initial configuration is unit volume, then the same material element after deformation occupies a volume equal to (det **F**). The second term can be written as follows:

$$\mathbf{F}^{-T} = (\mathbf{F}^{-1})^T$$

$$= \left(\begin{bmatrix} 1 + \dfrac{\partial u}{\partial x} & \dfrac{\partial u}{\partial y} & \dfrac{\partial u}{\partial z} \\[2mm] \dfrac{\partial v}{\partial x} & 1 + \dfrac{\partial v}{\partial y} & \dfrac{\partial v}{\partial z} \\[2mm] \dfrac{\partial w}{\partial x} & \dfrac{\partial w}{\partial y} & 1 + \dfrac{\partial w}{\partial z} \end{bmatrix}^{-1} \right)^T \tag{4.71}$$

$$= \begin{bmatrix} \tilde{i}_x & \tilde{j}_x & \tilde{k}_x \\ \tilde{i}_y & \tilde{j}_y & \tilde{k}_y \\ \tilde{i}_z & \tilde{j}_z & \tilde{k}_z \end{bmatrix}^{-T} = [\tilde{\mathbf{i}} \ \ \tilde{\mathbf{j}} \ \ \tilde{\mathbf{k}}]^{-T}$$

Matrix

$$\begin{bmatrix} \tilde{i}_x & \tilde{j}_x & \tilde{k}_x \\ \tilde{i}_y & \tilde{j}_y & \tilde{k}_y \\ \tilde{i}_z & \tilde{j}_z & \tilde{k}_z \end{bmatrix} \tag{4.72}$$

is not an orthogonal matrix, thus the inverse matrix of this matrix is a non-orthogonal matrix:

$$\left(\begin{bmatrix} 1 + \dfrac{\partial u}{\partial x} & \dfrac{\partial u}{\partial y} & \dfrac{\partial u}{\partial z} \\[2mm] \dfrac{\partial v}{\partial x} & 1 + \dfrac{\partial v}{\partial y} & \dfrac{\partial v}{\partial z} \\[2mm] \dfrac{\partial w}{\partial x} & \dfrac{\partial w}{\partial y} & 1 + \dfrac{\partial w}{\partial z} \end{bmatrix}^{-1} \right)^T = \begin{bmatrix} \hat{\mathbf{i}} \\ \hat{\mathbf{j}} \\ \hat{\mathbf{k}} \end{bmatrix}^T = \begin{bmatrix} \hat{i}_x & \hat{j}_x & \hat{k}_x \\ \hat{i}_y & \hat{j}_y & \hat{k}_y \\ \hat{i}_z & \hat{j}_z & \hat{k}_z \end{bmatrix} = [\hat{\mathbf{i}} \ \ \hat{\mathbf{j}} \ \ \hat{\mathbf{k}}]$$

$$\tag{4.73}$$

This inverse matrix represents the global components of a new triad of vectors

$$(\hat{\mathbf{i}}, \hat{\mathbf{j}}, \hat{\mathbf{k}}) \tag{4.74}$$

This triad of vectors is associated with a deformed configuration. Vectors of this triad have the following property:

$$\begin{array}{lll} \tilde{\mathbf{i}} \cdot \hat{\mathbf{i}} = 1; & \tilde{\mathbf{j}} \cdot \hat{\mathbf{i}} = 0; & \tilde{\mathbf{k}} \cdot \hat{\mathbf{i}} = 0 \\[1mm] \tilde{\mathbf{i}} \cdot \hat{\mathbf{j}} = 0; & \tilde{\mathbf{j}} \cdot \hat{\mathbf{j}} = 0; & \tilde{\mathbf{k}} \cdot \hat{\mathbf{j}} = 0 \\[1mm] \tilde{\mathbf{i}} \cdot \hat{\mathbf{k}} = 0; & \tilde{\mathbf{j}} \cdot \hat{\mathbf{k}} = 0; & \tilde{\mathbf{k}} \cdot \hat{\mathbf{k}} = 1 \end{array} \tag{4.75}$$

This means that, for instance, vector

$$\hat{\tilde{\mathbf{k}}} \text{ is parallel to } (\tilde{\mathbf{i}} \times \tilde{\mathbf{j}}) \tag{4.76}$$

i.e. it is orthogonal to the surface formed by these two vectors The volume of the deformed element of material is given by

$$(\det \mathbf{F}) = (\tilde{\mathbf{i}} \times \tilde{\mathbf{j}}) \cdot \hat{\tilde{\mathbf{k}}} \tag{4.77}$$

Since

$$\frac{\hat{\tilde{\mathbf{k}}} \cdot \tilde{\mathbf{k}}}{(\tilde{\mathbf{i}} \times \tilde{\mathbf{j}}) \cdot \hat{\tilde{\mathbf{k}}}} = \frac{\left|\hat{\tilde{\mathbf{k}}}\right|}{\left|(\tilde{\mathbf{i}} \times \tilde{\mathbf{j}})\right|} \tag{4.78}$$

and

$$\hat{\tilde{\mathbf{k}}} \cdot \tilde{\mathbf{k}} = 1; \quad (\tilde{\mathbf{i}} \times \tilde{\mathbf{j}}) \cdot \hat{\tilde{\mathbf{k}}} = \det \mathbf{F} \tag{4.79}$$

it follows that

$$\left|\hat{\tilde{\mathbf{k}}}\right| (\det \mathbf{F}) = \left|(\tilde{\mathbf{i}} \times \tilde{\mathbf{j}})\right| \tag{4.80}$$

Since vector

$$(\tilde{\mathbf{i}} \times \tilde{\mathbf{j}}) \text{ is parallel to the vector } \hat{\tilde{\mathbf{k}}} \tag{4.81}$$

it follows that

$$(\det \mathbf{F})\hat{\tilde{\mathbf{k}}} = (\tilde{\mathbf{i}} \times \tilde{\mathbf{j}}) \tag{4.82}$$

By analogy, the following expression for the other two vectors of the triad is obtained:

$$(\det \mathbf{F})\hat{\tilde{\mathbf{i}}} = (\tilde{\mathbf{j}} \times \tilde{\mathbf{k}}) \tag{4.83}$$

$$(\det \mathbf{F})\hat{\tilde{\mathbf{j}}} = (\tilde{\mathbf{k}} \times \tilde{\mathbf{i}})$$

In other words, these vectors simply represent the surface normals of the deformed material element. By substituting these into the defining formula for the first Piola-Kirchhoff stress, the following expressions are obtained:

$$\begin{aligned}
\mathbf{S} &= (\det \mathbf{F})\mathbf{T}\mathbf{F}^{-T} \\
&= \mathbf{T}[(\det \mathbf{F})\mathbf{F}^{-T}] \\
&= \mathbf{T}(\det \mathbf{F})[\hat{\tilde{\mathbf{i}}}, \hat{\tilde{\mathbf{j}}}, \hat{\tilde{\mathbf{k}}}] \\
&= \mathbf{T}[\hat{\tilde{\mathbf{i}}}(\det \mathbf{F}), \hat{\tilde{\mathbf{j}}}(\det \mathbf{F}), \hat{\tilde{\mathbf{k}}}(\det \mathbf{F})] \\
&= \mathbf{T}[\hat{\tilde{\mathbf{i}}}(\det \mathbf{F})) + \mathbf{T}\hat{\tilde{\mathbf{j}}}(\det \mathbf{F}) + \mathbf{T}\hat{\tilde{\mathbf{k}}}(\det \mathbf{F})
\end{aligned} \tag{4.84}$$

These represent calculation of the traction forces on the surfaces

$$\hat{\tilde{\mathbf{i}}}(\det \mathbf{F})), \quad \hat{\tilde{\mathbf{j}}}(\det \mathbf{F}) \text{ and } \hat{\tilde{\mathbf{k}}}(\det \mathbf{F}) \tag{4.85}$$

The matrix form of traction force calculation is given by

$$\mathbf{S} = \begin{bmatrix} S_{x\hat{\tilde{x}}} & S_{x\hat{\tilde{y}}} & S_{x\hat{\tilde{z}}} \\ S_{y\hat{\tilde{x}}} & S_{y\hat{\tilde{y}}} & S_{y\hat{\tilde{z}}} \\ S_{z\hat{\tilde{x}}} & S_{z\hat{\tilde{y}}} & S_{z\hat{\tilde{z}}} \end{bmatrix} = \begin{bmatrix} t_{xx} & t_{xy} & t_{xz} \\ t_{yx} & t_{yy} & t_{yz} \\ t_{zx} & t_{zy} & t_{zz} \end{bmatrix} \begin{bmatrix} \hat{\tilde{i}}_x & \hat{\tilde{j}}_x & \hat{\tilde{k}}_x \\ \hat{\tilde{i}}_y & \hat{\tilde{j}}_y & \hat{\tilde{k}}_y \\ \hat{\tilde{i}}_z & \hat{\tilde{j}}_z & \hat{\tilde{k}}_z \end{bmatrix} (\det \mathbf{F}) \tag{4.86}$$

Thus, the first index indicates the direction of the stress component (global x, y or z direction). The second index indicates the surface of the material element the stress component is associated with (the surface that was initially normal to the initial x, y or z direction). The stress components of the first Piola-Kirchhoff stress are shown in Figure 4.9.

The first Piola-Kirchhoff stress represents the traction force per unit area of the initial configuration. However, when viewed in a deformed configuration it does not represent stress per unit area. It represents traction force component per deformed area that was initially a unit area, but after deformation it has changed.

The same can be said for the Cauchy stress tensor, which truly represents stress per unit area of the deformed configuration. When the initial configuration is taken into account, Cauchy stress does not represent stress per unit area.

Thus, when a deformed configuration is considered, stress per unit area for any material point is best expressed using the Cauchy stress tensor and material element defined by the global triad, but on a deformed body (i.e. the surfaces of the material element are orthogonal to the unit vectors of global triad). The stress components associated with

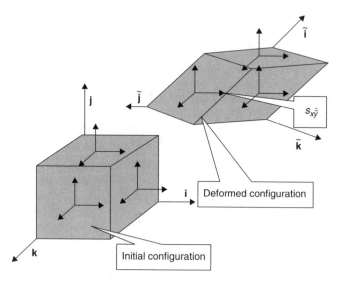

Figure 4.9 The first Piola-Kirchhoff stress components. Note that the material element is taken in the initial configuration in the directions of the vectors of the local triad.

the particular surfaces of the material element are orientated in the directions of vectors of the global triad. As these directions coincide with the edges of the material element, these components also represent normal and shear stress components. In the case of the first Piola-Kirchhoff stress, the material element is chosen to coincide with the local triad, which is chosen on the initial configuration. Thus, stress components represent surface traction per unit area of the initial configuration. The direction of the initial stress components corresponds with the directions of individual vectors of the global triad. Thus, when viewed on the initial configuration, they would appear as normal or tangential to the surfaces of the material element. However, this is not the case. The material element has deformed, and the surfaces of the material element are no longer orthogonal to the individual vectors of the global triad. Thus, the first Piola-Kirchhoff stress is best viewed as a surface traction per unit area of the initial configuration (undeformed configuration) with individual components expressed in the direction of the local triad. Thus, the only difference between these two stresses is in the surfaces to which individual stress components correspond. Traction force over a given surface can be obtained using the tensor

$$\int_{\partial P_t} \mathbf{T} \mathbf{m} \, dA = \int_{\partial P} (\det \mathbf{F}) \mathbf{T} \mathbf{F}^{-T} \mathbf{n} \, dA \tag{4.87}$$

where $\mathbf{m}$ is the outward unit normal field for a deformed configuration, expressed in some global coordinate system, and $\mathbf{n}$ is the corresponding outward unit normal field for the initial configuration expressed in the global coordinate system.

4.5.3   Second Piola-Kirchhoff stress tensor

As explained above, Cauchy stress components represent normal and tangential (shear) traction forces. This is not the case with the first Piola-Kirchhoff stress, where stress components are oriented at an arbitrary angle relative to the surface of the deformed elemental volume. Thus, the second Piola-Kirchhoff stress is introduced. It is defined by the expression

$$\sigma = \mathbf{F}^{-1} \mathbf{S} = (\det \mathbf{F}) \mathbf{F}^{-1} \mathbf{T} \mathbf{F}^{-T} \tag{4.88}$$

The same definition can be expressed using tensor matrices

$$\sigma = \begin{bmatrix} \hat{i}_x & \hat{j}_x & \hat{k}_x \\ \hat{i}_y & \hat{j}_y & \hat{k}_y \\ \hat{i}_z & \hat{j}_z & \hat{k}_z \end{bmatrix}^T \begin{bmatrix} S_{x\hat{x}} & S_{x\hat{y}} & S_{x\hat{z}} \\ S_{y\hat{x}} & S_{y\hat{y}} & S_{y\hat{z}} \\ S_{z\hat{x}} & S_{z\hat{y}} & S_{z\hat{z}} \end{bmatrix} = \begin{bmatrix} \hat{i} \\ \hat{j} \\ \hat{k} \end{bmatrix} \begin{bmatrix} S_{x\hat{x}} & S_{x\hat{y}} & S_{x\hat{z}} \\ S_{y\hat{x}} & S_{y\hat{y}} & S_{y\hat{z}} \\ S_{z\hat{x}} & S_{z\hat{y}} & S_{z\hat{z}} \end{bmatrix} \tag{4.89}$$

where, for instance, the component

$$\begin{bmatrix} \hat{i}_x & \hat{i}_y & \hat{i}_z \end{bmatrix} \begin{bmatrix} S_{x\hat{x}} \\ S_{y\hat{x}} \\ S_{z\hat{x}} \end{bmatrix} \tag{4.90}$$

represents the traction force components in the $\tilde{i}$ direction. In other words, the second Piola-Kirchhoff stress is obtained by expressing components of traction force on each surface of the deformed material element in terms of the base vectors of the deformed triad:

$$(\tilde{i}, \tilde{j}, \tilde{k}) \tag{4.91}$$

as shown in Figure 4.10.

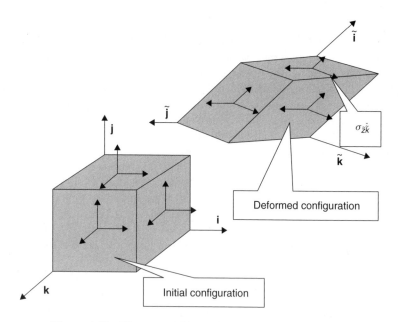

Figure 4.10   The second Piola-Kirchhoff stress components.

4.6 CONSTITUTIVE LAW

For an elastic body a constitutive law (physical equations) can be written as

$$\mathbf{T}(\mathbf{x}) = \hat{\mathbf{T}}(\mathbf{F}(\mathbf{p}), \mathbf{p}) \tag{4.92}$$

where $\mathbf{T}$ is Cauchy stress tensor, $\mathbf{x}$ represents deformed configuration and $\mathbf{p}$ represents initial configuration. A necessary and sufficient condition that the response is independent of the observer is that

$$\mathbf{R}\hat{\mathbf{T}}(\mathbf{F}(\mathbf{p}), \mathbf{p})\mathbf{R}^T = \hat{\mathbf{T}}(\mathbf{R}\mathbf{F}(\mathbf{p}), \mathbf{p}) \tag{4.93}$$

i.e. if rotation $\mathbf{R}$ is applied to the elastic body, the stress should not change. Actually, this rotation could also be viewed as rotation of the global triad. Rotation of an elastic body is equivalent to the rotation of the global coordinate system in the opposite direction by

$$\mathbf{R}^{-1} = \mathbf{R}^T \tag{4.94}$$

The body is called isotropic if

$$\hat{\mathbf{T}}(\mathbf{F}(\mathbf{p}), \mathbf{p}) = \hat{\mathbf{T}}(\mathbf{F}(\mathbf{p})\mathbf{R}, \mathbf{p}) \tag{4.95}$$

Many materials cannot undergo large (finite) strains, and often it is the case that only small strains are possible before fracture or failure occurs. In addition, in many problems of practical engineering importance the deformation gradients are also small, resulting in the deformed configuration being almost identical to the initial configuration.

The first case is the case of small strains, and the second case is the case of small displacements. In the case of small strains and small displacements, a suitable strain tensor is the so-called infinitesimal strain:

$$\mathbf{E} = \tfrac{1}{2}(\nabla \mathbf{u} + \nabla \mathbf{u}^T) \tag{4.96}$$

In the combined finite-discrete element method, the strain may be small in most problems of practical importance. However, the displacements are almost never small, thus (4.96) does not apply. Small strains only imply small stretches, while rotations and displacements are large. In such a case of small strains and large rotations, the deformation gradient can be decomposed into a stretch followed by rotation

$$\mathbf{F} = \mathbf{RU} \tag{4.97}$$

Stretch $\mathbf{U}$ is a result of displacements $\mathbf{u}$, i.e.

$$\mathbf{U} = \mathbf{I} + \nabla \mathbf{u} \tag{4.98}$$

Also,

$$\mathbf{C} = \mathbf{F}^T \mathbf{F} = \mathbf{U}^2 \tag{4.99}$$

where

$$\mathbf{U}^2 =
\begin{bmatrix}
1 + \dfrac{\partial u}{\partial x} & \dfrac{\partial v}{\partial x} & \dfrac{\partial w}{\partial x} \\[2mm]
\dfrac{\partial u}{\partial y} & 1 + \dfrac{\partial v}{\partial y} & \dfrac{\partial w}{\partial y} \\[2mm]
\dfrac{\partial u}{\partial z} & \dfrac{\partial v}{\partial z} & 1 + \dfrac{\partial w}{\partial z}
\end{bmatrix}
\begin{bmatrix}
1 + \dfrac{\partial u}{\partial x} & \dfrac{\partial u}{\partial y} & \dfrac{\partial u}{\partial z} \\[2mm]
\dfrac{\partial v}{\partial x} & 1 + \dfrac{\partial v}{\partial y} & \dfrac{\partial v}{\partial z} \\[2mm]
\dfrac{\partial w}{\partial x} & \dfrac{\partial w}{\partial y} & 1 + \dfrac{\partial w}{\partial z}
\end{bmatrix} \tag{4.100}$$

which after neglecting higher order terms yields

$$\mathbf{U}^2 =
\begin{bmatrix}
1 + 2\dfrac{\partial u}{\partial x} & \dfrac{\partial u}{\partial y} + \dfrac{\partial v}{\partial x} & \dfrac{\partial u}{\partial z} + \dfrac{\partial w}{\partial x} \\[2mm]
\dfrac{\partial u}{\partial y} + \dfrac{\partial v}{\partial x} & 1 + 2\dfrac{\partial v}{\partial y} & \dfrac{\partial v}{\partial z} + \dfrac{\partial w}{\partial y} \\[2mm]
\dfrac{\partial u}{\partial z} + \dfrac{\partial w}{\partial x} & \dfrac{\partial v}{\partial z} + \dfrac{\partial w}{\partial y} & 1 + 2\dfrac{\partial w}{\partial z}
\end{bmatrix} \tag{4.101}$$

This means that the small strain tensor can be approximated by

$$\mathbf{E} = \tfrac{1}{2}(\mathbf{U}^2 - \mathbf{I}) \tag{4.102}$$

In other words, if the strains are small, a small strain tensor (engineering strain) is obtained using the formula

$$\overline{\mathbf{E}} = \tfrac{1}{2}(\mathbf{F}^T \mathbf{F} - \mathbf{I}) = \tfrac{1}{2}[(\mathbf{RU})^T (\mathbf{RU}) - \mathbf{I}] = \tfrac{1}{2}(\mathbf{U}^T \mathbf{U} - \mathbf{I}) \tag{4.103}$$

This strain tensor is called a right Green–St. Venant strain tensor. It is worth mentioning that although strains are small, rotations in the combined finite-element method are always finite. With the right stretch tensor $\mathbf{U}$ a material is first stretched in the principal directions. This is followed by rotation. Thus, the right small strain tensor corresponds to the initial configuration in a sense that strain components expressed using a global triad are correct when applied to the initial configuration.

An equivalent small strain tensor is obtained using the left stretch tensor $\mathbf{V}$:

$$\tilde{\mathbf{E}} = \tfrac{1}{2}(\mathbf{F}\mathbf{F}^T - \mathbf{I}) = \tfrac{1}{2}[(\mathbf{V}\mathbf{R})(\mathbf{V}\mathbf{R})^T - \mathbf{I}] = \tfrac{1}{2}(\mathbf{V}\mathbf{V}^T - \mathbf{I}) \tag{4.104}$$

This strain tensor is called the left Green–St. Venant strain tensor. The left stretch tensor is defined in such a way that rotation occurs before stretching, i.e. stretching in three principal directions occurs on the rotated configuration. As rotation in the combined finite-discrete element method is always finite regardless of the strains, the strain tensor defined by equation (4.104) is generally different from the strain tensor obtained using equation (4.103).

The right Green–St. Venant strain tensor refers to the initial configuration. The left Green–St. Venant strain tensor refers to the deformed (current) configuration. Since strains are small, to obtain stresses from strains a small strain elasticity constitutive law can be employed. For homogeneous isotropic material the stress-strain relationship is given by Hooks law. Hooks law in terms of principal stresses and strains is given by

$$\sigma_1 = E \frac{1-v}{(1+v)(1-2v)} \left[\varepsilon_1 + \frac{v}{1-v}\varepsilon_2 + \frac{v}{1-v}\varepsilon_3 \right] \tag{4.105}$$

$$\sigma_2 = E \frac{1-v}{(1+v)(1-2v)} \left[\frac{v}{1-v}\varepsilon_1 + \varepsilon_2 + \frac{v}{1-v}\varepsilon_3 \right] \tag{4.106}$$

$$\sigma_3 = E \frac{1-v}{(1+v)(1-2v)} \left[\frac{v}{1-v}\varepsilon_1 + \frac{v}{1-v}\varepsilon_2 + \varepsilon_3 \right] \tag{4.107}$$

Additional decomposition of the small strain tensor in the form

$$\begin{bmatrix} \varepsilon_1 & 0 & 0 \\ 0 & \varepsilon_2 & 0 \\ 0 & 0 & \varepsilon_3 \end{bmatrix} = \begin{bmatrix} \varepsilon_1 - \varepsilon_s & 0 & 0 \\ 0 & \varepsilon_2 - \varepsilon_s & 0 \\ 0 & 0 & \varepsilon_3 - \varepsilon_s \end{bmatrix} + \begin{bmatrix} \varepsilon_s & 0 & 0 \\ 0 & \varepsilon_s & 0 \\ 0 & 0 & \varepsilon_s \end{bmatrix} \tag{4.108}$$

where

$$\varepsilon_s = \frac{1}{3}(\varepsilon_1 + \varepsilon_2 + \varepsilon_3) \tag{4.109}$$

separates change of volume of the material element from the change in shape of the material element. Hooks law expressed in terms of the strain components given by equation (4.108) is as follows:

$$\sigma_1 = \frac{1}{(1+v)}E(\varepsilon_1 - \varepsilon_s) + \frac{1}{(1-2v)}E\varepsilon_s \tag{4.110}$$

$$\sigma_2 = \frac{1}{(1+v)}E(\varepsilon_2 - \varepsilon_s) + \frac{1}{(1-2v)}E\varepsilon_s \tag{4.111}$$

$$\sigma_1 = \frac{1}{(1+v)}E(\varepsilon_3 - \varepsilon_s) + \frac{1}{(1-2v)}E\varepsilon_s \tag{4.112}$$

Unlike Lamé constants, the above formulation completely separates volumetric strains in a sense that the first part does not produce any change in the volume of the material element, while the second part does not produce any change in the shape of the material element.

To write the above constitutive law in terms of a Green–St. Venants strain tensor, homogeneous deformation is expressed as a composition of rotation $\mathbf{g}$, followed by the shape changing stretch s_d, followed by the volume changing stretch s_s:

$$\mathbf{f}(\mathbf{p}) = \mathbf{s}_s \circ \mathbf{s}_d \circ \mathbf{g} \tag{4.113}$$

The deformation gradient for this deformation is given by

$$\mathbf{F} = \mathbf{V}_s \mathbf{V}_d \mathbf{R} \quad \text{where} \quad \mathbf{R} = \nabla \mathbf{g}; \quad \mathbf{V}_d = \nabla \mathbf{s}_d; \quad \mathbf{V}_s = \nabla \mathbf{s}_s; \tag{4.114}$$

It is worth mentioning that by definition,

$$|\det \mathbf{F}| = \det \mathbf{V}_s; \quad \textbf{and} \quad \det \mathbf{V}_d = 1 \tag{4.115}$$

The left Green–St. Venant strain tensor is therefore given by

$$\tilde{\mathbf{E}} = \frac{1}{2}(\mathbf{F}\mathbf{F}^T - \mathbf{I}) = \frac{1}{2}[(\mathbf{V}_s \mathbf{V}_d \mathbf{R})(\mathbf{V}_s \mathbf{V}_d \mathbf{R})^T - \mathbf{I}] \tag{4.116}$$

$$= \frac{1}{2}[(\mathbf{V}_s \mathbf{V}_d \mathbf{R})\mathbf{R}^T(\mathbf{V}_s \mathbf{V}_d)^T - \mathbf{I}] = \frac{1}{2}[(\mathbf{V}_s \mathbf{V}_d \mathbf{R}\mathbf{R}^T \mathbf{V}_d^T \mathbf{V}_s^T - \mathbf{I}]$$

$$= \frac{1}{2}[(\mathbf{V}_s \mathbf{V}_d \mathbf{V}_d^T \mathbf{V}_s^T - \mathbf{I}] = \frac{1}{2}[(\mathbf{V}_d \mathbf{V}_d^T (|\det \mathbf{F}|)^{2/3} - \mathbf{I}]$$

The last term in the above equation comes from the fact that the volumetric stretch carries all volume change with it, and can therefore be written as three identical successive stretches in any three mutually orthogonal directions:

$$\mathbf{V}_s = \mathbf{I}\sqrt[3]{|\det \mathbf{F}|} \tag{4.117}$$

Thus

$$\mathbf{V}_s \mathbf{V}_s^T = \mathbf{I}\,(|\det \mathbf{F}|)^{2/3} \tag{4.118}$$

The Green–St. Venant strain tensor due to the shape changing stretch is therefore given by

$$\tilde{\mathbf{E}}_d = \frac{1}{2}(\mathbf{V}_d \mathbf{V}_d^T - \mathbf{I}) = \frac{1}{2}\left(\frac{\mathbf{F}\mathbf{F}^T}{(|\det \mathbf{F}|)^{2/3}} - \mathbf{I}\right) \tag{4.119}$$

while the Green–St. Venant strain tensor due to volume changing stretch is given by

$$\tilde{\mathbf{E}}_s = \frac{1}{2}(\mathbf{V}_s \mathbf{V}_s^T - \mathbf{I}) = \frac{1}{2}(\mathbf{I}(|\det \mathbf{F}|)^{2/3} - \mathbf{I}) = \mathbf{I}\left(\frac{(|\det \mathbf{F}|)^{2/3} - 1}{2}\right) \tag{4.120}$$

The Green–St. Venant strain tensor due to the shape changing stretch has no volume changing component in it. In a similar way, the Green–St. Venant strain tensor due to the volume changing stretch has no shape changing component in it. In other words, by using these two strain tensors, change in the shape of the material element is completely separated from change in the volume of the material element. A constitutive law for homogeneous isotropic material is derived by analogy with the constitutive law described above in terms of small strains. The physical equations obtained are as follows:

$$\mathbf{T} = \frac{E}{(1+v)}\tilde{\mathbf{E}}_d + \frac{E}{(1-2v)}\tilde{\mathbf{E}}_s \tag{4.121}$$

It was explained above that the strains are small because the material fails or fractures as significant strains are reached. This is not the case with all materials; for instance, rubber can undergo very large strains before failure. However, under hydrostatic pressure many other materials can undergo significant change in volume without being damaged. To cater for such materials that undergo small shape change but finite volume change, the first part of the above constitutive law should be independent of the current volume of the material element. As **T** is the Cauchy stress tensor, this is not the case, for if the volume decreases the surface tractions on the surfaces of material element will decrease as well. The change in surface area of each of the six surfaces of the material element with change in volume is proportional to

$$(|\det \mathbf{F}|)^{2/3} \tag{4.122}$$

Thus, to take into account the change in the surface area of the surfaces of the material element with a change in the volume of the material element, the following modification of the above constitutive law is adopted:

$$\mathbf{T} = \frac{E}{(1+v)}\frac{1}{(|\det\mathbf{F}|)^{2/3}}\tilde{\mathbf{E}}_d + \frac{E}{(1-2v)}\frac{1}{(|\det\mathbf{F}|)^{2/3}}\tilde{\mathbf{E}}_s \tag{4.123}$$

Practical implementation of the above concepts is described in detail in the following sections.

It is worth noting that this constitutive law applies only to linear elastic homogeneous and isotropic materials. However, in principle, it is possible to implement any constitutive law. For instance, for rubber-like materials undergoing large (finite) strains, it is easier to express the constitutive law in terms of logarithmic strains. However, in such a case, base vectors coinciding with principal stretch directions must be employed. These base vectors are orthogonal to each other. The matrix of the left stretch tensor is therefore diagonal, i.e.

$$\mathbf{V} = \begin{bmatrix} s_1 & 0 & 0 \\ 0 & s_2 & 0 \\ 0 & 0 & s_3 \end{bmatrix} \tag{4.124}$$

The matrix of the left Cauchy–Green strain tensor is also diagonal:

$$\mathbf{B} = \mathbf{V}\mathbf{V}^T = \begin{bmatrix} s_1^2 & 0 & 0 \\ 0 & s_2^2 & 0 \\ 0 & 0 & s_3^2 \end{bmatrix} \tag{4.125}$$

The matrix of the Green–St. Venant strain tensor is therefore given as follows:

$$\mathbf{E} = \frac{1}{2}(\mathbf{B} - \mathbf{I}) = \begin{bmatrix} (s_1^2 - 1)/2 & 0 & 0 \\ 0 & (s_2^2 - 1)/2 & 0 \\ 0 & 0 & (s_3^2 - 1)/2 \end{bmatrix} \qquad (4.126)$$

In a similar way, the matrix of the logarithmic strain tensor strain tensor is given by

$$\mathbf{E}_o = \begin{bmatrix} \ln(s_1) & 0 & 0 \\ 0 & \ln(s_2) & 0 \\ 0 & 0 & \ln(s_3) \end{bmatrix} \qquad (4.127)$$

The left stretch tensor can be expressed as a composition of shape changing stretch followed by volume changing stretch:

$$\mathbf{V} = \mathbf{V}_s \mathbf{V}_d = \begin{bmatrix} s & 0 & 0 \\ 0 & s & 0 \\ 0 & 0 & s \end{bmatrix} \begin{bmatrix} s_1/s & 0 & 0 \\ 0 & s_2/s & 0 \\ 0 & 0 & s_3/s \end{bmatrix} \qquad (4.128)$$

This represents the stretching of elemental volume, as shown in Figure 4.11, where the elemental volume of edge a (infinitesimally small number) is considered. It can be observed from the figure that the three edges of the initial volume have stretched by a factor s_1, s_2 and s_3, respectively. Thus, the elemental volume is subject to strain in each direction. The Green–St. Venant strain tensor quantifies this strain as given by equation (4.126).

The same elemental volume has changed its volume. The volume change can be expressed as the ratio

$$s_v = \frac{s_1 a \cdot s_2 a \cdot s_3 a}{a^3} = s_1 s_2 s_3 \qquad (4.129)$$

The matrix of the logarithmic strain tensor is calculated from stretch components as follows:

$$\mathbf{E}_o = \begin{bmatrix} \ln(s_1/s) & 0 & 0 \\ 0 & \ln(s_2/s) & 0 \\ 0 & 0 & \ln(s_3/s) \end{bmatrix} + \begin{bmatrix} \ln(s) & 0 & 0 \\ 0 & \ln(s) & 0 \\ 0 & 0 & \ln(s) \end{bmatrix} \qquad (4.130)$$

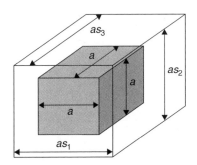

Figure 4.11   Stretching of elemental volume in three orthogonal directions.

Using these logarithmic strain components, the constitutive law can be formulated for any material, thus yielding the Cauchy stress components.

4.7 CONSTANT STRAIN TRIANGLE FINITE ELEMENT

In the combined finite-discrete element method, finite element discretisation of discrete elements is also used to process contact interaction. To arrive at efficient contact interaction algorithms, it is important to employ the simplest possible geometry of finite elements. In 2D space this is a three noded triangle (Figure 4.12). The deformation of this element is shown in Figure 4.13.

To describe this deformation, two frames are introduced:

- *Initial frame:* this frame corresponds to the initial (undeformed) configuration. The base vectors of this frame are identical in magnitude and orientation with the two edges of the finite element, as shown in Figure 4.14. Thus, base vectors are neither unit vectors nor orthogonal to each other:

$$(\hat{\mathbf{i}}, \hat{\mathbf{j}}) \tag{4.131}$$

- *Deformed initial frame:* this frame corresponds to the deformed (current) configuration. The base vectors of this frame are identical in magnitude and orientation with two edges of the deformed triangle, as shown in Figure 4.14. The base vectors are not unit vectors. They are not orthogonal to each other either:

$$(\check{\mathbf{i}}, \check{\mathbf{j}}) \tag{4.132}$$

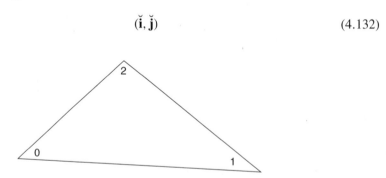

Figure 4.12   Geometry of the constant strain triangle finite element.

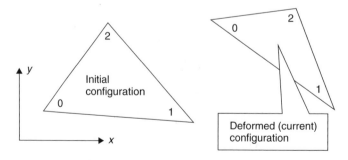

Figure 4.13   Left: Initial configuration, right; deformed configuration.

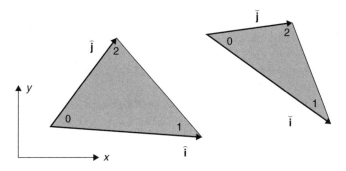

Figure 4.14   Left: initial frame, right; deformed initial frame.

The base vectors of the initial frame can be expressed using the base vectors of the deformed initial frame:

$$\hat{\mathbf{i}} = \hat{i}_{\breve{x}}\breve{\mathbf{i}} + \hat{i}_{\breve{x}}\breve{\mathbf{j}}$$

$$\hat{\mathbf{j}} = \hat{j}_{\breve{x}}\breve{\mathbf{i}} + \hat{j}_{\breve{y}}\breve{\mathbf{j}}$$
(4.133)

Also, the base vectors of the deformed frame can be expressed using the base vectors of the initial frame:

$$\breve{\mathbf{i}} = \breve{i}_{\hat{x}}\hat{\mathbf{i}} + \breve{i}_{\hat{x}}\hat{\mathbf{j}}$$

$$\breve{\mathbf{j}} = \breve{j}_{\hat{x}}\hat{\mathbf{i}} + \breve{j}_{\hat{y}}\hat{\mathbf{j}}$$
(4.134)

Any vectors **a** or **b** can be written using either base, as shown in Figure 4.15. Vector **a** is, for instance, given by

$$\mathbf{a} = 0.3\hat{\mathbf{i}} + 0.9\hat{\mathbf{j}}$$
(4.135)

and vector **b** by

$$\mathbf{b} = 0.4\breve{\mathbf{i}} + 0.3\breve{\mathbf{j}}$$
(4.136)

In implementation of the combined finite-discrete element method, both vectors are defined using the global base:

$$(\mathbf{i}, \mathbf{j})$$
(4.137)

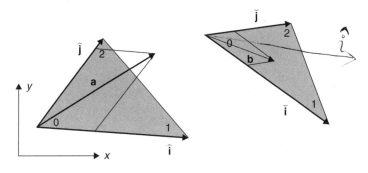

Figure 4.15   Left: initial frame, right; deformed initial frame.

Components of vectors **a** and **b** in the global frame are as follows:

$$\mathbf{a} = a_x \mathbf{i} + a_y \mathbf{j}$$
$$\mathbf{b} = b_x \mathbf{i} + b_y \mathbf{j}$$

(4.138)

In a similar way, components of the base vectors of the initial frame written in the global frame are given by

$$\hat{\mathbf{i}} = \hat{i}_x \mathbf{i} + \hat{i}_y \mathbf{j}$$
$$\hat{\mathbf{j}} = \hat{j}_x \mathbf{i} + \hat{j}_y \mathbf{j}$$

(4.139)

The components of the base vectors of the deformed initial frame written in the global frame are

$$\breve{\mathbf{i}} = \breve{i}_x \mathbf{i} + \breve{i}_y \mathbf{j}$$
$$\breve{\mathbf{j}} = \breve{j}_x \mathbf{i} + \breve{j}_y \mathbf{j}$$

(4.140)

It follows that vector **a** can be written as

$$\mathbf{a} = a_x \mathbf{i} + a_y \mathbf{j}$$

(4.141)

$$= a_{\hat{x}} \hat{\mathbf{i}} + a_{\hat{y}} \hat{\mathbf{j}}$$
$$= a_{\hat{x}}(\hat{i}_x \mathbf{i} + \hat{i}_y \mathbf{j}) + a_{\hat{y}}(\hat{j}_x \mathbf{i} + \hat{j}_y \mathbf{j})$$
$$= (a_{\hat{x}} \hat{i}_x + a_{\hat{y}} \hat{j}_x)\mathbf{i} + (a_{\hat{x}} \hat{i}_y + a_{\hat{y}} \hat{j}_y)\mathbf{j}$$

Thus

$$\begin{bmatrix} a_x \\ a_y \end{bmatrix} = \begin{bmatrix} \hat{i}_x & \hat{j}_x \\ \hat{i}_y & \hat{j}_y \end{bmatrix} \begin{bmatrix} a_{\hat{x}} \\ a_{\hat{y}} \end{bmatrix}$$

(4.142)

The matrix

$$\begin{bmatrix} \hat{i}_x & \hat{j}_x \\ \hat{i}_y & \hat{j}_y \end{bmatrix}$$

(4.143)

is called the initial transformation matrix.

Vector **b** can be written as

$$\mathbf{b} = b_x \mathbf{i} + b_y \mathbf{j}$$

(4.144)

$$= b_{\breve{x}} \breve{\mathbf{i}} + b_{\breve{y}} \breve{\mathbf{j}}$$
$$= b_{\breve{x}}(\breve{i}_x \mathbf{i} + \breve{i}_y \mathbf{j}) + b_{\breve{y}}(\breve{j}_x \mathbf{i} + \breve{j}_y \mathbf{j})$$
$$= (b_{\breve{x}} \breve{i}_x + b_{\breve{y}} \breve{j}_x)\mathbf{i} + (b_{\breve{x}} \breve{i}_y + b_{\breve{y}} \breve{j}_y)\mathbf{j}$$

Thus

$$\begin{bmatrix} b_x \\ b_y \end{bmatrix} = \begin{bmatrix} \breve{i}_x & \breve{j}_x \\ \breve{i}_y & \breve{j}_y \end{bmatrix} \begin{bmatrix} b_{\breve{x}} \\ b_{\breve{y}} \end{bmatrix}$$

(4.145)

The matrix

$$\begin{bmatrix} \breve{i}_x & \breve{j}_x \\ \breve{i}_y & \breve{j}_y \end{bmatrix} \tag{4.146}$$

is called the deformed initial transformation matrix.

Transformation of vector components from the global frame into the initial frame is obtained using an inverse initial transformation matrix:

$$\begin{bmatrix} a_{\hat{x}} \\ a_{\hat{y}} \end{bmatrix} = \begin{bmatrix} \hat{i}_x & \hat{j}_x \\ \hat{i}_y & \hat{j}_y \end{bmatrix}^{-1} \begin{bmatrix} a_x \\ a_y \end{bmatrix} \tag{4.147}$$

Transformation of vector components from the global frame into the deformed initial frame is obtained using an inverse deformed initial frame matrix:

$$\begin{bmatrix} b_{\breve{x}} \\ b_{\breve{y}} \end{bmatrix} = \begin{bmatrix} \breve{i}_x & \breve{j}_x \\ \breve{i}_y & \breve{j}_y \end{bmatrix}^{-1} \begin{bmatrix} b_x \\ b_y \end{bmatrix} \tag{4.148}$$

Transformation of vector components from the initial into the deformed initial frame is obtained as follows:

$$\mathbf{a} = a_{\hat{x}}\hat{\mathbf{i}} + a_{\hat{y}}\hat{\mathbf{j}} \tag{4.149}$$

$$= a_{\breve{x}}\breve{\mathbf{i}} + a_{\breve{y}}\breve{\mathbf{j}}$$

$$= a_{\hat{x}}(\hat{i}_x\breve{\mathbf{i}} + \hat{i}_y\breve{\mathbf{j}}) + a_{\hat{y}}(\hat{j}_x\breve{\mathbf{i}} + \hat{j}_y\breve{\mathbf{j}})$$

$$= (a_{\hat{x}}\hat{i}_{\breve{x}} + a_{\hat{y}}\hat{j}_{\breve{x}})\breve{\mathbf{i}} + (a_{\hat{x}}\hat{i}_{\breve{y}} + a_{\hat{y}}\hat{j}_{\breve{y}})\breve{\mathbf{j}}$$

Thus

$$\begin{bmatrix} a_{\breve{x}} \\ a_{\breve{y}} \end{bmatrix} = \begin{bmatrix} \hat{i}_{\breve{x}} & \hat{j}_{\breve{x}} \\ \hat{i}_{\breve{y}} & \hat{j}_{\breve{y}} \end{bmatrix} \begin{bmatrix} a_{\hat{x}} \\ a_{\hat{y}} \end{bmatrix} \tag{4.150}$$

Transformation of vector components from the deformed initial frame into the initial frame is obtained as follows:

$$\mathbf{a} = a_{\breve{x}}\breve{\mathbf{i}} + a_{\breve{y}}\breve{\mathbf{j}} \tag{4.151}$$

$$= a_{\hat{x}}\hat{\mathbf{i}} + a_{\hat{y}}\hat{\mathbf{j}}$$

$$= a_{\breve{x}}(\breve{i}_x\hat{\mathbf{i}} + \breve{i}_y\hat{\mathbf{j}}) + a_{\breve{y}}(\breve{j}_x\hat{\mathbf{i}} + \breve{j}_y\hat{\mathbf{j}})$$

$$= (a_{\breve{x}}\breve{i}_{\hat{x}} + a_{\breve{y}}\breve{j}_{\hat{x}})\hat{\mathbf{i}} + (a_{\breve{x}}\breve{i}_{\hat{y}} + a_{\breve{y}}\breve{j}_{\hat{y}})\hat{\mathbf{j}}$$

Thus

$$\begin{bmatrix} a_{\hat{x}} \\ a_{\hat{y}} \end{bmatrix} = \begin{bmatrix} \breve{i}_{\hat{x}} & \breve{j}_{\hat{x}} \\ \breve{i}_{\hat{y}} & \breve{j}_{\hat{y}} \end{bmatrix} \begin{bmatrix} a_{\breve{x}} \\ a_{\breve{y}} \end{bmatrix} = \begin{bmatrix} \hat{i}_{\breve{x}} & \hat{j}_{\breve{x}} \\ \hat{i}_{\breve{y}} & \hat{j}_{\breve{y}} \end{bmatrix}^{-1} \begin{bmatrix} a_{\breve{x}} \\ a_{\breve{y}} \end{bmatrix} \tag{4.152}$$

Use of three noded finite element results in the deformation gradient being constant over the domain of the finite element. This is because the deformation over the domain of finite element is a linear function of type

$$x_c = \alpha_x x_i + \beta_x y_i$$
$$y_c = \alpha_y x_i + \beta_y y_i$$

(4.153)

where x_c and y_c represent current coordinates (deformed configuration), and x_i and y_i represent the initial coordinates (undeformed configuration).

Deformation over the domain of the finite element is expressed in terms of the deformation at nodes of the triangle. Deformation at nodes of a triangle is defined by the initial nodal coordinates corresponding to the initial configuration, and current nodal coordinates corresponding to the deformed configuration.

The easiest way to calculate the deformation gradient for a three noded triangle is to use the deformed initial frame

$$(\check{\mathbf{i}}, \check{\mathbf{j}})$$

(4.154)

Using this frame, the matrix of the deformation gradient tensor is

$$\mathbf{F} = \begin{bmatrix} \dfrac{\partial x_c}{\partial \widehat{x}_i} & \dfrac{\partial x_c}{\partial \widehat{y}_i} \\ \dfrac{\partial y_c}{\partial \widehat{x}_i} & \dfrac{\partial y_c}{\partial \widehat{y}_i} \end{bmatrix}$$

(4.155)

where x_c and y_c are the current global coordinates given in the global frame

$$(\mathbf{i}, \mathbf{j})$$

(4.156)

while

$$\widehat{x} \text{ and } \widehat{y}$$

(4.157)

are the current coordinates expressed using the initial frame. In other words,

$$x_c = x_c(\widehat{x}_i, \widehat{y}_i)$$
$$y_c = y_c(\widehat{x}_i, \widehat{y}_i)$$

(4.158)

and the physical meaning of operators

$$\frac{\partial x_c}{\partial \widehat{x}_i} \text{ and } \frac{\partial x_c}{\partial \widehat{y}_i}$$

(4.159)

is the change in global current coordinate when from point **p** in the initial configuration one moves towards point **q**, which is in the direction of the initial base vectors

$$(\widehat{\mathbf{i}}, \widehat{\mathbf{j}})$$

(4.160)

As this base vector coincides in both direction and magnitude with two edges of the triangle in the initial configuration, this means that the matrix of the deformation gradient tensor for the constant strain triangle is simply

$$
\mathbf{F} = \begin{bmatrix} \dfrac{\partial x_c}{\partial \widehat{x}_i} & \dfrac{\partial x_c}{\partial \widehat{y}_i} \\ \dfrac{\partial y_c}{\partial \widehat{x}_i} & \dfrac{\partial y_c}{\partial \widehat{y}_i} \end{bmatrix} = \begin{bmatrix} x_{1c} - x_{0c} & x_{2c} - x_{0c} \\ y_{1c} - y_{0c} & y_{2c} - y_{0c} \end{bmatrix} \tag{4.161}
$$

where x_{1c} and y_{1c} are the current global coordinates of node 1; x_{2c} and y_{2c} are the current global coordinates of node 2 and x_{0c} and y_{0c} are the current global coordinates of node 0. It is worth noting that node 0 in the initial configuration coincides with the origin of the initial frame.

The matrix of the velocity gradient tensor is obtained in the same way:

$$
\mathbf{L} = \begin{bmatrix} \dfrac{\partial v_{xc}}{\partial \widehat{x}_i} & \dfrac{\partial v_{xc}}{\partial \widehat{y}_i} \\ \dfrac{\partial v_{yc}}{\partial \widehat{x}_i} & \dfrac{\partial v_{yc}}{\partial \widehat{y}_i} \end{bmatrix} = \begin{bmatrix} v_{1xc} - v_{0xc} & v_{2xc} - v_{0xc} \\ v_{1yc} - v_{0yc} & v_{2yc} - v_{0yc} \end{bmatrix} \tag{4.162}
$$

where v_{ixc} is the current velocity of node i in the direction of the global base vector $\mathbf{i}$, while is v_{iyc} is the current velocity of node i in the direction of the global base vector $\mathbf{j}$.

The matrix of the deformation gradient tensor in the form

$$
\mathbf{F} = \begin{bmatrix} \dfrac{\partial x_c}{\partial x_i} & \dfrac{\partial x_c}{\partial y_i} \\ \dfrac{\partial y_c}{\partial x_i} & \dfrac{\partial y_c}{\partial y_i} \end{bmatrix} \tag{4.163}
$$

is obtained by using the initial base, which is given by

$$
\begin{bmatrix} \widehat{i}_x & \widehat{j}_x \\ \widehat{i}_y & \widehat{j}_y \end{bmatrix} = \begin{bmatrix} x_{1i} - x_{0i} & x_{2i} - x_{0i} \\ y_{1i} - y_{0i} & y_{2i} - y_{0i} \end{bmatrix} \tag{4.164}
$$

where x_{1i} and y_{1i} are the initial global coordinates of node 1; x_{2i} and y_{2i} are the initial global coordinates of node 2, and x_{0i} and y_{0i} are the initial global coordinates of node 0. Using coordinate transformations explained above, the following expression for the matrix of the deformation gradient tensor is obtained:

$$
\mathbf{F} = \begin{bmatrix} \dfrac{\partial x_c}{\partial x_i} & \dfrac{\partial x_c}{\partial y_i} \\ \dfrac{\partial y_c}{\partial x_i} & \dfrac{\partial y_c}{\partial y_i} \end{bmatrix} = \begin{bmatrix} \dfrac{\partial x_c}{\partial \widehat{x}_i} & \dfrac{\partial x_c}{\partial \widehat{y}_i} \\ \dfrac{\partial y_c}{\partial \widehat{x}_i} & \dfrac{\partial y_c}{\partial \widehat{y}_i} \end{bmatrix} \begin{bmatrix} \widehat{i}_x & \widehat{j}_x \\ \widehat{i}_y & \widehat{j}_y \end{bmatrix}^{-1} \tag{4.165}
$$

The physical meaning of this transformation is best understood by considering the inverse transformation

$$
\mathbf{F} = \begin{bmatrix} \dfrac{\partial x_c}{\partial \widehat{x}_i} & \dfrac{\partial x_c}{\partial \widehat{y}_i} \\[3mm] \dfrac{\partial y_c}{\partial \widehat{x}_i} & \dfrac{\partial y_c}{\partial \widehat{y}_i} \end{bmatrix} = \begin{bmatrix} \dfrac{\partial x_c}{\partial x_i} & \dfrac{\partial x_c}{\partial y_i} \\[3mm] \dfrac{\partial y_c}{\partial x_i} & \dfrac{\partial y_c}{\partial y_i} \end{bmatrix} \begin{bmatrix} \widehat{i}_x & \widehat{j}_x \\[2mm] \widehat{i}_y & \widehat{j}_y \end{bmatrix} \tag{4.166}
$$

The first line of $\mathbf{F}$ matrix is the vector

$$
\begin{bmatrix} \dfrac{\partial x_c}{\partial \widehat{x}_i} & \dfrac{\partial x_c}{\partial \widehat{y}_i} \end{bmatrix} \tag{4.167}
$$

the components of which are given in global frame $(\mathbf{i}, \mathbf{j})$. It represents the gradient of the function $x_c(x_i, y_i)$. Column

$$
\widehat{\mathbf{i}} = \begin{bmatrix} \widehat{i}_x \\[2mm] \widehat{i}_y \end{bmatrix} \tag{4.168}
$$

is the base vector of the initial frame

$$
(\widehat{\mathbf{i}}, \widehat{\mathbf{j}}) \tag{4.169}
$$

expressed using the global base $(\mathbf{i}, \mathbf{j})$. Thus,

$$
\begin{bmatrix} \dfrac{\partial x_c}{\partial x_i} & \dfrac{\partial x_c}{\partial y_i} \end{bmatrix} \begin{bmatrix} \widehat{i}_x \\[2mm] \widehat{i}_y \end{bmatrix} = \dfrac{\partial x_c}{\partial \widehat{x}_i} \tag{4.170}
$$

The same applies to

$$
\begin{bmatrix} \dfrac{\partial y_c}{\partial x_i} & \dfrac{\partial y_c}{\partial y_i} \end{bmatrix} \begin{bmatrix} \widehat{i}_x \\[2mm] \widehat{i}_y \end{bmatrix} = \dfrac{\partial y_c}{\partial \widehat{x}_i} \tag{4.171}
$$

Having understood the physical meaning behind the components of the deformation gradient, it is evident that the same applies to the velocity gradient:

$$
\mathbf{L} = \begin{bmatrix} \dfrac{\partial v_{xc}}{\partial x_i} & \dfrac{\partial v_{xc}}{\partial y_i} \\[3mm] \dfrac{\partial v_{yc}}{\partial x_i} & \dfrac{\partial v_{yc}}{\partial y_i} \end{bmatrix} = \begin{bmatrix} \dfrac{\partial v_{xc}}{\partial \widehat{x}_i} & \dfrac{\partial v_{xc}}{\partial \widehat{y}_i} \\[3mm] \dfrac{\partial v_{yc}}{\partial \widehat{x}_i} & \dfrac{\partial v_{yc}}{\partial \widehat{y}_i} \end{bmatrix} \begin{bmatrix} \widehat{i}_x & \widehat{j}_x \\[2mm] \widehat{i}_y & \widehat{j}_y \end{bmatrix}^{-1} \tag{4.172}
$$

The deformation gradient is comprised of either rotation followed by stretch

$$
\mathbf{F} = \mathbf{VR} \tag{4.173}
$$

or stretch followed by rotation

$$
\mathbf{F} = \mathbf{RU} \tag{4.174}
$$

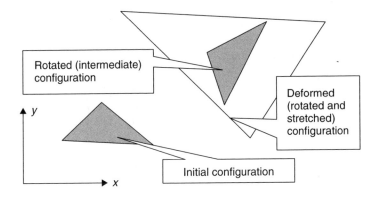

Figure 4.16   Rotation followed by stretch.

Either can be employed. However, employing the left stretch $\mathbf{V}$ is computationally more efficient. The physical meaning of this is that the triangle is first rotated and then stretched (Figure 4.16).

Using the left stretch tensor, the matrix of the left Cauchy–Green strain tensor is calculated as follows:

$$\mathbf{B} = \mathbf{FF}^T = \mathbf{VV}^T = \begin{bmatrix} \dfrac{\partial x_c}{\partial x_i} & \dfrac{\partial x_c}{\partial y_i} \\ \dfrac{\partial y_c}{\partial x_i} & \dfrac{\partial y_c}{\partial y_i} \end{bmatrix} \begin{bmatrix} \dfrac{\partial x_c}{\partial x_i} & \dfrac{\partial y_c}{\partial x_i} \\ \dfrac{\partial x_c}{\partial y_i} & \dfrac{\partial y_c}{\partial y_i} \end{bmatrix} \tag{4.175}$$

In a similar way, the matrix of the rate of deformation tensor is obtained from the velocity gradient:

$$\mathbf{D} = \frac{1}{2}(\mathbf{L} + \mathbf{L})^{\mathcal{K}} = \frac{1}{2} \left(\begin{bmatrix} \dfrac{\partial v_{xc}}{\partial x_i} & \dfrac{\partial v_{xc}}{\partial y_i} \\ \dfrac{\partial v_{yc}}{\partial x_i} & \dfrac{\partial v_{yc}}{\partial y_i} \end{bmatrix} + \begin{bmatrix} \dfrac{\partial v_{xc}}{\partial x_i} & \dfrac{\partial v_{yc}}{\partial x_i} \\ \dfrac{\partial v_{xc}}{\partial y_i} & \dfrac{\partial v_{yc}}{\partial y_i} \end{bmatrix} \right) \tag{4.176}$$

From the matrix of the left Cauchy–Green strain tensor, for small strains, the matrix of the Green–St. Venant strain tensor is obtained as follows:

$$\check{\mathbf{E}} = \frac{1}{2}(\mathbf{V}^2 - I) = \frac{1}{2}(\mathbf{B} - \mathbf{I}) \tag{4.177}$$

$$\frac{1}{2} \left(\begin{bmatrix} \dfrac{\partial x_c}{\partial x_i} & \dfrac{\partial x_c}{\partial y_i} \\ \dfrac{\partial y_c}{\partial x_i} & \dfrac{\partial y_c}{\partial y_i} \end{bmatrix} \begin{bmatrix} \dfrac{\partial x_c}{\partial x_i} & \dfrac{\partial y_c}{\partial x_i} \\ \dfrac{\partial x_c}{\partial y_i} & \dfrac{\partial y_c}{\partial y_i} \end{bmatrix} - \begin{bmatrix} 1 & 0 \\ 0 & 1 \end{bmatrix} \right)$$

This is represented as a shape changing part:

$$\check{\mathbf{E}}_d = \frac{1}{2}(\mathbf{V}^2/(|\det \mathbf{F}|) - \mathbf{I}) = \frac{1}{2}(\mathbf{B}/(|\det \mathbf{F}|) - \mathbf{I}) \tag{4.178}$$

$$\frac{1}{2}\left(\frac{1}{(|\det \mathbf{F}|)} \begin{bmatrix} \dfrac{\partial x_c}{\partial x_i} & \dfrac{\partial x_c}{\partial y_i} \\ \dfrac{\partial y_c}{\partial x_i} & \dfrac{\partial y_c}{\partial y_i} \end{bmatrix} \begin{bmatrix} \dfrac{\partial x_c}{\partial x_i} & \dfrac{\partial y_c}{\partial x_i} \\ \dfrac{\partial x_c}{\partial y_i} & \dfrac{\partial y_c}{\partial y_i} \end{bmatrix} - \begin{bmatrix} 1 & 0 \\ 0 & 1 \end{bmatrix} \right)$$

and volume changing part

$$\check{\mathbf{E}}_s = \frac{1}{2}(\mathbf{V}_s \mathbf{V}_s^T - \mathbf{I}) = \frac{1}{2}\left(\mathbf{I}\left[(|\det \mathbf{F}|)^{1/2} \right]^2 - \mathbf{I} \right) = \frac{1}{2}(|\det \mathbf{F}| - 1)\begin{bmatrix} 1 & 0 \\ 0 & 1 \end{bmatrix} \tag{4.179}$$

From the matrix of the strain tensor, the matrix of the Cauchy stress tensor is obtained using the constitutive law. For instance, for plane stress homogeneous isotropic material, using the plane stress equations for Hooks law, is given by

$$\sigma_1 = \frac{E}{(1 - v^2)}(\varepsilon_1 + v\varepsilon_2) \tag{4.180}$$

$$\sigma_2 = \frac{E}{(1 - v^2)}(v\varepsilon_1 + \varepsilon_2)$$

The same constitutive law can be written as follows:

$$\sigma_1 = \frac{E}{(1 + v)}\left[\varepsilon_1 - \frac{1}{2}(\varepsilon_1 + \varepsilon_2) \right] + \frac{E}{(1 - v)}\left[\frac{1}{2}(\varepsilon_1 + \varepsilon_2) \right] \tag{4.181}$$

$$\sigma_2 = \frac{E}{(1 + v)}\left[\varepsilon_2 - \frac{1}{2}(\varepsilon_1 + \varepsilon_2) \right] + \frac{E}{(1 - v)}\left[\frac{1}{2}(\varepsilon_1 + \varepsilon_2) \right]$$

where therefore 'volume' change

$$\left[\frac{1}{2}(\varepsilon_1 + \varepsilon_2) \right] \tag{4.182}$$

is separated from the change in the shape of the material element. By analogy and using the Green–St. Venant strain tensor, the following constitutive law is arrived at:

$$\mathbf{T} = \frac{E}{(1 + v)}\tilde{\mathbf{E}}_d + \frac{E}{(1 - v)}\tilde{\mathbf{E}}_s + 2\bar{\mu}\mathbf{D} \tag{4.183}$$

where the last term is due to the rate of deformation.

To make the first part independent of volume change, further modification by dividing the stress tensor by the ratio of the increase of edge length of the material element due to the volume change ($\sqrt{|\det F|}$) is adopted, which yields

$$\mathbf{T} = \frac{1}{\sqrt{|\det \mathbf{F}|}} \frac{E}{(1+v)} \tilde{\mathbf{E}}_d + \tag{4.184}$$

$$\frac{1}{\sqrt{|\det \mathbf{F}|}} \frac{E}{(1-v)} \tilde{\mathbf{E}}_s + \frac{2\overline{\mu}}{\sqrt{|\det \mathbf{F}|}}$$

These stress components apply to the deformed (current) configuration. The components of stress tensor $\mathbf{T}$ represent stress in the global frame per unit area of the deformed triangle. Thus traction force over the each of the edges of the triangle can be calculated using the normal on the edge of the deformed configuration, with components of the normal given in the global frame (Figure 4.17). Edge traction is given by the expression

$$\mathbf{s} = \mathbf{Tm} = \begin{bmatrix} s_x \\ s_y \end{bmatrix} = \begin{bmatrix} t_{xx} & t_{xy} \\ t_{yx} & t_{yy} \end{bmatrix} \begin{bmatrix} m_x \\ m_y \end{bmatrix} \tag{4.185}$$

Edge traction for each of the tree edges of the deformed configuration is distributed in equal proportion to each of the nodes belonging to a particular edge. To each node, equivalent nodal force

$$\mathbf{f} = \frac{1}{2}\mathbf{s} = \frac{1}{2} \begin{bmatrix} s_x \\ s_y \end{bmatrix} = \frac{1}{2} \begin{bmatrix} t_{xx} & t_{xy} \\ t_{yx} & t_{yy} \end{bmatrix} \begin{bmatrix} m_x \\ m_y \end{bmatrix} \tag{4.186}$$

is assigned.

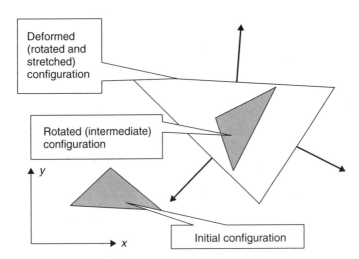

Figure 4.17   Normal vectors used in calculation of traction forces; magnitude of each normal is equal to the length of the corresponding edge.

4.8 CONSTANT STRAIN TETRAHEDRON FINITE ELEMENT

To arrive at an efficient discretised distributed potential force contact interaction algorithm in 3D, as discussed in Chapter 2, it is important to employ the simplest possible geometry of finite elements. In 3D space this is a four nodded tetrahedron (Figure 4.18).

These simple elements are now also widely used in the finite element method. Special techniques to avoid problems such as locking are employed. Locking occurs when the kinematics of nodes comprising the finite element mesh is over-constrained due to the presence of volumetric rigidity. In elastic problems, locking occurs when incompressible materials (materials with a Poisson ratio close to 0.5) are analysed. It can also occur when material nonlinearity such as plasticity is considered, in which case, very often by default, volumetric plastic strains are zero, which is to say that the material does not change volume in plastic deformations. The most widely used technique to avoid locking is to integrate volumetric strain over two or more neighbouring finite elements. With these modifications, the four node tetrahedron solid finite element can be used for a wide range of 3D problems, including nonlinear problems of elasticity and plasticity. The deformation of this element is shown in Figure 4.19.

To describe the deformation of the tetrahedron, two frames are introduced:

- The first frame is called the initial frame, because it corresponds to the initial (undeformed) configuration. The base vectors of this frame are identical in magnitude and orientation with three undeformed edges of the tetrahedron, as shown in Figure 4.20.

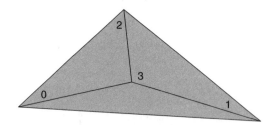

Figure 4.18   Solid tetrahedron finite element.

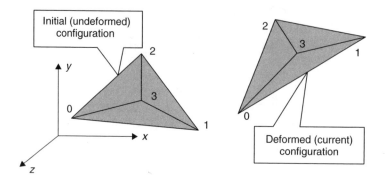

Figure 4.19   Left: Initial configuration, right; deformed configuration.

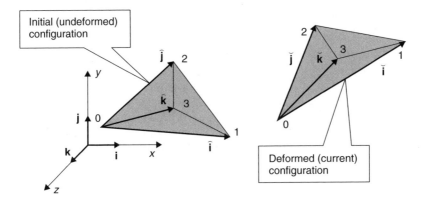

Figure 4.20 Global, initial and deformed initial frames of reference.

The base vectors of this frame are not unit vectors. These vectors are not mutually orthogonal to each other either:

$$(\widehat{\mathbf{i}}, \widehat{\mathbf{j}}, \widehat{\mathbf{k}}) \tag{4.187}$$

- The second frame is called the deformed initial frame, because it corresponds to the deformed (current configuration), while at instances when the deformed configuration is identical to the initial configuration it is identical to the initial frame. The base vectors of the deformed initial frame are identical in magnitude and orientation with three deformed edges of the tetrahedron, as shown in Figure 4.20. The base vectors are therefore not unit vectors, and they are not orthogonal to each other either:

$$(\widecheck{\mathbf{i}}, \widecheck{\mathbf{j}}, \widecheck{\mathbf{k}}) \tag{4.188}$$

Vector component transformation rules explained with the three noded triangle finite element are also valid here. The base vectors of the initial frame can be expressed using the base vectors of the deformed initial frame:

$$\widehat{\mathbf{i}} = \widehat{i}_{\widecheck{x}}\widecheck{\mathbf{i}} + \widehat{i}_{\widecheck{x}}\widecheck{\mathbf{j}} + \widehat{i}_{\widecheck{z}}\widecheck{\mathbf{k}} \tag{4.189}$$

$$\widehat{\mathbf{j}} = \widehat{j}_{\widecheck{x}}\widecheck{\mathbf{i}} + \widehat{j}_{\widecheck{y}}\widecheck{\mathbf{j}} + \widehat{j}_{\widecheck{z}}\widecheck{\mathbf{k}}$$

$$\widehat{\mathbf{k}} = \widehat{k}_{\widecheck{x}}\widecheck{\mathbf{i}} + \widehat{k}_{\widecheck{y}}\widecheck{\mathbf{j}} + \widehat{k}_{\widecheck{z}}\widecheck{\mathbf{k}}$$

Also, the base vectors of the deformed frame can be expressed using the base vectors of the initial frame:

$$\widecheck{\mathbf{i}} = \widecheck{i}_{\widehat{x}}\widehat{\mathbf{i}} + \widecheck{i}_{\widehat{x}}\widehat{\mathbf{j}} + \widecheck{i}_{\widehat{x}}\widehat{\mathbf{k}} \tag{4.190}$$

$$\widecheck{\mathbf{j}} = \widecheck{j}_{\widehat{x}}\widehat{\mathbf{i}} + \widecheck{j}_{\widehat{y}}\widehat{\mathbf{j}} + \widecheck{j}_{\widehat{z}}\widehat{\mathbf{k}}$$

$$\widehat{\mathbf{k}} = \widecheck{k}_{\widehat{x}}\widehat{\mathbf{i}} + \widecheck{k}_{\widehat{y}}\widehat{\mathbf{j}} + \widecheck{k}_{\widehat{z}}\widehat{\mathbf{k}}$$

In a similar way, components of the base vectors of the initial frame written in the global frame are given by

$$\hat{\mathbf{i}} = \hat{i}_x \mathbf{i} + \hat{i}_y \mathbf{j} + \hat{i}_z \mathbf{k} \tag{4.191}$$

$$\hat{\mathbf{j}} = \hat{j}_x \mathbf{i} + \hat{j}_y \mathbf{j} + \hat{j}_z \mathbf{k}$$

$$\hat{\mathbf{k}} = \hat{k}_x \mathbf{i} + \hat{k}_y \mathbf{j} + \hat{k}_z \mathbf{k}$$

Components of the base vectors of the deformed initial frame written in the global frame are

$$\check{\mathbf{i}} = \check{i}_x \mathbf{i} + \check{i}_y \mathbf{j} + \check{i}_z \mathbf{k} \tag{4.192}$$

$$\check{\mathbf{j}} = \check{j}_x \mathbf{i} + \check{j}_y \mathbf{j} + \check{j}_z \mathbf{k}$$

$$\check{\mathbf{k}} = \check{k}_x \mathbf{i} + \check{k}_y \mathbf{j} + \check{k}_z \mathbf{k}$$

Any vector **a** can be written using either frame

$$
\begin{aligned}
\mathbf{a} &= a_x \mathbf{i} + a_y \mathbf{j} + a_z \mathbf{k} \\
&= a_{\hat{x}} \hat{\mathbf{i}} + a_{\hat{y}} \hat{\mathbf{j}} + a_{\hat{z}} \hat{\mathbf{k}} \\
&= a_{\hat{x}}(\hat{i}_x \mathbf{i} + \hat{i}_y \mathbf{j} + \hat{i}_z \mathbf{k}) \\
&\quad + a_{\hat{y}}(\hat{j}_x \mathbf{i} + \hat{j}_y \mathbf{j} + \hat{j}_z \mathbf{k}) \\
&\quad + a_{\hat{z}}(\hat{k}_x \mathbf{i} + \hat{k}_y \mathbf{j} + \hat{k}_z \mathbf{k}) \\
&= (a_{\hat{x}} \hat{i}_x + a_{\hat{y}} \hat{j}_x + a_{\hat{z}} \hat{k}_x)\mathbf{i} \\
&\quad + (a_{\hat{x}} \hat{i}_y + a_{\hat{y}} \hat{j}_y + a_{\hat{z}} \hat{k}_y)\mathbf{j} \\
&\quad + (a_{\hat{x}} \hat{i}_z + a_{\hat{y}} \hat{j}_z + a_{\hat{z}} \hat{k}_z)\mathbf{k}
\end{aligned}
\tag{4.193}
$$

The transformation of vector components from one frame to another frame is given by

$$
\begin{bmatrix} a_x \\ a_y \\ a_z \end{bmatrix} =
\begin{bmatrix} \hat{i}_x & \hat{j}_x & \hat{k}_x \\ \hat{i}_y & \hat{j}_y & \hat{k}_y \\ \hat{i}_z & \hat{j}_z & \hat{k}_z \end{bmatrix}
\begin{bmatrix} a_{\hat{x}} \\ a_{\hat{y}} \\ a_{\hat{z}} \end{bmatrix}
\tag{4.194}
$$

The transformation matrix

$$
\begin{bmatrix} \hat{i}_x & \hat{j}_x & \hat{k}_x \\ \hat{i}_y & \hat{j}_y & \hat{k}_y \\ \hat{i}_z & \hat{j}_z & \hat{k}_z \end{bmatrix}
\tag{4.195}
$$

is called the initial transformation matrix. In a similar way, for any vector **b**

$$
\begin{aligned}
\mathbf{b} &= b_x \mathbf{i} + b_y \mathbf{j} + b_z \mathbf{k} \\
&= b_{\hat{x}} \check{\mathbf{i}} + b_{\hat{y}} \check{\mathbf{j}} + b_{\hat{z}} \mathbf{k}
\end{aligned}
\tag{4.196}
$$

transformation of vector components from the deformed initial frame into the global frame is given by

$$
\begin{bmatrix} b_x \\ b_y \\ b_z \end{bmatrix} = \begin{bmatrix} \breve{i}_x & \breve{j}_x & \breve{k}_x \\ \breve{i}_y & \breve{j}_y & \breve{k}_y \\ \breve{i}_z & \breve{j}_z & \breve{k}_z \end{bmatrix} \begin{bmatrix} b_{\breve{x}} \\ b_{\breve{y}} \\ b_{\breve{z}} \end{bmatrix}
\tag{4.197}
$$

The transformation matrix

$$
\begin{bmatrix} \breve{i}_x & \breve{j}_x & \breve{k}_x \\ \breve{i}_y & \breve{j}_y & \breve{k}_y \\ \breve{i}_z & \breve{j}_z & \breve{k}_z \end{bmatrix}
\tag{4.198}
$$

is called the deformed initial transformation matrix.

Transformation of vector components from the global frame into the initial and deformed initial frames is obtained using inverse initial transformation matrices:

$$
\begin{bmatrix} a_{\hat{x}} \\ a_{\hat{y}} \\ a_{\hat{z}} \end{bmatrix} = \begin{bmatrix} \hat{i}_x & \hat{j}_x & \hat{k}_x \\ \hat{i}_y & \hat{j}_y & \hat{k}_y \\ \hat{i}_z & \hat{j}_z & \hat{k}_z \end{bmatrix}^{-1} \begin{bmatrix} a_x \\ a_y \\ a_z \end{bmatrix} \quad \text{and} \quad \begin{bmatrix} b_{\breve{x}} \\ b_{\breve{y}} \\ b_{\breve{z}} \end{bmatrix} = \begin{bmatrix} \breve{i}_x & \breve{j}_x & \breve{k}_x \\ \breve{i}_y & \breve{j}_y & \breve{k}_y \\ \breve{i}_z & \breve{j}_z & \breve{k}_z \end{bmatrix}^{-1} \begin{bmatrix} b_x \\ b_y \\ b_z \end{bmatrix}
\tag{4.199}
$$

Deformation over the domain of a four noded tetrahedron is approximated using deformation at four nodes of the tetrahedron. This approximation is therefore of the type

$$
x_c = \alpha_x x_i + \beta_x y_i + \gamma_x z_i
\tag{4.200}
$$

$$
y_c = \alpha_y x_i + \beta_y y_i + \gamma_y z_i
$$

$$
z_c = \alpha_z x_i + \beta_z y_i + \gamma_z z_i
$$

where x_c, y_c and z_c represent the current coordinates corresponding to the deformed configuration, while x_i, y_i and z_i represent the initial coordinates corresponding to the undeformed configuration. Both sets of coordinates refer to the global frame:

$$
(\mathbf{i}, \mathbf{j}, \mathbf{k})
\tag{4.201}
$$

It follows from (4.200) that the deformation gradient over the domain of the tetrahedron finite element is constant. The easiest way to calculate this deformation gradient is to use the deformed initial frame:

$$
(\breve{\mathbf{i}}, \breve{\mathbf{j}}, \breve{\mathbf{k}})
\tag{4.202}
$$

Using this frame, the following matrix of the deformation gradient is obtained:

$$
\mathbf{F} = \begin{bmatrix} \dfrac{\partial x_c}{\partial \hat{x}_i} & \dfrac{\partial x_c}{\partial \hat{y}_i} & \dfrac{\partial x_c}{\partial \hat{z}_i} \\[2mm] \dfrac{\partial y_c}{\partial \hat{x}_i} & \dfrac{\partial y_c}{\partial \hat{y}_i} & \dfrac{\partial y_c}{\partial \hat{z}_i} \\[2mm] \dfrac{\partial z_c}{\partial \hat{x}_i} & \dfrac{\partial z_c}{\partial \hat{y}_i} & \dfrac{\partial z_c}{\partial \hat{z}_i} \end{bmatrix}
\tag{4.203}
$$

where x_c and y_c are current global coordinates given in the global frame

$$(\mathbf{i}, \mathbf{j}, \mathbf{k}) \tag{4.204}$$

while

$$\hat{x}, \hat{y} \text{ and } \hat{z} \tag{4.205}$$

are initial coordinates of material points over the domain of the tetrahedron expressed using the initial frame.

In other words, deformation is expressed in terms

$$x_c = x_c(\hat{x}_i, \hat{y}_i, \hat{z}_i) \tag{4.206}$$

$$y_c = y_c(\hat{x}_i, \hat{y}_i, \hat{z}_i)$$

$$z_c = z_c(\hat{x}_i, \hat{y}_i, \hat{z}_i)$$

As the base vectors of the initial frame coincide in both direction and magnitude with three edges of the tetrahedron in the initial configuration (undeformed edges of the tetrahedron), the deformation gradient for a constant strain triangle is simply

$$\mathbf{F} = \begin{bmatrix} \dfrac{\partial x_c}{\partial \hat{x}_i} & \dfrac{\partial x_c}{\partial \hat{y}_i} & \dfrac{\partial x_c}{\partial \hat{z}_i} \\[2mm] \dfrac{\partial y_c}{\partial \hat{x}_i} & \dfrac{\partial y_c}{\partial \hat{y}_i} & \dfrac{\partial y_c}{\partial \hat{z}_i} \\[2mm] \dfrac{\partial z_c}{\partial \hat{x}_i} & \dfrac{\partial z_c}{\partial \hat{y}_i} & \dfrac{\partial z_c}{\partial \hat{z}_i} \end{bmatrix} = \begin{bmatrix} x_{1c} - x_{0c} & x_{2c} - x_{0c} & x_{3c} - x_{0c} \\ y_{1c} - y_{0c} & y_{2c} - y_{0c} & y_{3c} - y_{0c} \\ z_{1c} - z_{0c} & z_{2c} - z_{0c} & z_{3c} - z_{0c} \end{bmatrix} \tag{4.207}$$

where x_{1c}, y_{1c} and z_{1c} are the current global coordinates of node 1; x_{2c}, y_{2c} and z_{2c} are the global coordinates of node 2 of the deformed tetrahedron; x_{3c}, y_{3c} and z_{3c} are the global coordinates of node 3 of the deformed tetrahedron; and x_{0c}, y_{0c} and z_{0c} are the global coordinates of node 0 of the deformed tetrahedron. Also, x_{1i}, y_{1i} and z_{1i} are the global coordinates of node 1 of the undeformed tetrahedron (initial configuration); x_{2i}, y_{2i} and z_{2i} are the global coordinates of node 2 of the undeformed tetrahedron (initial configuration); x_{3i}, y_{3i} and z_{3i} are the global coordinates of node 3 of the undeformed tetrahedron (initial configuration); x_{0i}, y_{0i} and z_{0i} are the global coordinates of node 0 of the undeformed tetrahedron (initial configuration).

The velocity gradient is obtained in the same way:

$$L = \begin{bmatrix} \dfrac{\partial v_{xc}}{\partial \hat{x}_i} & \dfrac{\partial v_{xc}}{\partial \hat{y}_i} & \dfrac{\partial v_{xc}}{\partial \hat{z}_i} \\[2mm] \dfrac{\partial v_{yc}}{\partial \hat{x}_i} & \dfrac{\partial v_{yc}}{\partial \hat{y}_i} & \dfrac{\partial v_{yc}}{\partial \hat{z}_i} \\[2mm] \dfrac{\partial v_{zc}}{\partial \hat{x}_i} & \dfrac{\partial v_{zc}}{\partial \hat{y}_i} & \dfrac{\partial v_{zc}}{\partial \hat{z}_i} \end{bmatrix} = \begin{bmatrix} v_{1xc} - v_{0xc} & v_{2xc} - v_{0xc} & v_{3xc} - v_{0xc} \\ v_{1yc} - v_{0yc} & v_{2yc} - v_{0yc} & v_{3yc} - v_{0yc} \\ v_{1zc} - v_{0zc} & v_{2zc} - v_{0zc} & v_{3zc} - v_{0zc} \end{bmatrix} \tag{4.208}$$

where, for instance, v_{ixc} is global (i.e. in the direction of the global base vector $\mathbf{i}$) velocity of node i of the deformed tetrahedron.

A deformation gradient in the form

$$
\mathbf{F} = \begin{bmatrix}
\dfrac{\partial x_c}{\partial x_i} & \dfrac{\partial x_c}{\partial y_i} & \dfrac{\partial x_c}{\partial z_i} \\[2ex]
\dfrac{\partial y_c}{\partial x_i} & \dfrac{\partial y_c}{\partial y_i} & \dfrac{\partial y_c}{\partial z_i} \\[2ex]
\dfrac{\partial z_c}{\partial x_i} & \dfrac{\partial z_c}{\partial y_i} & \dfrac{\partial z_c}{\partial z_i}
\end{bmatrix}
\tag{4.209}
$$

is obtained using the initial base. This base is given by the matrix

$$
\begin{bmatrix}
\widehat{i}_x & \widehat{j}_x & \widehat{k}_x \\
\widehat{i}_y & \widehat{j}_y & \widehat{k}_y \\
\widehat{i}_z & \widehat{j}_z & \widehat{k}_z
\end{bmatrix}
=
\begin{bmatrix}
x_{1i} - x_{0i} & x_{2i} - x_{0i} & x_{3i} - x_{0i} \\
y_{1i} - y_{0i} & y_{2i} - y_{0i} & y_{3i} - y_{0i} \\
z_{1i} - z_{0i} & z_{2i} - z_{0i} & z_{3i} - z_{0i}
\end{bmatrix}
\tag{4.210}
$$

It is worth noting that node 0 in the initial configuration coincides with the origin of the initial frame.

Using coordinate transformations explained above, the following expression for the deformation gradient is obtained:

$$
\mathbf{F} = \begin{bmatrix}
\dfrac{\partial x_c}{\partial x_i} & \dfrac{\partial x_c}{\partial y_i} & \dfrac{\partial x_c}{\partial z_i} \\[2ex]
\dfrac{\partial y_c}{\partial x_i} & \dfrac{\partial y_c}{\partial y_i} & \dfrac{\partial y_c}{\partial z_i} \\[2ex]
\dfrac{\partial z_c}{\partial x_i} & \dfrac{\partial z_c}{\partial y_i} & \dfrac{\partial z_c}{\partial z_i}
\end{bmatrix}
=
\begin{bmatrix}
\dfrac{\partial x_c}{\partial \widehat{x}_i} & \dfrac{\partial x_c}{\partial \widehat{y}_i} & \dfrac{\partial x_c}{\partial \widehat{z}_i} \\[2ex]
\dfrac{\partial y_c}{\partial \widehat{x}_i} & \dfrac{\partial y_c}{\partial \widehat{y}_i} & \dfrac{\partial y_c}{\partial \widehat{z}_i} \\[2ex]
\dfrac{\partial z_c}{\partial \widehat{x}_i} & \dfrac{\partial z_c}{\partial \widehat{y}_i} & \dfrac{\partial z_c}{\partial \widehat{z}_i}
\end{bmatrix}
\begin{bmatrix}
\widehat{i}_x & \widehat{j}_x & \widehat{k}_x \\
\widehat{i}_y & \widehat{j}_y & \widehat{k}_y \\
\widehat{i}_z & \widehat{j}_z & \widehat{k}_z
\end{bmatrix}^{-1}
\tag{4.211}
$$

In a similar way, the velocity gradient is given by

$$
L = \begin{bmatrix}
\dfrac{\partial v_{xc}}{\partial x_i} & \dfrac{\partial v_{xc}}{\partial y_i} & \dfrac{\partial v_{xc}}{\partial z_i} \\[2ex]
\dfrac{\partial v_{yc}}{\partial x_i} & \dfrac{\partial v_{yc}}{\partial y_i} & \dfrac{\partial v_{yc}}{\partial z_i} \\[2ex]
\dfrac{\partial v_{zc}}{\partial x_i} & \dfrac{\partial v_{zc}}{\partial y_i} & \dfrac{\partial v_{zc}}{\partial z_i}
\end{bmatrix}
=
\begin{bmatrix}
\dfrac{\partial v_{xc}}{\partial \widehat{x}_i} & \dfrac{\partial v_{xc}}{\partial \widehat{y}_i} & \dfrac{\partial v_{xc}}{\partial \widehat{z}_i} \\[2ex]
\dfrac{\partial v_{yc}}{\partial \widehat{x}_i} & \dfrac{\partial v_{yc}}{\partial \widehat{y}_i} & \dfrac{\partial v_{yc}}{\partial \widehat{z}_i} \\[2ex]
\dfrac{\partial v_{zc}}{\partial \widehat{x}_i} & \dfrac{\partial v_{zc}}{\partial \widehat{y}_i} & \dfrac{\partial v_{zc}}{\partial \widehat{z}_i}
\end{bmatrix}
\begin{bmatrix}
\widehat{i}_x & \widehat{j}_x & \widehat{k}_x \\
\widehat{i}_y & \widehat{j}_y & \widehat{k}_y \\
\widehat{i}_z & \widehat{j}_z & \widehat{k}_z
\end{bmatrix}^{-1}
\tag{4.212}
$$

As explained before, the deformation gradient is comprised of either rotation followed by stretch

$$
\mathbf{F} = \mathbf{VR}
\tag{4.213}
$$

or stretch followed by rotation

$$
\mathbf{F} = \mathbf{RU}
\tag{4.214}
$$

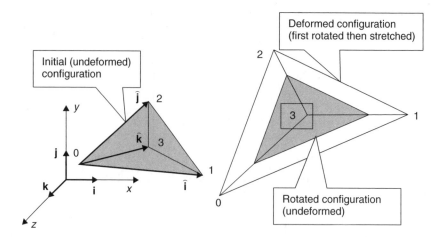

Figure 4.21   Left stretch: rotation of tetrahedron followed by stretch.

Either can be employed. However, employing the left stretch **V** is computationally more efficient. The physical meaning of this is that the tetrahedron is first rotated and then stretched (Figure 4.21).

Using left stretch tensor, the left Cauchy–Green strain tensor is calculated as follows:

$$
\mathbf{B} = \mathbf{F}\mathbf{F}^T = \mathbf{V}\mathbf{V}^T =
\begin{bmatrix}
\dfrac{\partial x_c}{\partial x_i} & \dfrac{\partial x_c}{\partial y_i} & \dfrac{\partial x_c}{\partial z_i} \\[2mm]
\dfrac{\partial y_c}{\partial x_i} & \dfrac{\partial y_c}{\partial y_i} & \dfrac{\partial y_c}{\partial z_i} \\[2mm]
\dfrac{\partial z_c}{\partial x_i} & \dfrac{\partial z_c}{\partial y_i} & \dfrac{\partial z_c}{\partial z_i}
\end{bmatrix}
\begin{bmatrix}
\dfrac{\partial x_c}{\partial x_i} & \dfrac{\partial y_c}{\partial x_i} & \dfrac{\partial z_c}{\partial x_i} \\[2mm]
\dfrac{\partial x_c}{\partial y_i} & \dfrac{\partial y_c}{\partial y_i} & \dfrac{\partial z_c}{\partial y_i} \\[2mm]
\dfrac{\partial x_c}{\partial z_i} & \dfrac{\partial y_c}{\partial z_i} & \dfrac{\partial z_c}{\partial z_i}
\end{bmatrix}
\tag{4.215}
$$

In a similar way, the rate of deformation is obtained from velocity gradient

$$
\mathbf{D} = \frac{1}{2}(\mathbf{L} + \mathbf{L}^T) = \frac{1}{2}\left(
\begin{bmatrix}
\dfrac{\partial v_{xc}}{\partial x_i} & \dfrac{\partial v_{xc}}{\partial y_i} & \dfrac{\partial v_{xc}}{\partial z_i} \\[2mm]
\dfrac{\partial v_{yc}}{\partial x_i} & \dfrac{\partial v_{yc}}{\partial y_i} & \dfrac{\partial v_{yc}}{\partial z_i} \\[2mm]
\dfrac{\partial v_{zc}}{\partial x_i} & \dfrac{\partial v_{zc}}{\partial y_i} & \dfrac{\partial v_{zc}}{\partial z_i}
\end{bmatrix}
+
\begin{bmatrix}
\dfrac{\partial v_{xc}}{\partial x_i} & \dfrac{\partial v_{yc}}{\partial x_i} & \dfrac{\partial v_{zc}}{\partial x_i} \\[2mm]
\dfrac{\partial v_{xc}}{\partial y_i} & \dfrac{\partial v_{yc}}{\partial y_i} & \dfrac{\partial v_{zc}}{\partial y_i} \\[2mm]
\dfrac{\partial v_{xc}}{\partial z_i} & \dfrac{\partial v_{yc}}{\partial z_i} & \dfrac{\partial v_{zc}}{\partial z_i}
\end{bmatrix}
\right)
\tag{4.216}
$$

From the left Cauchy–Green strain tensor, for small strains a Green–St. Venant strain tensor is obtained as follows:

$$
\check{\mathbf{E}} = \frac{1}{2}\mathbf{V}^2 - \mathbf{I} = \frac{1}{2}\mathbf{B} - \mathbf{I}
\tag{4.217}
$$

The Green–St. Venant strain tensor due to the shape changing stretch is therefore given by

$$
\tilde{\mathbf{E}}_d = \frac{1}{2}(\mathbf{V}_d \mathbf{V}_d^T - \mathbf{I}) = \frac{1}{2}\left(\frac{\mathbf{F}\mathbf{F}^T}{(|\det \mathbf{F}|)^{2/3}} - \mathbf{I}\right)
\tag{4.218}
$$

while the Green–St. Venant strain tensor due to volume changing stretch is given by

$$\tilde{\mathbf{E}}_s = \frac{1}{2}(\mathbf{V}_s \mathbf{V}_s^T - \mathbf{I}) = \frac{1}{2}\left(\mathbf{I}(|\det \mathbf{F}|)^{2/3} - \mathbf{I}\right) = \mathbf{I}\left(\frac{(|\det \mathbf{F}|)^{2/3} - 1}{2}\right) \tag{4.219}$$

From the strain tensors, the Cauchy stress is calculated using the constitutive law. Different constitutive laws can be employed at this point. For instance, a constitutive law for homogeneous isotropic material derived by analogy with the Hooks constitutive law is given by

$$\mathbf{T} = \frac{E}{(1 + \nu)} \frac{1}{(|\det \mathbf{F}|)^{2/3}} \tilde{\mathbf{E}}_d + \tag{4.220}$$

$$\frac{E}{(1 - 2\nu)} \frac{1}{(|\det \mathbf{F}|)^{2/3}} \tilde{\mathbf{E}}_s + \frac{2\overline{\mu}}{(|\det \mathbf{F}|)^{2/3}} \mathbf{D}$$

The last component is the viscous stress, which is responsible for most of the material damping due to deformation.

Cauchy stress components apply to the deformed (current) configuration. Components of stress tensor $\mathbf{T}$ represent stress in the global frame per unit area of the deformed tetrahedron. Thus, traction force over each of the surfaces of the tetrahedron is calculated using the normal on the edge of the deformed configuration with components of the normal given in global frame (Figure 4.22). Surface traction is given by the following expression:

$$\mathbf{s} = \mathbf{Tm} = \begin{bmatrix} s_x \\ s_y \\ s_z \end{bmatrix} = \begin{bmatrix} t_{xx} & t_{xy} & t_{xz} \\ t_{yx} & t_{yy} & t_{yz} \\ t_{zx} & t_{zy} & t_{zz} \end{bmatrix} \begin{bmatrix} m_x \\ m_y \\ m_z \end{bmatrix} \tag{4.221}$$

Surface traction for each of the four surfaces of the deformed tetrahedron is distributed in equal proportion to each of the nodes that belong to a particular surface. To each node

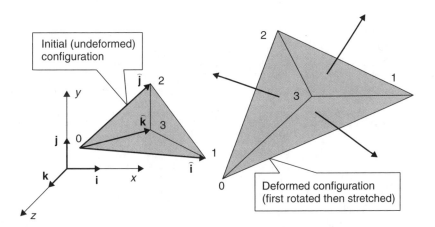

Figure 4.22   Vectors normal to the surfaces of the tetrahedron; the magnitude of each vector is equal to the area of the corresponding surface.

an equivalent nodal force

$$\mathbf{f} = \frac{1}{3}\mathbf{s}$$

(4.222)

is assigned.

4.9 NUMERICAL DEMONSTRATION OF FINITE ROTATION ELASTICITY IN THE COMBINED FINITE-DISCRETE ELEMENT METHOD

In Figure 4.23 a combined finite-discrete element problem comprised of two discrete elements is shown. Each discrete element is discretised into constant strain triangular finite elements. The discrete elements move towards each other with initial velocity. The initial motion sequence involves the simplest form of deformation, namely translation. It is evident from Figure 4.23 that in this translation no stress is produced.

As discrete elements move closer towards each other (Figure 4.24), the discrete elements start impacting each other. Discretised distributed potential contact force is used to resolve contacts. In this process, discretisation coinciding with the finite element mesh is used on both discrete elements, thus the same grid is used to process interaction and deformability. No energy dissipation either at contact or due to material damping (dissipation due to stretching of discrete elements) is present.

Due to impact, contact forces are generated on the edges of both discrete elements, resulting in material of discrete elements both stretching and rotating (Figure 4.25).

In Figures 4.26–4.28, the discrete elements continue to deform and move away from each other, performing a sort of 'dancing motion', which is the reason why the discrete elements were called 'dancing triangles'. It is worth mentioning that although straining

Figure 4.23   Dancing triangles:translation towards each other.

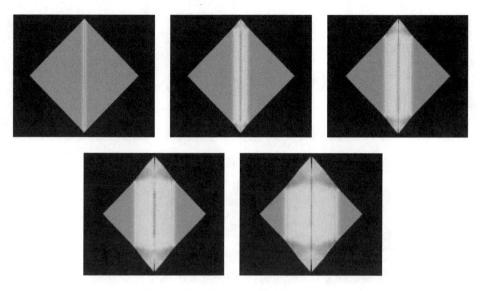

Figure 4.24 Dancing triangles: Impact.

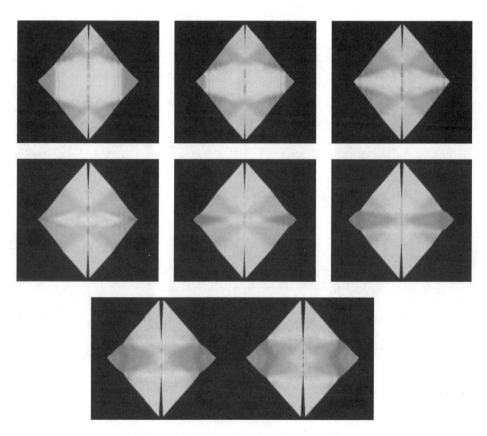

Figure 4.25 Dancing triangles: Separation.

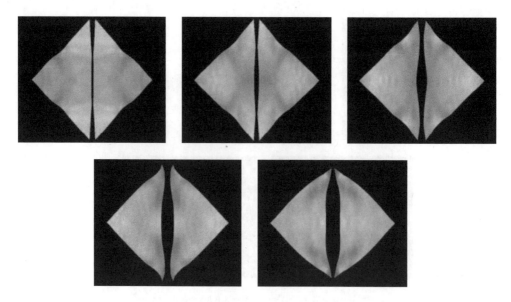

Figure 4.26   Dancing triangles: Bouncing back.

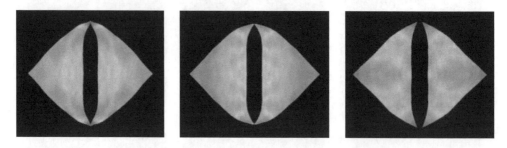

Figure 4.27   Dancing triangles: Touching again.

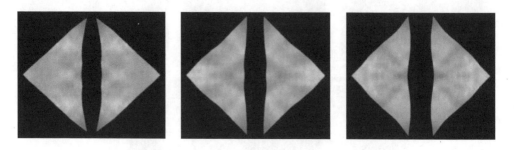

Figure 4.28   Dancing triangles: Deforming freely.

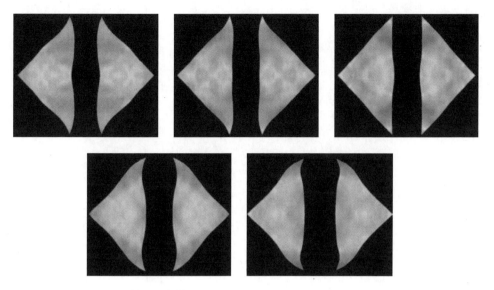

Figure 4.29   Dancing triangles: Dancing away.

of the material is not large, the rotations of individual finite elements is significant. Thus, assumptions employed in formulation of finite-rotation elasticity are valid, including the approximate validity of the strain tensor definition and stress-strain relationship. For simulations involving large stretches (say over 20%), the Green–St. Venant strain tensor is not applicable, nor is the constitutive law described earlier in this chapter. Thus, in such cases both strain and stress tensors and the constitutive law need modification. One possible solution is to employ a logarithmic strain tensor.

Unlike the Cauchy–St. Venant strain tensor, a logarithmic strain tensor involves logarithmic functions. The only way to resolve these is to employ the frame of reference coinciding with the principal directions of the stretch tensor in either the initial or deformed configurations. The stress-strain relationship is also best resolved using these principal directions. This requires spectral decomposition of the stretch tensor. The CPU overheads are not significant, and implementation is relatively simple.

5

Temporal Discretisation

5.1 THE CENTRAL DIFFERENCE TIME INTEGRATION SCHEME

Contact between discrete elements together with the deformability of discrete elements is described in terms of nodal forces and nodal displacements. Since each discrete element is discretised into finite elements, the shape of each discrete element and its position in space at any time instance is given by the current coordinates of the finite element nodes, i.e. nodal coordinates.

$$\mathbf{x} = \begin{bmatrix} x_1 \\ x_2 \\ x_3 \\ \dots \\ x_i \\ \dots \\ x_n \end{bmatrix} \qquad (5.1)$$

where n is the total number of degrees of freedom for a particular discrete element.

In a similar way, the velocity field over the discrete element is defined by nodal velocities $\mathbf{v}$:

$$\mathbf{v} = \dot{\mathbf{x}} = \begin{bmatrix} \dot{x}_1 \\ \dot{x}_2 \\ \dot{x}_3 \\ \dots \\ \dot{x}_i \\ \dots \\ \dot{x}_n \end{bmatrix} \qquad (5.2)$$

The acceleration field over the discrete element is given by

$$\mathbf{a} = \dot{\mathbf{v}} = \ddot{\mathbf{x}} = \begin{bmatrix} \ddot{x}_1 \\ \ddot{x}_2 \\ \ddot{x}_3 \\ \dots \\ \ddot{x}_i \\ \dots \\ \ddot{x}_n \end{bmatrix} \qquad (5.3)$$

The Combined Finite-Discrete Element Method   A. Munjiza
© 2004 John Wiley & Sons, Ltd   ISBN: 0-470-84199-0

The inertia of the discrete element is defined by the mass of the discrete element, which is obtained by integration of density over the volume of the discrete element, i.e.

$$dm = \rho dV \tag{5.4}$$

Discretisation of the discrete element into finite elements also results in discretisation of the mass. The most convenient way of discretisation of the mass used in the combined finite-discrete element method is a so-called lumped mass approach. In essence, instead of considering the mass being distributed over the discrete element, it is assumed that the mass is lumped into the nodes of the finite element mesh. Thus, the mass associated with each degree of freedom is given by

$$\mathbf{m} = \begin{bmatrix} m_1 \\ m_2 \\ m_3 \\ \ldots \\ m_i \\ \ldots \\ m_n \end{bmatrix} \tag{5.5}$$

It is worth noting that in the finite element literature, discretisation of mass is done through the mass matrix, which is in general non-diagonal. However, elimination of non-diagonal terms leads to a diagonal lumped mass matrix:

$$\mathbf{M} = \begin{bmatrix} m_1 & & & & & \\ & m_2 & & & & \\ & & m_3 & & & \\ & & & \ldots & & \\ & & & & m_i & \\ & & & & & \ldots \\ & & & & & & m_n \end{bmatrix} \tag{5.6}$$

where all non-diagonal terms are zero. This approach, in conjunction with the stiffness matrix for dynamic problems, is suitable for both implicit and explicit direct integration in the time domain.

In the context of the combined finite-discrete element method, thousands or even millions of finite element meshes are present. Deformability together with rigid rotation and translation is considered, and contact interaction is resolved together with fracture and fragmentation. Assembling a stiffness matrix and a non-diagonal mass matrix would lead nowhere, for any available implicit time integration scheme could not be used without significant modifications.

Thus in the context of the combined finite-discrete element method, no stiffness matrices are calculated. A time integration scheme is applied on element-by-element, node-by-node and degree of freedom by degree of freedom bases in an explicit form.

Nodal forces from:

- contact interaction,
- deformation of a discrete element,

- external loads, and
- damping forces (due to either 'external' damping or 'internal' damping: external damping is, for instance, drag on discrete elements due to interaction with fluid, while internal damping is, for instance, due to elastic or plastic deformation of a discrete element)

are all added together, and a vector of nodal forces is obtained

$$\mathbf{f} = \begin{bmatrix} f_1 \\ f_2 \\ f_3 \\ \dots \\ f_i \\ \dots \\ f_n \end{bmatrix} \tag{5.7}$$

The dynamic equilibrium of the discrete element is therefore given by

$$\begin{bmatrix} m_1 & & & & & \\ & m_2 & & & & \\ & & m_3 & & & \\ & & & \dots & & \\ & & & & m_i & \\ & & & & & \dots \\ & & & & & & m_n \end{bmatrix} \begin{bmatrix} \ddot{x}_1 \\ \ddot{x}_2 \\ \ddot{x}_3 \\ \dots \\ \ddot{x}_i \\ \dots \\ \ddot{x}_n \end{bmatrix} = \begin{bmatrix} f_1 \\ f_2 \\ f_3 \\ \dots \\ f_i \\ \dots \\ f_n \end{bmatrix} \tag{5.8}$$

The mass matrix may be constant provided no fracture occurs. However, the vector of nodal forces is a function of nodal velocities and nodal coordinates.

For integration of the above equations, the central difference time integration scheme has been traditionally employed. It is an explicit scheme resulting in no need for stiffness matrices to be assembled or stored. In addition, it is conditionally stable, meaning that the stability of the scheme is achieved through reducing the size of the time step. The accuracy of the scheme is also controlled by the size of the time step.

The essence of the central difference time integration scheme is the explicit integration of the governing equation for each degree of freedom separately. The scheme can be formulated as follows:

$$v_{next} = v_{current} + a_{current} \frac{h_{current} + h_{next}}{2} \tag{5.9}$$

$$x_{next} = x_{current} + v_{next} h_{next} \tag{5.10}$$

where

$$a_{current} = \frac{f_{current}}{m} \tag{5.11}$$

is the sum of body forces, contact forces and external loads, together with any damping forces (friction, viscous drag, material viscous damping), and m is the mass associated with the particular degree of freedom.

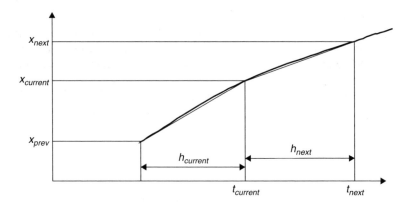

Figure 5.1   The central difference time integration scheme.

A graphical interpretation of the central difference time integration scheme is given in Figure 5.1.

In the case of the constant time step, formulation of the central difference time integration scheme is given as follows:

$$v_{next} = v_{current} + a_{current}h \tag{5.12}$$

$$x_{next} = x_{current} + v_{next}h \tag{5.13}$$

5.1.1   Stability of the central difference time integration scheme

For the zero external load and internal and contact forces being proportional to the displacement with no damping present, the force can be written as a linear function of displacement:

$$f_{current} = -kx_{current} \tag{5.14}$$

In such a case, the central difference time integration scheme is reduced to

$$v_{next} = v_{current} + a_{current}h = v_{current} - \frac{kx_{current}}{m}h \tag{5.15}$$

$$x_{next} = x_{current} + v_{next}h \tag{5.16}$$

$$= v_{current}h + \left(x_{current} - \frac{kx_{current}}{m}h^2 \right)$$

$$= v_{current}h + \left(1 - \frac{k}{m}h^2 \right) x_{current}$$

After multiplication with h

$$hv_{next} = hv_{current} - \frac{k}{m}h^2 x_{current} \tag{5.17}$$

$$x_{next} = v_{current}h + \left(1 - \frac{k}{m}h^2 \right) x_{current} \tag{5.18}$$

which can be written in matrix form:

$$\begin{bmatrix} vh \\ x \end{bmatrix}_{next} = \begin{bmatrix} 1 & -h^2k/m \\ 1 & 1 - h^2k/m \end{bmatrix} \begin{bmatrix} vh \\ x \end{bmatrix}_{current} \tag{5.19}$$

Thus the central difference scheme for damping a free linear system can be reduced to a recursive formula. For the scheme to be stable, it is necessary that the spectral radius of the recursive linear operator

$$\mathbf{A} = \begin{bmatrix} 1 & -h^2k/m \\ 1 & 1 - h^2k/m \end{bmatrix} \tag{5.20}$$

is not greater than one. The associated eigenvalue problem yields the following equation:

$$\begin{vmatrix} 1 - \lambda & -h^2k/m \\ 1 & 1 - h^2k/m - \lambda \end{vmatrix} = 0 \tag{5.21}$$

which results in the following characteristic equation:

$$(1 - \lambda)(1 - h^2k/m - \lambda) + h^2k/m = 0 \tag{5.22}$$

or

$$\lambda^2 + \lambda(h^2k/m - 2) + 1 = 0 \tag{5.23}$$

Solution of this equation is given by

$$\lambda_{1,2} = \frac{-(h^2k/m - 2) \pm \sqrt{(h^2k/m - 2)^2 - 4}}{2} \tag{5.24}$$

which for

$$h^2k/m = 4 \tag{5.25}$$

results in

$$\lambda_{1,2} = -1 \tag{5.26}$$

and the spectral radius is equal to 1. For

$$h^2k/m < 4 \tag{5.27}$$

equation (5.24) yields

$$\lambda_{1,2} = \frac{-(h^2k/m - 2) \pm i\sqrt{4 - (h^2k/m - 2)^2}}{2} \tag{5.28}$$

and the spectral radius is given by

$$\max |\lambda_{1,2}| = \tfrac{1}{2}\sqrt{(h^2k/m - 2)^2 + [4 - (h^2k/m - 2)^2]} = 1 \tag{5.29}$$

i.e. is again equal to 1. For

$$h^2 k/m > 4 \tag{5.30}$$

Equation (5.24) yields

$$\lambda_{1,2} = \frac{-(h^2 k/m - 2) \pm \sqrt{(h^2 k/m - 2)^2 - 4}}{2} \tag{5.31}$$

and the spectral radius is given by

$$_{\max}|\lambda_{1,2}| = \frac{(h^2 k/m - 2) + \sqrt{(h^2 k/m - 2)^2 - 4}}{2} > 1 \tag{5.32}$$

and is always greater than 1. A graph of spectral radius is shown in Figure 5.2.
It is evident that the spectral radius is equal to 1 for

$$h^2 k/m \le 4 \tag{5.33}$$

i.e. the central difference time integration scheme is stable for the time step smaller than

$$h \le \frac{2}{\sqrt{k/m}} \tag{5.34}$$

For the time steps

$$h > \frac{2}{\sqrt{k/m}} \tag{5.35}$$

the central difference time integration scheme is always numerically unstable.

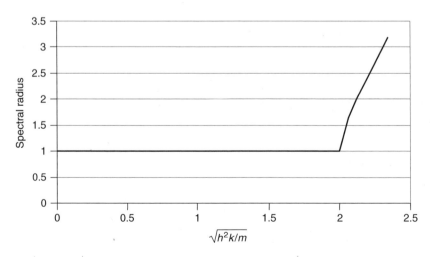

Figure 5.2   Spectral radius of the recursive operator **A** and stability of the central difference time integration scheme.

5.2 DYNAMICS OF IRREGULAR DISCRETE ELEMENTS SUBJECT TO FINITE ROTATIONS IN 3D

Very often a combined finite-discrete element system also comprises rigid discrete elements. In such cases, no forces due to deformation are present, while discretisation of rigid discrete elements is necessary only for a description of the geometry of discrete elements and processing of contact interaction. In these types of problems, once the contact forces and external loads are known, the governing equations can be integrated.

For systems comprising rigid bodies in 2D, in general this is achieved through solving equations for translation and rotation about the centre of mass, i.e. assigning to each discrete element three degrees of freedom – two translations in the direction of the coordinate axes and one rotation about the z-axis. Thus, in 2D problems the central difference time integration scheme as described in the previous section is directly applicable.

In 3D problems this situation is complicated by the presence of finite rotations about the centre of mass of the discrete element. The problem with simply extending 2D algorithms into 3D is that the angular velocity describing such rotation in general does not coincide with the principal axes of the discrete element. Simple extension of 2D algorithms into 3D would therefore not work. For instance, in 3D problems a description of spatial orientation is not a trivial task.

5.2.1 Frames of reference

To describe the motion of a particular rigid discrete element, two reference frames are introduced. The first is an inertial frame which does not move with the discrete element, and in the following text it is referred to as the 'inertial frame' or 'fixed frame'. The orientation of the inertial frame is defined by a triad of unit vectors parallel to the respective axes of a Cartesian coordinate system, (Figure 5.3).

$$(\tilde{\mathbf{i}}, \tilde{\mathbf{j}}, \tilde{\mathbf{k}}) \tag{5.36}$$

The second frame is fixed to the discrete element. The origin of this frame coincides with the centre of mass of the discrete element. This frame is assumed to move (translate and

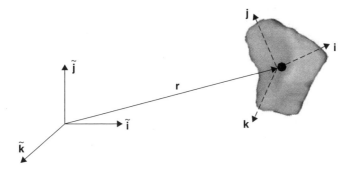

Figure 5.3 Moving and fixed frames of reference.

rotate) together with the discrete element. In the following text, this frame is referred to as the 'element frame' or 'moving frame'. The axes of the Cartesian coordinate system associated with the element frame are assumed to coincide with the principal axes of inertia of the discrete element. The orientation of the element frame is defined by a triad of unit vectors that are orthogonal to each other, (Figure 5.3).

$$(\mathbf{i}, \mathbf{j}, \mathbf{k}) \tag{5.37}$$

5.2.2   Kinematics of the discrete element in general motion

The general motion of the discrete element in 3D is described by the translation superposed by rotation about the centre of mass. Translation of the discrete element is described by the position $\mathbf{r}$ of its centre of mass, (Figure 5.3). From the position of the centre of mass, the velocity $\mathbf{v}$ of the centre of mass is obtained:

$$\mathbf{v} = \frac{d\mathbf{r}}{dt} = (\mathbf{v} \cdot \tilde{\mathbf{i}})\tilde{\mathbf{i}} + (\mathbf{v} \cdot \tilde{\mathbf{j}})\tilde{\mathbf{j}} + (\mathbf{v} \cdot \tilde{\mathbf{k}})\tilde{\mathbf{k}} = v_{\tilde{x}}\tilde{\mathbf{i}} + v_{\tilde{y}}\tilde{\mathbf{j}} + v_{\tilde{z}}\tilde{\mathbf{k}} = \begin{bmatrix} v_{\tilde{x}} \\ v_{\tilde{y}} \\ v_{\tilde{z}} \end{bmatrix} \tag{5.38}$$

It should be noted that in equation (5.38), both the position and velocity of the centre of mass are given in the inertial frame.

A change in orientation of the discrete element occurs due to the presence of angular velocity. In the combined finite-discrete element method, it is convenient to express angular velocity $\boldsymbol{\omega}$ in the inertial frame:

$$\boldsymbol{\omega} = (\boldsymbol{\omega} \cdot \tilde{\mathbf{i}})\tilde{\mathbf{i}} + (\boldsymbol{\omega} \cdot \tilde{\mathbf{j}})\tilde{\mathbf{j}} + (\boldsymbol{\omega} \cdot \tilde{\mathbf{k}})\tilde{\mathbf{k}} = \omega_{\tilde{x}}\tilde{\mathbf{i}} + \omega_{\tilde{y}}\tilde{\mathbf{j}} + \omega_{\tilde{z}}\tilde{\mathbf{k}} = \begin{bmatrix} \omega_{\tilde{x}} \\ \omega_{\tilde{y}} \\ \omega_{\tilde{z}} \end{bmatrix} \tag{5.39}$$

5.2.3   Spatial orientation of the discrete element

Representation of the spatial orientation of the discrete element using approaches such as Euler or Cardan angles of different kinds is difficult to implement in the context of the combined finite-discrete element method, because of the presence of singularities associated with specific values of these angles. To avoid these singularities, in the combined finite-discrete element method, spatial orientation of the discrete element is described by the orientation of the element frame, i.e. by the triad of unit vectors

$$(\mathbf{i}, \mathbf{j}, \mathbf{k}) \tag{5.40}$$

The components of these unit vectors are expressed in the inertial frame of reference

$$\mathbf{i} = i_{\tilde{x}}\tilde{\mathbf{i}} + i_{\tilde{y}}\tilde{\mathbf{j}} + i_{\tilde{z}}\tilde{\mathbf{k}} \tag{5.41}$$

$$\mathbf{j} = j_{\tilde{x}}\tilde{\mathbf{i}} + j_{\tilde{y}}\tilde{\mathbf{j}} + j_{\tilde{z}}\tilde{\mathbf{k}}$$

$$\mathbf{k} = k_{\tilde{x}}\tilde{\mathbf{i}} + k_{\tilde{y}}\tilde{\mathbf{j}} + k_{\tilde{z}}\tilde{\mathbf{k}}$$

In a similar way,

$$\tilde{\mathbf{i}} = \tilde{i}_x \mathbf{i} + \tilde{i}_y \mathbf{j} + \tilde{i}_z \mathbf{k} \tag{5.42}$$

$$\tilde{\mathbf{j}} = \tilde{j}_x \mathbf{i} + \tilde{j}_y \mathbf{j} + \tilde{j}_z \mathbf{k}$$

$$\tilde{\mathbf{k}} = \tilde{k}_x \mathbf{i} + \tilde{k}_y \mathbf{j} + \tilde{k}_z \mathbf{k}$$

5.2.4 Transformation matrices

Representation of the spatial orientation by using equation (5.41) also provides an efficient way of transforming vector components from one reference frame into another (much used in contact procedures). Any particular vector written in the element frame is easily written in the inertial frame:

$$\mathbf{a} = a_x \mathbf{i} + a_y \mathbf{j} + a_z \mathbf{k} \tag{5.43}$$

$$= a_x (i_{\tilde{x}} \tilde{\mathbf{i}} + i_{\tilde{y}} \tilde{\mathbf{j}} + i_{\tilde{z}} \tilde{\mathbf{k}})$$

$$+ a_y (j_{\tilde{x}} \tilde{\mathbf{i}} + j_{\tilde{y}} \tilde{\mathbf{j}} + j_{\tilde{z}} \tilde{\mathbf{k}})$$

$$+ a_z (k_{\tilde{x}} \tilde{\mathbf{i}} + k_{\tilde{y}} \tilde{\mathbf{j}} + k_{\tilde{z}} \tilde{\mathbf{k}})$$

$$= a_{\tilde{x}} \tilde{\mathbf{i}} + a_{\tilde{y}} \tilde{\mathbf{j}} + a_{\tilde{z}} \tilde{\mathbf{k}}$$

The vector components in the inertial frame are therefore calculated from the vector components in the element frame using the following transformation:

$$\begin{bmatrix} a_{\tilde{x}} \\ a_{\tilde{y}} \\ a_{\tilde{z}} \end{bmatrix} = \begin{bmatrix} i_{\tilde{x}} & j_{\tilde{x}} & k_{\tilde{x}} \\ i_{\tilde{y}} & j_{\tilde{y}} & k_{\tilde{y}} \\ i_{\tilde{z}} & j_{\tilde{z}} & k_{\tilde{z}} \end{bmatrix} \begin{bmatrix} a_x \\ a_y \\ a_z \end{bmatrix} \tag{5.44}$$

The components of the same vector in the element frame can be calculated from the components given in the inertial frame of reference as follows:

$$\begin{bmatrix} a_x \\ a_y \\ a_z \end{bmatrix} = \begin{bmatrix} \tilde{i}_x & \tilde{j}_x & \tilde{k}_x \\ \tilde{i}_y & \tilde{j}_y & \tilde{k}_y \\ \tilde{i}_z & \tilde{j}_z & \tilde{k}_z \end{bmatrix} \begin{bmatrix} a_{\tilde{x}} \\ a_{\tilde{y}} \\ a_{\tilde{z}} \end{bmatrix} = \begin{bmatrix} i_{\tilde{x}} & j_{\tilde{x}} & k_{\tilde{x}} \\ i_{\tilde{y}} & j_{\tilde{y}} & k_{\tilde{y}} \\ i_{\tilde{z}} & j_{\tilde{z}} & k_{\tilde{z}} \end{bmatrix}^{-1} \begin{bmatrix} a_{\tilde{x}} \\ a_{\tilde{y}} \\ a_{\tilde{z}} \end{bmatrix} \tag{5.45}$$

As the triad of unit vectors are orthogonal to each other, these transformation matrices are orthogonal, and

$$\begin{bmatrix} \tilde{i}_x & \tilde{j}_x & \tilde{k}_x \\ \tilde{i}_y & \tilde{j}_y & \tilde{k}_y \\ \tilde{i}_z & \tilde{j}_z & \tilde{k}_z \end{bmatrix} = \begin{bmatrix} i_{\tilde{x}} & j_{\tilde{x}} & k_{\tilde{x}} \\ i_{\tilde{y}} & j_{\tilde{y}} & k_{\tilde{y}} \\ i_{\tilde{z}} & j_{\tilde{z}} & k_{\tilde{z}} \end{bmatrix}^{-1} = \begin{bmatrix} i_{\tilde{x}} & j_{\tilde{x}} & k_{\tilde{x}} \\ i_{\tilde{y}} & j_{\tilde{y}} & k_{\tilde{y}} \\ i_{\tilde{z}} & j_{\tilde{z}} & k_{\tilde{z}} \end{bmatrix}^{T} \tag{5.46}$$

5.2.5 The inertia of the discrete element

The inertia of the discrete element in translational motion is defined by the mass of the discrete element

$$m = \int_{vol} \rho \, dV \tag{5.47}$$

where ρ is mass per unit volume of the discrete element.

The inertia of the discrete element in rotational motion is defined by the inertia tensor:

$$I = \begin{bmatrix} Ixx & Ixy & Ixz \\ Iyx & Iyy & Iyz \\ Izx & Izy & Izz \end{bmatrix} \tag{5.48}$$

where the moments of inertia are given by

$$Ixx = \int_{vol} (y^2 + z^2)\rho \, dV \tag{5.49}$$

$$Iyy = \int_{vol} (z^2 + x^2)\rho \, dV \tag{5.50}$$

$$Izz = \int_{vol} (x^2 + y^2)\rho \, dV \tag{5.51}$$

while the products of inertia are

$$Ixy = Iyx = -\int_{vol} (xy)\rho \, dV \tag{5.52}$$

$$Iyz = Izy = -\int_{vol} (yz)\rho \, dV \tag{5.53}$$

$$Ixz = Izx = -\int_{vol} (xz)\rho \, dV \tag{5.54}$$

As the axes of the element frame are assumed to coincide with principal axes of inertia, the products of inertia are zero, and the inertia tensor is represented by a diagonal matrix:

$$I = \begin{bmatrix} Ixx & 0 & 0 \\ 0 & Iyy & 0 \\ 0 & 0 & Izz \end{bmatrix} = \begin{bmatrix} Ix & 0 & 0 \\ 0 & Iy & 0 \\ 0 & 0 & Iz \end{bmatrix} \tag{5.55}$$

5.2.6 Governing equation of motion

In the inertial reference frame, the motion of the discrete element is governed by the Euler equations:

$$\mathbf{F} = \frac{d\mathbf{L}}{dt} = m\frac{d\mathbf{v}}{dt} \tag{5.56}$$

$$\mathbf{M} = \frac{d\mathbf{H}}{dt} \tag{5.57}$$

where $\mathbf{v}$ is the velocity of the centre of mass, $\mathbf{H}$ is moment of momentum about the centre of mass, $\mathbf{F}$ is the resultant force acting at the centre of mass, and $\mathbf{M}$ is resultant moment about the centre of mass.

Equation (5.56) governs translational motion of the discrete element, and is best resolved in the inertial reference frame, as the velocity components of the mass centre are given in the inertial reference frame.

Equation (5.57) governs the rotational motion about the centre of mass of the discrete element. The discrete element is rigid, and therefore the moment of momentum is expressible in terms of the inertia properties of the discrete element and angular velocity. Thus, it is simply referred to as the 'angular momentum'. The angular momentum about the centre of mass is given as follows:

$$\mathbf{H} = \int_{vol} (\mathbf{p} \times (\boldsymbol{\omega} \times \mathbf{p}))\rho \, dV = (\mathbf{H} \cdot \mathbf{i})\mathbf{i} + (\mathbf{H} \cdot \mathbf{j})\mathbf{j} + (\mathbf{H} \cdot \mathbf{k})\mathbf{k} \qquad (5.58)$$

$$= H_x\mathbf{i} + H_y\mathbf{j} + H_z\mathbf{k}$$

$$= \begin{bmatrix} H_x \\ H_y \\ H_z \end{bmatrix} = \begin{bmatrix} I_x & 0 & 0 \\ 0 & I_y & 0 \\ 0 & 0 & I_z \end{bmatrix} \begin{bmatrix} \omega_x \\ \omega_y \\ \omega_z \end{bmatrix}$$

The time derivative of angular momentum in equation (5.57) is taken in the inertial reference frame, while the angular momentum in equation (5.58) is expressed in the element reference frame, which is fixed to the body and rotates at angular velocity $\boldsymbol{\omega}$ relative to the inertial reference frame. Thus, the time derivative of angular momentum is as follows:

$$\frac{d\mathbf{H}}{dt} = \dot{\mathbf{H}} \qquad (5.59)$$

$$= \dot{H}_x\mathbf{i} + \dot{H}_y\mathbf{j} + \dot{H}_z\mathbf{k} + \boldsymbol{\omega} \times \mathbf{H}$$

$$= I_x\dot{\omega}_x\mathbf{i} + I_y\dot{\omega}_y\mathbf{j} + I_z\dot{\omega}_z\mathbf{k}$$

$$+ (I_z - I_y)\omega_z\omega_y\mathbf{i} + (I_x - I_z)\omega_x\omega_z\mathbf{j} + (I_y - I_x)\omega_y\omega_x\mathbf{k}$$

$$= \begin{bmatrix} I_x\dot{\omega}_x + (I_z - I_y)\omega_z\omega_y \\ I_y\dot{\omega}_y + (I_x - I_z)\omega_x\omega_z \\ I_z\dot{\omega}_z + (I_y - I_x)\omega_y\omega_x \end{bmatrix}$$

After substitution into equation (5.58), the Euler equations governing rotational motion of the discrete element are obtained:

$$\begin{bmatrix} M_x \\ M_y \\ M_z \end{bmatrix} = \begin{bmatrix} I_x\dot{\omega}_x + (I_z - I_y)\omega_z\omega_y \\ I_y\dot{\omega}_y + (I_x - I_z)\omega_x\omega_z \\ I_z\dot{\omega}_z + (I_y - I_x)\omega_y\omega_x \end{bmatrix} \qquad (5.60)$$

Combined finite-discrete element systems usually comprise very large numbers of discrete elements, all interacting with each other. Contact forces between discrete elements change rapidly with the motion of the discrete elements. For such a general motion of a particular discrete element, a closed form solution to the Euler equations is not available. It follows that direct integration is necessary to obtain angular velocity and spatial orientation of the discrete element.

5.2.7   Change in spatial orientation during a single time step

By employing equations (5.40) and (5.41), the spatial orientation of the discrete element at time t (i.e. at the beginning of the time step) can be obtained:

$$(_t\mathbf{i}, _t\mathbf{j}, _t\mathbf{k}) \tag{5.61}$$

In a similar way, the spatial orientation of the discrete element at time $t + h$ (i.e. at the end of the time step) is

$$(_{t+h}\mathbf{i}, _{t+h}\mathbf{j}, _{t+h}\mathbf{k}) \tag{5.62}$$

In equations (5.61) and (5.62) h is the time step, while the unit vectors are described using the inertial frame of reference:

$$_t\mathbf{i} = {_t i_{\tilde{x}}}\tilde{\mathbf{i}} + {_t i_{\tilde{y}}}\tilde{\mathbf{j}} + {_t i_{\tilde{z}}}\tilde{\mathbf{k}} \tag{5.63}$$

$$_t\mathbf{j} = {_t j_{\tilde{x}}}\tilde{\mathbf{i}} + {_t j_{\tilde{y}}}\tilde{\mathbf{j}} + {_t j_{\tilde{z}}}\tilde{\mathbf{k}}$$

$$_t\mathbf{k} = {_t k_{\tilde{x}}}\tilde{\mathbf{i}} + {_t k_{\tilde{y}}}\tilde{\mathbf{j}} + {_t k_{\tilde{z}}}\tilde{\mathbf{k}}$$

$$_{t+h}\mathbf{i} = {_{t+h} i_{\tilde{x}}}\tilde{\mathbf{i}} + {_{t+h} i_{\tilde{y}}}\tilde{\mathbf{j}} + {_{t+h} i_{\tilde{z}}}\tilde{\mathbf{k}} \tag{5.64}$$

$$_{t+h}\mathbf{j} = {_{t+h} j_{\tilde{x}}}\tilde{\mathbf{i}} + {_{t+h} j_{\tilde{y}}}\tilde{\mathbf{j}} + {_{t+h} j_{\tilde{z}}}\tilde{\mathbf{k}}$$

$$_{t+h}\mathbf{k} = {_{t+h} k_{\tilde{x}}}\tilde{\mathbf{i}} + {_{t+h} k_{\tilde{y}}}\tilde{\mathbf{j}} + {_{t+h} k_{\tilde{z}}}\tilde{\mathbf{k}}$$

The difference between the two triads is due to the rotation of the discrete element. This rotation is uniquely described by the angular velocity:

$$_{t+h}\mathbf{i} = {_t\mathbf{i}} + \int_t^{t+h} \boldsymbol{\omega}(t) \times \mathbf{i}(t)dt \tag{5.65}$$

$$_{t+h}\mathbf{j} = {_t\mathbf{j}} + \int_t^{t+h} \boldsymbol{\omega}(t) \times \mathbf{j}(t)dt$$

$$_{t+h}\mathbf{k} = {_t\mathbf{k}} + \int_t^{t+h} \boldsymbol{\omega}(t) \times \mathbf{k}(t)dt$$

If the time step h is small, it is reasonable to assume that the angular velocity is constant during the time step, so that

$$\overline{\boldsymbol{\omega}} = \begin{vmatrix} \overline{\omega}_{\tilde{x}} \\ \overline{\omega}_{\tilde{y}} \\ \overline{\omega}_{\tilde{z}} \end{vmatrix} \tag{5.66}$$

The assumption of a constant angular velocity means that the discrete element rotates about a fixed axis during this time interval. The total rotation angle is

$$\overline{\boldsymbol{\psi}} = h\overline{\boldsymbol{\omega}}$$

$$\overline{\psi} = \sqrt{\overline{\psi}_{\tilde{x}}^2 + \overline{\psi}_{\tilde{y}}^2 + \overline{\psi}_{\tilde{z}}^2}$$

(5.67)

while the rotated triad of unit vectors is as follows:

$$
_{t+h}\mathbf{i} = \frac{(\overline{\boldsymbol{\psi}} \cdot_t \mathbf{i})}{\overline{\psi}^2}\overline{\boldsymbol{\psi}} + \left[{}_t\mathbf{i} - \frac{(\overline{\boldsymbol{\psi}} \cdot_t \mathbf{i})}{\overline{\psi}^2}\overline{\boldsymbol{\psi}} \right]\cos(\overline{\psi})
$$

(5.68)

$$
+ \frac{1}{\overline{\psi}}(\overline{\boldsymbol{\psi}} \times_t \mathbf{i})\sin(\overline{\psi})
$$

$$
_{t+h}\mathbf{j} = \frac{(\overline{\boldsymbol{\psi}} \cdot_t \mathbf{j})}{\overline{\psi}^2}\overline{\boldsymbol{\psi}} + \left[{}_t\mathbf{j} - \frac{(\overline{\boldsymbol{\psi}} \cdot_t \mathbf{j})}{\overline{\psi}^2}\overline{\boldsymbol{\psi}} \right]\cos(\overline{\psi})
$$

$$
+ \frac{1}{\overline{\psi}}(\overline{\boldsymbol{\psi}} \times_t \mathbf{j})\sin(\overline{\psi})
$$

$$
_{t+h}\mathbf{k} = \frac{(\overline{\boldsymbol{\psi}} \cdot_t \mathbf{k})}{\overline{\psi}^2}\overline{\boldsymbol{\psi}} + \left[{}_t\mathbf{k} - \frac{(\overline{\boldsymbol{\psi}} \cdot_t \mathbf{k})}{\overline{\psi}^2}\overline{\boldsymbol{\psi}} \right]\cos(\overline{\psi})
$$

$$
+ \frac{1}{\overline{\psi}}(\overline{\boldsymbol{\psi}} \times_t \mathbf{k})\sin(\overline{\psi})
$$

The above described rotation preserves the orthogonal relationship between the unit vectors of the triad and also the magnitude of these vectors, Figure 5.4.

5.6.8 Change in angular momentum due to external loads

As mentioned earlier, for each discrete element the components of angular velocity at time t are conveniently expressed using the inertial frame of reference:

$$
_t\boldsymbol{\omega} = {}_t\,\omega_{\tilde{x}}\tilde{\mathbf{i}} + {}_t\,\omega_{\tilde{y}}\tilde{\mathbf{j}} + {}_t\,\omega_{\tilde{z}}\tilde{\mathbf{k}} = \begin{bmatrix} {}_t\omega_{\tilde{x}} \\ {}_t\omega_{\tilde{y}} \\ {}_t\omega_{\tilde{z}} \end{bmatrix}
$$

(5.69)

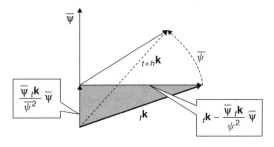

Figure 5.4 Rotation of a base vector.

However, the angular momentum at time t is easier to calculate using the element frame of reference:

$$_{-t}\mathbf{H} = \begin{bmatrix} -_tH_x \\ -_tH_y \\ -_tH_z \end{bmatrix} = \begin{bmatrix} I_x & 0 & 0 \\ 0 & I_y & 0 \\ 0 & 0 & I_z \end{bmatrix} \begin{bmatrix} _t\omega_x \\ _t\omega_y \\ _t\omega_z \end{bmatrix} \tag{5.70}$$

Thus the transformation matrix (5.46) is employed to obtain the components of angular velocity, which after substitution in (5.70) yields

$$_{-t}\mathbf{H} = \begin{bmatrix} -_tH_x \\ -_tH_y \\ -_tH_z \end{bmatrix} = \begin{bmatrix} I_x & 0 & 0 \\ 0 & I_y & 0 \\ 0 & 0 & I_z \end{bmatrix} \begin{bmatrix} _t\tilde{i}_x & _t\tilde{j}_x & _t\tilde{k}_x \\ _t\tilde{i}_y & _t\tilde{j}_y & _t\tilde{k}_y \\ _t\tilde{i}_z & _t\tilde{j}_z & _t\tilde{k}_z \end{bmatrix} \begin{bmatrix} _t\omega_{\tilde{x}} \\ _t\omega_{\tilde{y}} \\ _t\omega_{\tilde{z}} \end{bmatrix} \tag{5.71}$$

The same angular momentum in the inertial frame of reference is given by

$$_{-t}\mathbf{H} = \begin{bmatrix} -_tH_{\tilde{x}} \\ -_tH_{\tilde{y}} \\ -_tH_{\tilde{z}} \end{bmatrix} = \begin{bmatrix} _ti_{\tilde{x}} & _tj_{\tilde{x}} & _tk_{\tilde{x}} \\ _ti_{\tilde{y}} & _tj_{\tilde{y}} & _tk_{\tilde{y}} \\ _ti_{\tilde{z}} & _tj_{\tilde{z}} & _tk_{\tilde{z}} \end{bmatrix} \begin{bmatrix} -_tH_x \\ -_tH_y \\ -_tH_z \end{bmatrix} \tag{5.72}$$

$$= \begin{bmatrix} _ti_{\tilde{x}} & _tj_{\tilde{x}} & _tk_{\tilde{x}} \\ _ti_{\tilde{y}} & _tj_{\tilde{y}} & _tk_{\tilde{y}} \\ _ti_{\tilde{z}} & _tj_{\tilde{z}} & _tk_{\tilde{z}} \end{bmatrix} \begin{bmatrix} I_x & 0 & 0 \\ 0 & I_y & 0 \\ 0 & 0 & I_z \end{bmatrix} \begin{bmatrix} _t\tilde{i}_x & _t\tilde{j}_x & _t\tilde{k}_x \\ _t\tilde{i}_y & _t\tilde{j}_y & _t\tilde{k}_y \\ _t\tilde{i}_z & _t\tilde{j}_z & _t\tilde{k}_z \end{bmatrix} \begin{bmatrix} _t\omega_{\tilde{x}} \\ _t\omega_{\tilde{y}} \\ _t\omega_{\tilde{z}} \end{bmatrix}$$

In the inertial frame of reference, rotational motion of the discrete element is governed by the Euler second law, equation (5.57). At time t the total forces and moments acting on the discrete element are calculated using the usual contact routines, which include detection of contacts and solving contact interaction between each couple of interacting discrete elements. These are expressed as the resultant force and moment acting on the centre of mass of the discrete element. At time t the latter is given in the inertial reference frame as

$$_t\mathbf{M} =_t M_{\tilde{x}}\tilde{\mathbf{i}} +_t M_{\tilde{y}}\tilde{\mathbf{j}} +_t M_{\tilde{z}}\tilde{\mathbf{k}} = \begin{bmatrix} _tM_{\tilde{x}} \\ _tM_{\tilde{y}} \\ _tM_{\tilde{z}} \end{bmatrix} \tag{5.73}$$

At this point an assumption that the change in angular momentum due to the external moment from equation (5.73) can be approximated by

$$\Delta\mathbf{H} = \mathbf{M}h \tag{5.74}$$

is introduced. The physical meaning of this approximation is equivalent to the assumption that the external moment is applied as impulse at time t. Thus the total external moment on the discrete element is approximated as a sum of impulse loads acting at specified time intervals, i.e. at the beginning of each time step (Figure 5.5).

The updated angular momentum at time t is therefore given as follows:

$$_t\mathbf{H} = \begin{bmatrix} _tH_{\tilde{x}} \\ _tH_{\tilde{y}} \\ _tH_{\tilde{z}} \end{bmatrix} = \begin{bmatrix} -_tH_{\tilde{x}} \\ -_tH_{\tilde{y}} \\ -_tH_{\tilde{z}} \end{bmatrix} + \begin{bmatrix} _tM_{\tilde{x}}h \\ _tM_{\tilde{y}}h \\ _tM_{\tilde{z}}h \end{bmatrix} \tag{5.75}$$

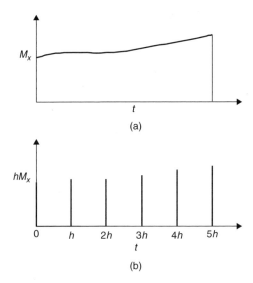

Figure 5.5 Approximation of external load. (a) Continuous load, (b) equivalent series of impulse loads.

5.6.9 Change in angular velocity during a single time step

The assumption that the change of angular momentum is instantaneous implies that the motion of the discrete element during the time step is free of external moments. In other words, the angular momentum at the beginning of the time step (before the discrete element has changed its spatial orientation, but after the external impulse load has been taken into account) is equal to the angular momentum at the end of the time step, i.e.

$$_{t+h}\mathbf{H} =_t \mathbf{H} \tag{5.76}$$

The angular momentum at time $t + h$ is a function of the angular velocity at time $t + h$, i.e.

$$_{t+h}\mathbf{H} = \begin{bmatrix} _{t+h}H_{\tilde{x}} \\ _{t+h}H_{\tilde{y}} \\ _{t+h}H_{\tilde{z}} \end{bmatrix} \tag{5.77}$$

$$= \begin{bmatrix} _{t+h}i_{\tilde{x}} & _{t+h}j_{\tilde{x}} & _{t+h}k_{\tilde{x}} \\ _{t+h}i_{\tilde{y}} & _{t+h}j_{\tilde{y}} & _{t+h}k_{\tilde{y}} \\ _{+ht}i_{\tilde{z}} & _{t+h}j_{\tilde{z}} & _{t+h}k_{\tilde{z}} \end{bmatrix} \begin{bmatrix} I_x & 0 & 0 \\ 0 & I_y & 0 \\ 0 & 0 & I_z \end{bmatrix} \begin{bmatrix} _{t+h}\tilde{i}_x & _{t+h}\tilde{j}_x & _{t+h}\tilde{k}_x \\ _{t+h}\tilde{i}_y & _{t+h}\tilde{j}_y & _{t+h}\tilde{k}_y \\ _{t+h}\tilde{i}_z & _{t+h}\tilde{j}_z & _{t+h}\tilde{k}_z \end{bmatrix} \begin{bmatrix} _{t+h}\omega_{\tilde{x}} \\ _{t+h}\omega_{\tilde{y}} \\ _{t+h}\omega_{\tilde{z}} \end{bmatrix}$$

Equations (5.76) and (5.77), when combined, yield angular velocity at time $t + h$

$$\begin{bmatrix} _{t+h}\omega_{\tilde{x}} \\ _{t+h}\omega_{\tilde{y}} \\ _{t+h}\omega_{\tilde{z}} \end{bmatrix} \tag{5.78}$$

$$= \left(\begin{bmatrix} _{t+h}i_{\tilde{x}} & _{t+h}j_{\tilde{x}} & _{t+h}k_{\tilde{x}} \\ _{t+h}i_{\tilde{y}} & _{t+h}j_{\tilde{y}} & _{t+h}k_{\tilde{y}} \\ _{t+h}i_{\tilde{z}} & _{t+h}j_{\tilde{z}} & _{t+h}k_{\tilde{z}} \end{bmatrix} \begin{bmatrix} I_x & 0 & 0 \\ 0 & I_y & 0 \\ 0 & 0 & I_z \end{bmatrix}^{-1} \begin{bmatrix} _{t+h}\tilde{i}_x & _{t+h}\tilde{j}_x & _{t+h}\tilde{k}_x \\ _{t+h}\tilde{i}_y & _{t+h}\tilde{j}_y & _{t+h}\tilde{k}_y \\ _{t+h}\tilde{i}_z & _{t+h}\tilde{j}_z & _{t+h}\tilde{k}_z \end{bmatrix} \right) \begin{bmatrix} _t H_{\tilde{x}} \\ _t H_{\tilde{y}} \\ _t H_{\tilde{z}} \end{bmatrix}$$

5.6.10 Munjiza direct time integration scheme

The temporal discretisation of the governing equations described above is used to formu-
late the integration algorithm. This is achieved through the average angular velocity in
equation (5.66) being defined using formulae for the fourth-order Runge–Kutta method.
The resulting direct time integration scheme was first formulated by Munjiza, and is called
the Munjiza time integration scheme. It can be summarised as follows:

Step 1: set the first approximation of the average angular velocity:

$$_1\overline{\omega} = \begin{bmatrix} {}_t\omega_{\bar{x}} \\ {}_t\omega_{\bar{y}} \\ {}_t\omega_{\bar{z}} \end{bmatrix} \tag{5.79}$$

the total angle of rotation:

$$_1\overline{\psi} = \frac{h}{2}{}_1\overline{\omega} \tag{5.80}$$

$$_1\overline{\psi} = \sqrt{{}_1\overline{\psi}_{\bar{x}}^2 + {}_1\overline{\psi}_{\bar{y}}^2 + {}_1\overline{\psi}_{\bar{z}}^2}$$

and the first approximation of the intermediate spatial orientation:

$$_{t+h/2}^1\mathbf{i} = \frac{({}_1\overline{\psi} \cdot {}_t\mathbf{i})}{{}_1\overline{\psi}^2}{}_1\overline{\psi} + \left[{}_t\mathbf{i} - \frac{({}_1\overline{\psi} \cdot {}_t\mathbf{i})}{{}_1\overline{\psi}^2}{}_1\overline{\psi} \right]\cos({}_1\overline{\psi}) + \frac{1}{{}_1\overline{\psi}}({}_1\overline{\psi} \times_t \mathbf{i})\sin({}_1\overline{\psi}) \tag{5.81}$$

$$_{t+h/2}^1\mathbf{j} = \frac{({}_1\overline{\psi} \cdot {}_t\mathbf{j})}{{}_1\overline{\psi}^2}{}_1\overline{\psi} + \left[{}_t\mathbf{j} - \frac{({}_1\overline{\psi} \cdot {}_t\mathbf{j})}{{}_1\overline{\psi}^2}{}_1\overline{\psi} \right]\cos({}_1\overline{\psi}) + \frac{1}{{}_1\overline{\psi}}({}_1\overline{\psi} \times_t \mathbf{j})\sin({}_1\overline{\psi})$$

$$_{t+h/2}^1\mathbf{k} = \frac{({}_1\overline{\psi} \cdot {}_t\mathbf{k})}{{}_1\overline{\psi}^2}{}_1\overline{\psi} + \left[{}_t\mathbf{k} - \frac{({}_1\overline{\psi} \cdot {}_t\mathbf{k})}{{}_1\overline{\psi}^2}{}_1\overline{\psi} \right]\cos({}_1\overline{\psi}) + \frac{1}{{}_1\overline{\psi}}({}_1\overline{\psi} \times_t \mathbf{k})\sin({}_1\overline{\psi})$$

Step 2: calculate the second approximation of the average angular velocity

$$_2\overline{\omega} = \begin{bmatrix} {}_2\overline{\omega}_{\bar{x}} \\ {}_2\overline{\omega}_{\bar{y}} \\ {}_2\overline{\omega}_{\bar{z}} \end{bmatrix} = \tag{5.82}$$

$$\left(\begin{bmatrix} {}_{t+h/2}^1 i_{\bar{x}} & {}_{t+h/2}^1 j_{\bar{x}} & {}_{t+h/2}^1 k_{\bar{x}} \\ {}_{t+h/2}^1 i_{\bar{y}} & {}_{t+h/2}^1 j_{\bar{y}} & {}_{t+h/2}^1 k_{\bar{y}} \\ {}_{t+h/2}^1 i_{\bar{z}} & {}_{t+h/2}^1 j_{\bar{z}} & {}_{t+h/2}^1 k_{\bar{z}} \end{bmatrix} \begin{bmatrix} I_x & 0 & 0 \\ 0 & I_y & 0 \\ 0 & 0 & I_z \end{bmatrix}^{-1} \begin{bmatrix} {}_{t+h/2}^1 \tilde{i}_x & {}_{t+h/2}^1 \tilde{j}_x & {}_{t+h/2}^1 \tilde{k}_x \\ {}_{t+h/2}^1 \tilde{i}_y & {}_{t+h/2}^1 \tilde{j}_y & {}_{t+h/2}^1 \tilde{k}_y \\ {}_{t+h/2}^1 \tilde{i}_z & {}_{t+h/2}^1 \tilde{j}_z & {}_{t+h/2}^1 \tilde{k}_z \end{bmatrix} \right.$$

$$\left(\begin{bmatrix} {}_t i_{\bar{x}} & {}_t j_{\bar{x}} & {}_t k_{\bar{x}} \\ {}_t i_{\bar{y}} & {}_t j_{\bar{y}} & {}_t k_{\bar{y}} \\ {}_t i_{\bar{z}} & {}_t j_{\bar{z}} & {}_t k_{\bar{z}} \end{bmatrix} \begin{bmatrix} I_x & 0 & 0 \\ 0 & I_y & 0 \\ 0 & 0 & I_z \end{bmatrix} \begin{bmatrix} {}_t \tilde{i}_x & {}_t \tilde{j}_x & {}_t \tilde{k}_x \\ {}_t \tilde{i}_y & {}_t \tilde{j}_y & {}_t \tilde{k}_y \\ {}_t \tilde{i}_z & {}_t \tilde{j}_z & {}_t \tilde{k}_z \end{bmatrix} \begin{bmatrix} {}_t \omega_{\bar{x}} \\ {}_t \omega_{\bar{y}} \\ {}_t \omega_{\bar{z}} \end{bmatrix} + \begin{bmatrix} {}_t M_{\bar{x}} h \\ {}_t M_{\bar{y}} h \\ {}_t M_{\bar{z}} h \end{bmatrix} \right)$$

the total angle of rotation:

$$_2\overline{\boldsymbol{\psi}} = \frac{h}{2}{}_2\overline{\boldsymbol{\omega}} \tag{5.83}$$

$$_2\overline{\psi} = \sqrt{_2\overline{\psi}_{\tilde{x}}^2 + _2\overline{\psi}_{\tilde{y}}^2 + _2\overline{\psi}_{\tilde{z}}^2}$$

and the second approximation of the intermediate spatial orientation:

$$_{t+h/2}{}^2\mathbf{i} = \frac{(_2\overline{\boldsymbol{\psi}} \cdot_t \mathbf{i})}{_2\overline{\psi}^2}{}_2\overline{\boldsymbol{\psi}} + \left[{}_t\mathbf{i} - \frac{(_2\overline{\boldsymbol{\psi}} \cdot_t \mathbf{i})}{_2\overline{\psi}^2}{}_2\overline{\boldsymbol{\psi}}\right]\cos(_2\overline{\psi}) \tag{5.84}$$

$$+ \frac{1}{_2\overline{\psi}}(_2\overline{\boldsymbol{\psi}} \times_t \mathbf{i})\sin(_2\overline{\psi})$$

$$_{t+h/2}{}^2\mathbf{j} = \frac{(_2\overline{\boldsymbol{\psi}} \cdot_t \mathbf{j})}{_2\overline{\psi}^2}{}_2\overline{\boldsymbol{\psi}} + \left[{}_t\mathbf{j} - \frac{(_1\overline{\boldsymbol{\psi}} \cdot_t \mathbf{j})}{_2\overline{\psi}^2}{}_2\overline{\boldsymbol{\psi}}\right]\cos(_2\overline{\psi})$$

$$+ \frac{1}{_2\overline{\psi}}(_2\overline{\boldsymbol{\psi}} \times_t \mathbf{j})\sin(_2\overline{\psi})$$

$$_{t+h/2}{}^2\mathbf{k} = \frac{(_2\overline{\boldsymbol{\psi}} \cdot_t \mathbf{k})}{_2\overline{\psi}^2}{}_2\overline{\boldsymbol{\psi}} + \left[{}_t\mathbf{k} - \frac{(_2\overline{\boldsymbol{\psi}} \cdot_t \mathbf{k})}{_2\overline{\psi}^2}{}_2\overline{\boldsymbol{\psi}}\right]\cos(_1\overline{\psi})$$

$$+ \frac{1}{_2\overline{\psi}}(_2\overline{\boldsymbol{\psi}} \times_t \mathbf{k})\sin(_2\overline{\psi})$$

Step 3: calculate the third approximation of the average angular velocity:

$$_3\overline{\boldsymbol{\omega}} = \begin{bmatrix} _3\overline{\omega}_{\tilde{x}} \\ _3\overline{\omega}_{\tilde{y}} \\ _3\overline{\omega}_{\tilde{z}} \end{bmatrix} =$$

$$\left(\begin{bmatrix} _{t+h/2}{}^2 i_{\tilde{x}} & _{t+h/2}{}^2 j_{\tilde{x}} & _{t+h/2}{}^2 k_{\tilde{x}} \\ _{t+h/2}{}^2 i_{\tilde{y}} & _{t+h/2}{}^2 i_{\tilde{y}} & _{t+h/2}{}^2 k_{\tilde{y}} \\ _{t+h/2}{}^2 i_{\tilde{z}} & _{t+h/2}{}^2 j_{\tilde{z}} & _{t+h/2}{}^2 k_{\tilde{z}} \end{bmatrix} \begin{bmatrix} I_x & 0 & 0 \\ 0 & I_y & 0 \\ 0 & 0 & I_z \end{bmatrix}^{-1} \begin{bmatrix} _{t+h/2}{}^2 \tilde{j}_x & _{t+h/2}{}^2 \tilde{j}_x & _{t+h/2}{}^2 \tilde{k}_x \\ _{t+h/2}{}^2 \tilde{i}_y & _{t+h/2}{}^2 \tilde{j}_y & _{t+h/2}{}^2 \tilde{k}_y \\ _{t+h/2}{}^2 \tilde{i}_z & _{t+h/2}{}^2 \tilde{j}_z & _{t+h/2}{}^2 \tilde{k}_z \end{bmatrix}\right)$$

$$\left(\begin{bmatrix} _t i_{\tilde{x}} & _t j_{\tilde{x}} & _t k_{\tilde{x}} \\ _t i_{\tilde{y}} & j_{\tilde{y}} & _t k_{\tilde{y}} \\ _t i_{\tilde{z}} & _t j_{\tilde{z}} & _t k_{\tilde{z}} \end{bmatrix} \begin{bmatrix} I_x & 0 & 0 \\ 0 & I_y & 0 \\ 0 & 0 & I_z \end{bmatrix} \begin{bmatrix} _t \tilde{i}_x & _t \tilde{j}_x & _t \tilde{k}_x \\ _t \tilde{i}_y & _t \tilde{j}_y & _t \tilde{k}_y \\ _t \tilde{i}_z & _t \tilde{j}_z & _t \tilde{k}_z \end{bmatrix} \begin{bmatrix} _t \omega_{\tilde{x}} \\ _t \omega_{\tilde{y}} \\ _t \omega_{\tilde{z}} \end{bmatrix} + \begin{bmatrix} _t M_{\tilde{x}} h \\ _t M_{\tilde{y}} h \\ _t M_{\tilde{z}} h \end{bmatrix}\right)$$

$$\tag{5.85}$$

the total angle of rotation:

$$_3\overline{\boldsymbol{\psi}} = h_3\overline{\boldsymbol{\omega}}$$

$$_3\overline{\psi} = \sqrt{_3\overline{\psi}_{\tilde{x}}^2 + _3\overline{\psi}_{\tilde{y}}^2 + _3\overline{\psi}_{\tilde{z}}^2} \tag{5.86}$$

and the third approximation of the intermediate spatial orientation:

$$
_{t+h}^{3}\mathbf{i} = \frac{(_3\overline{\boldsymbol{\psi}} \cdot_t \mathbf{i})}{_3\overline{\psi}^2}{}_3\overline{\boldsymbol{\psi}} + \left[_t\mathbf{i} - \frac{(_3\overline{\boldsymbol{\psi}} \cdot_t \mathbf{i})}{_3\overline{\psi}^2}{}_3\overline{\boldsymbol{\psi}} \right]\cos(_3\overline{\psi})
$$

$$
+ \frac{1}{_3\overline{\psi}}(_3\overline{\boldsymbol{\psi}} \times_t \mathbf{i})\sin(_3\overline{\psi})
$$

$$
_{t+h}^{3}\mathbf{j} = \frac{(_3\overline{\boldsymbol{\psi}} \cdot_t \mathbf{j})}{_3\overline{\psi}^2}{}_3\overline{\boldsymbol{\psi}} + \left[_t\mathbf{j} - \frac{(_1\overline{\boldsymbol{\psi}} \cdot_t \mathbf{j})}{_3\overline{\psi}^2}{}_3\overline{\boldsymbol{\psi}} \right]\cos(_3\overline{\psi})
$$

$$
+ \frac{1}{_3\overline{\psi}}(_3\overline{\boldsymbol{\psi}} \times_t \mathbf{j})\sin(_3\overline{\psi})
$$

$$
_{t+h}^{3}\mathbf{k} = \frac{(_3\overline{\boldsymbol{\psi}} \cdot_t \mathbf{k})}{_3\overline{\psi}^2}{}_3\overline{\boldsymbol{\psi}} + \left[_t\mathbf{k} - \frac{(_3\overline{\boldsymbol{\psi}} \cdot_t \mathbf{k})}{_3\overline{\psi}^2}{}_3\overline{\boldsymbol{\psi}} \right]\cos(_3\overline{\psi})
$$

$$
+ \frac{1}{_3\overline{\psi}}(_2\overline{\boldsymbol{\psi}} \times_t \mathbf{k})\sin(_3\overline{\psi})
$$

(5.87)

Step 4: calculate the fourth approximation of the average angular velocity:

$$
_4\overline{\boldsymbol{\omega}} = \begin{bmatrix} _4\overline{\omega}_{\tilde{x}} \\ _4\overline{\omega}_{\tilde{y}} \\ _4\overline{\omega}_{\tilde{z}} \end{bmatrix}
$$

$$
= \left(\begin{bmatrix} _{t+h}^{3}i_{\tilde{x}} & _{t+h}^{3}j_{\tilde{x}} & _{t+h}^{3}k_{\tilde{x}} \\ _{t+h}^{3}i_{\tilde{y}} & _{t+h}^{3}j_{\tilde{y}} & _{t+h}^{3}k_{\tilde{y}} \\ _{t+h}^{3}i_{\tilde{z}} & _{t+h}^{3}j_{\tilde{z}} & _{t+h}^{3}k_{\tilde{z}} \end{bmatrix} \begin{bmatrix} I_x & 0 & 0 \\ 0 & I_y & 0 \\ 0 & 0 & I_z \end{bmatrix}^{-1} \begin{bmatrix} _{t+h}^{3}\tilde{i}_x & _{t+h}^{3}\tilde{j}_x & _{t+h}^{3}\tilde{k}_x \\ _{t+h}^{3}\tilde{i}_y & _{t+h}^{3}\tilde{j}_y & _{t+h}^{3}\tilde{k}_y \\ _{t+h}^{3}\tilde{i}_z & _{t+h}^{3}\tilde{j}_z & _{t+h}^{3}\tilde{k}_z \end{bmatrix} \right)
$$

$$
\left(\begin{bmatrix} _t i_{\tilde{x}} & _t j_{\tilde{x}} & _t k_{\tilde{x}} \\ _t i_{\tilde{y}} & _t j_{\tilde{y}} & _t k_{\tilde{y}} \\ _t i_{\tilde{z}} & _t j_{\tilde{z}} & _t k_{\tilde{z}} \end{bmatrix} \begin{bmatrix} I_x & 0 & 0 \\ 0 & I_y & 0 \\ 0 & 0 & I_z \end{bmatrix} \begin{bmatrix} _t\tilde{i}_x & _t\tilde{j}_x & _t\tilde{k}_x \\ _t\tilde{i}_y & _t\tilde{j}_y & _t\tilde{k}_y \\ _t\tilde{i}_z & _t\tilde{j}_z & _t\tilde{k}_z \end{bmatrix} \begin{bmatrix} _t\omega_{\tilde{x}} \\ _t\omega_{\tilde{y}} \\ _t\omega_{\tilde{z}} \end{bmatrix} + \begin{bmatrix} _tM_{\tilde{x}}h \\ _tM_{\tilde{y}}h \\ _tM_{\tilde{z}}h \end{bmatrix} \right)
$$

(5.88)

Step 5: calculate the average angular velocity:

$$
\overline{\boldsymbol{\omega}} = \begin{bmatrix} \overline{\omega}_{\tilde{x}} \\ \overline{\omega}_{\tilde{y}} \\ \overline{\omega}_{\tilde{z}} \end{bmatrix} = \frac{1}{6}\left(\begin{bmatrix} _1\overline{\omega}_{\tilde{x}} \\ _1\overline{\omega}_{\tilde{y}} \\ _1\overline{\omega}_{\tilde{z}} \end{bmatrix} + 2\begin{bmatrix} _2\overline{\omega}_{\tilde{x}} \\ _2\overline{\omega}_{\tilde{y}} \\ _2\overline{\omega}_{\tilde{z}} \end{bmatrix} + 2\begin{bmatrix} _3\overline{\omega}_{\tilde{x}} \\ _3\overline{\omega}_{\tilde{y}} \\ _3\overline{\omega}_{\tilde{z}} \end{bmatrix} + \begin{bmatrix} _4\overline{\omega}_{\tilde{x}} \\ _4\overline{\omega}_{\tilde{y}} \\ _4\overline{\omega}_{\tilde{z}} \end{bmatrix} \right)
$$

(5.89)

Step 6: calculate the total angle of rotation:

$$
\overline{\boldsymbol{\psi}} = h\overline{\boldsymbol{\omega}}
$$

$$
\overline{\psi} = \sqrt{\overline{\psi}_x^2 + \overline{\psi}_y^2 + \overline{\psi}_z^2}
$$

(5.90)

and the spatial orientation of the discrete element at time $t + h$:

$$_{t+h}\mathbf{i} = \frac{(\overline{\boldsymbol{\psi}} \cdot_t \mathbf{i})}{\overline{\psi}^2}\overline{\boldsymbol{\psi}} + \left[_t\mathbf{i} - \frac{(\overline{\boldsymbol{\psi}} \cdot_t \mathbf{i})}{\overline{\psi}^2}\overline{\boldsymbol{\psi}}\right]\cos(\overline{\psi}) \tag{5.91}$$

$$+ \frac{1}{\overline{\psi}}(\overline{\boldsymbol{\psi}} \times_t \mathbf{i})\sin(\overline{\psi})$$

$$_{t+h}\mathbf{j} = \frac{(\overline{\boldsymbol{\psi}} \cdot_t \mathbf{j})}{\overline{\psi}^2}\overline{\boldsymbol{\psi}} + \left[_t\mathbf{j} - \frac{(\overline{\boldsymbol{\psi}} \cdot_t \mathbf{j})}{\overline{\psi}^2}\overline{\boldsymbol{\psi}}\right]\cos(\overline{\psi})$$

$$+ \frac{1}{\overline{\psi}}(\overline{\boldsymbol{\psi}} \times_t \mathbf{j})\sin(\overline{\psi})$$

$$_{t+h}\mathbf{k} = \frac{(\overline{\boldsymbol{\psi}} \cdot_t \mathbf{k})}{\overline{\psi}^2}\overline{\boldsymbol{\psi}} + \left[_t\mathbf{k} - \frac{(\overline{\boldsymbol{\psi}} \cdot_t \mathbf{k})}{\overline{\psi}^2}\overline{\boldsymbol{\psi}}\right]\cos(\overline{\psi})$$

$$+ \frac{1}{\overline{\psi}}(\overline{\boldsymbol{\psi}} \times_t \mathbf{k})\sin(\overline{\psi})$$

Step 7: set

$$\begin{bmatrix} _t\omega_{\tilde{x}} \\ _t\omega_{\tilde{y}} \\ _t\omega_{\tilde{z}} \end{bmatrix} = \begin{bmatrix} _{t+h}\omega_{\tilde{x}} \\ _{t+h}\omega_{\tilde{y}} \\ _{t+h}\omega_{\tilde{z}} \end{bmatrix} \tag{5.92}$$

$$_t\mathbf{i} =_{t+h} \mathbf{i} \tag{5.93}$$

$$_t\mathbf{j} =_{t+h} \mathbf{j}$$

$$_t\mathbf{k} =_{t+h} \mathbf{k}$$

$$t = t + h$$

and return to Step 1.

The above described direct integration scheme is best demonstrated using numerical examples. In Figure 5.6 a single rigid discrete element with one axis of symmetry is shown.

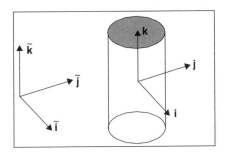

Figure 5.6 Axisymmetric discrete element subject to initial angular velocity.

The z-axis is one of symmetry. The inertia tensor of the element and initial angular velocity at time $t = 0$ are as follows:

$$\mathbf{I}_c = \begin{bmatrix} I_x & 0 & 0 \\ 0 & I_y & 0 \\ 0 & 0 & I_z \end{bmatrix} = \begin{bmatrix} J & 0 & 0 \\ 0 & J & 0 \\ 0 & 0 & I \end{bmatrix} = \begin{bmatrix} 2I & 0 & 0 \\ 0 & 2I & 0 \\ 0 & 0 & I \end{bmatrix} \qquad (5.94)$$

$$\boldsymbol{\omega}_o = \begin{bmatrix} \omega_{xo} \\ \omega_{yo} \\ \omega_{zo} \end{bmatrix} = \begin{bmatrix} 0 \\ 1 \\ 100 \end{bmatrix} \qquad (5.95)$$

The analytical solution for this problem is obtained using the Euler equations:

$$\begin{bmatrix} M_x \\ M_y \\ M_z \end{bmatrix} = \begin{bmatrix} I_x\dot{\omega}_x + (I_z - I_y)\omega_z\omega_y \\ I_y\dot{\omega}_y + (I_x - I_z)\omega_x\omega_z \\ I_z\dot{\omega}_z + (I_y - I_x)\omega_y\omega_x \end{bmatrix} = \begin{bmatrix} I\dot{\omega}_x - (I - J)\omega_z\omega_y \\ I\dot{\omega}_y - (I - J)\omega_x\omega_z \\ J\dot{\omega}_z \end{bmatrix} = \begin{bmatrix} 0 \\ 0 \\ 0 \end{bmatrix} \qquad (5.96)$$

The third equation yields

$$\omega_z = \text{constant} = \omega_{z_o} \qquad (5.97)$$

Differentiation of the second equation and substitution into the second equation yields

$$\ddot{\omega}_y + p^2\omega^2{}_y = 0; \, p = \left(\frac{J}{I} - 1\right)\omega_{z_o} = \left(\frac{2I}{I} - 1\right)100 = 100 \qquad (5.98)$$

A similar equation is obtained for ω_x, which after substitution of the initial conditions, yields

$$\omega_y = \omega_{y_o} \cos pt = \cos 100t \qquad (5.99)$$

$$\omega_x = -\omega_{y_o} \sin pt = -\sin 100t \qquad (5.100)$$

The same problem has been solved using direct integration. A comparison of the analytical and numerical result for the time step $h = 0.0005$ seconds is given in Figure 5.7. The analytical and numerical results for this time step are almost indistinguishable.

A similar comparison of the numerical and analytical results for ω_y, using time step $h = 0.001$ seconds, is shown in Figure 5.8. Again, very good agreement between the numerical and analytical results is obtained.

In Figure 5.9 a comparison of analytical results ω_y and numerical results obtained using time step $h = 0.005$ seconds is shown. The results show relatively good agreement during the first 0.05 seconds. Afterwards, the difference in results is increasing due to the numerical result lagging behind the analytical results.

Application of the Munjiza direct integration scheme in a problem comprising irregular discrete elements is shown in Figure 5.10, where numerical simulation of the motion of two very irregular discrete elements inside a rigid box is presented. The discrete elements represent pebbles subject to initial velocity and acceleration of gravity towards the bottom of the box. Initially, the pebbles move at different velocities, impact against each other and bounce apart in mid-air (Figure 5.10), while at the same time falling towards the

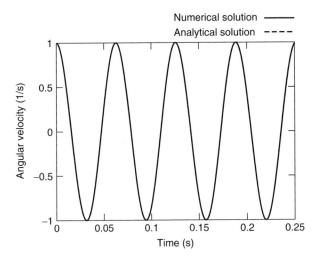

Figure 5.7   Comparison of analytical and numerical results for ω_y and $h = 0.0005$ seconds.

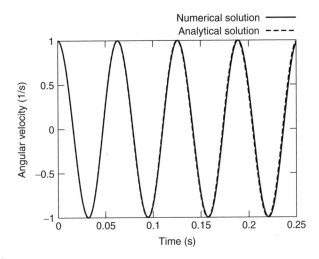

Figure 5.8   Comparison of analytical and numerical results for ω_y and $h = 0.001$ seconds.

bottom of the box (Figure 5.11). The pebbles are then seen to bounce on the bottom of the box, (Figure 5.12).

Subsequent motion is shown in Figure 5.13, with pebbles again moving towards each other and hitting each other. It is noticeable that the right-hand side pebble moves towards the right wall of the box under the impact from the left-hand side pebble.

In Figure 5.14 the discrete elements again move away from each other. In fact, the right-hand side discrete element is mostly stationary, while most of the motion is done by the left-hand side element.

In Figure 5.14 the left-hand side pebble is again falling under gravity and hitting the bottom of the box. The resulting motion can be described as 'pirouetting', or spinning

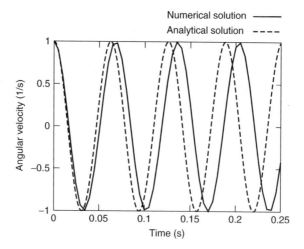

Figure 5.9   Comparison of analytical and numerical results for ω_y and $h = 0.005$ seconds.

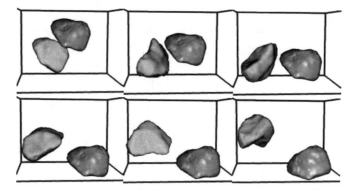

Figure 5.10   Pebbles in a box: initial motion sequence.

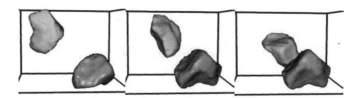

Figure 5.11   Pebbles in a box: motion sequence leading to contact.

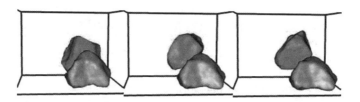

Figure 5.12   Pebbles in a box: motion sequence immediately after the contact.

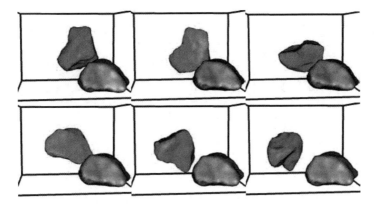

Figure 5.13 Pebbles in a box: motion sequence after the first contact.

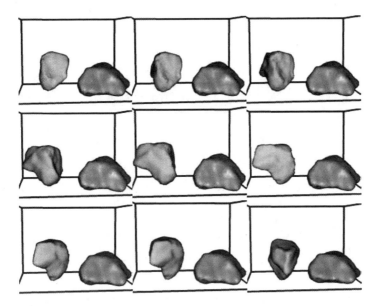

Figure 5.14 Pebbles in a box: 'Pirouetting'.

on a point. Towards the end of the motion sequence shown in this figure, the pebble, although attached to the bottom of the box, starts slowly overturning toward the floor of the box. At the end of this sequence it is sitting flat on the box floor. However, it is still moving, as shown in Figure 5.15, where the left-hand side pebble hits the right-hand side pebble, which then moves towards the wall of the box.

Subsequent motion occurs with both pebbles being 'glued' to the box floor, but still moving. This motion is shown in Figure 5.16, and can be best described as 'rocking' back and forth.

All the time the energy of the pebbles is dissipated through both friction and material and contact damping. No damping in the form of external forces has been employed.

Figure 5.15 Pebbles in a box: overturning towards the floor.

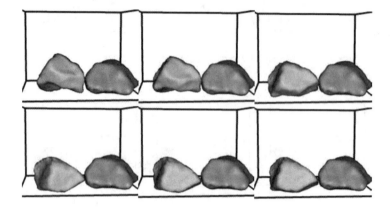

Figure 5.16 Pebbles in a box: 'rocking'.

Eventually, the total kinetic energy is reduced to a negligible level that it can be said that the pebbles are at state of rest. The final state of rest is shown in Figure 5.17.

It is evident from this example that the Munjiza direct integration scheme is clearly able to predict complex behaviour. The virtual pebbles presented were created from 3D laser scans of rock pebbles. The sequence, when animated at speed, gives a remarkably

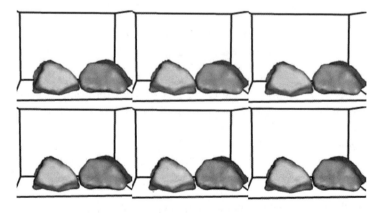

Figure 5.17 Pebbles in a box: state of rest.

convincing representation of this highly complex motion. It stands up to sensory perception of collisions between spinning and bouncing fragments of rock and their container walls and floor. The predicted rocking and pirouetting behaviour of such irregular shaped pebbles is extremely realistic.

5.3   ALTERNATIVE EXPLICIT TIME INTEGRATION SCHEMES

5.3.1   The Central Difference time integration scheme (CD)

CD as described in detail in this chapter is a second order time integration scheme, originally developed in the context of structural dynamics. Apart from CD, a whole range of explicit direct time integration schemes is available for the combined finite-discrete element simulations. Recursive formulae for some of the schemes are as follows.

- *The Leap Frog or Position Verlet Time Integration Scheme (PV):* PV is very similar to CD. It appears to be in essence the same as CD, except that velocities and positions are shifted by $h/2$. The recursive formulae are given by

$$x_{t+h/2} = x_{t-h/2} + h v_t \tag{5.101}$$

$$v_{t+h} = v_t + \frac{h}{m} f_{t+h/2}$$

- *The T-1/12 Time Integration Scheme (T-1/12):* T-1/12 is a third order scheme. The expressions for the position and the velocity at time $(t + h)$ are as follows:

$$x_{t+h} = x_t + \frac{v_t h + \dfrac{1}{12m}(7 f_t - f_{t-h})h^2}{1 - \dfrac{1}{12}\tilde{b}_t h^2} \tag{5.102}$$

where $\tilde{b}_t = \left(\dfrac{1}{m}\dfrac{df}{dx}\right)_t$

$$v_{t+h} = v_t + \frac{h}{12m}(8 f_t + 5 f_{t+h} - f_{t-h}) \tag{5.103}$$

- *The T-1/6 Time Integration Scheme (T-1/6):* T-1/6 is a third order algorithm. The position and the velocity at time $t + h$ are given by

$$x_{t+h} = x_t + v_t h + \frac{f_t}{2m}h^2 + \frac{1}{6}\tilde{b}_t v_t h^3 \tag{5.104}$$

$$v_{t+h} = \frac{v_t + \dfrac{f_t}{m}h + \dfrac{1}{3}\tilde{b}_t v_t h^2}{1 - \dfrac{1}{6}\tilde{b}_{t+h}h^2} \quad \text{where} \tag{5.105}$$

$$\tilde{b}_t = \left(\frac{1}{m}\frac{df}{dx}\right)_t \quad \text{and} \quad \tilde{b}_{t+h} = \left(\frac{1}{m}\frac{df}{dx}\right)_{t+h}$$

- *The D-1/12 Time Integration Scheme (D-1/12):* The recursive formula for D-1/12 is as follows:

$$x_{t+h} = x_{t-h} + \frac{x_t - x_{t-h} + v_t h + \frac{f_t}{2m} h^2}{1 - \frac{1}{12}\tilde{b}_t \Delta t^2} \quad \text{where} \quad \tilde{b}_t = \left(\frac{1}{m}\frac{df}{dx}\right)_t \quad (5.106)$$

$$v_{t+h} = v_t + \frac{h}{12m}(8f_t + 5f_{t+h} - f_{t-h}) \quad (5.107)$$

5.3.2 Gear's predictor-corrector time integration schemes (PC-3, PC-4, and PC-5)

The third order scheme is referred to as PC-3, while the fourth and fifth order schemes are referred to as PC-4 and PC-5, respectively. All three schemes use the same recursive formula, which comprises three stages:

(a) *Prediction*: in the prediction stage, positions are calculated at time $(t + \Delta t)$ by means of a Taylor series based on positions and their derivatives:

$$x_{t+h,p} = x_t + \dot{x}_t h + \ddot{x}_t \frac{h^2}{2!} + \dddot{x}_t \frac{h^3}{3!} + x_t^{iv} \frac{t^4}{4!} + x_t^{v} \frac{h^5}{5!} \quad (5.108)$$

$$\dot{x}_{t+h,p} = \dot{x}_t + \ddot{x}_t h + \dddot{x}_t \frac{h^2}{2!} + x_t^{iv} \frac{h^3}{3!} + x_t^{v} \frac{h^4}{4!}$$

$$\ddot{x}_{t+h,p} = \ddot{x}_t + \dddot{x}_t h + x_t^{iv} \frac{x^2}{2!} + x_t^{v} \frac{x^3}{3!}$$

$$\dddot{x}_{t+h,p} = \dddot{x}_t + x_t^{iv} h + x_t^{v} \frac{h^2}{2!}$$

$$x_{t+h,p}^{iv} = x_t^{iv} + x_t^{v} h$$

$$x_{t+\Delta t,p}^{v} = x_t^{v}$$

(b) *Evaluation*: in the evaluation stage, the force f_{t+h} at time $(t + h)$ is evaluated using the $x_{t+h,p}$ position from the prediction stage. From this force, acceleration $\ddot{x}_{t+h}$ at time $(t + h)$ is calculated, and the discrepancy between this acceleration and the predicted acceleration $\ddot{x}_{t+h,p}$ is evaluated, i.e.

$$\Delta\ddot{x} = (\ddot{x}_{t+h} - \ddot{x}_{t+h,p}) \quad (5.109)$$

(c) *Correction*: in the correction stage, the predicted positions and time derivatives are corrected using the following formulae:

$$x_{t+h} = x_{t+h,p} + \alpha_0 \frac{\Delta\ddot{x}h^2}{2!} \quad (5.110)$$

$$\dot{x}_{t+h}h = \dot{x}_{t+h,p}h + \alpha_1 \frac{\Delta\ddot{x}h^2}{2!}$$

Table 5.1   Values of α_i

	PC-3 (3rd Order)	PC4 (4th Order)	PC5 (5th Order)
α_0	1/6	19/120	3/16
α_1	5/6	3/4	251/360
α_2	1	1	1
α_3	1/3	1/2	11/18
α_4	–	1/12	1/6
α_5	–	–	1/60

$$\ddot{x}_{t+\Delta t}\frac{h^2}{2!} = \ddot{x}_{t+h,p}\frac{h^2}{2!} + \alpha_2\frac{\Delta\ddot{x}h^2}{2!}$$

$$\dddot{x}_{t+h}\frac{h^3}{3!} = \dddot{x}_{t+h,p}\frac{h^3}{3!} + \alpha_3\frac{\Delta\ddot{x}h^2}{2!}$$

$$x_{t+h}^{iv}\frac{h^4}{4!} = x_{t+h,p}^{iv}\frac{h^4}{4!} + \alpha_4\frac{\Delta\ddot{x}h^2}{2!}$$

$$x_{t+h}^{v}\frac{h^5}{5!} = x_{t+h,p}^{v}\frac{h^5}{5!} + \alpha_5\frac{\Delta\ddot{x}h^2}{2!}$$

The PC-3 scheme uses only first, second and third derivatives. PC-4 also uses a fourth derivative, and PC5 uses all five derivatives – the coefficients α for all three schemes are given in Table 5.1.

5.3.3   CHIN integration scheme

The recursive formula for the CHIN integration scheme is given by

$$v_1 = v_t + \frac{h}{6}\frac{f_t}{m} \tag{5.111}$$

$$x_1 = x_t + \frac{h}{2}v_1$$

$$v_2 = v_1 + \frac{2}{3}h\left(\frac{f_1}{m} + \frac{h^2}{48}G_1\right)$$

$$x_{t+h} = x_1 + \frac{h}{2}v_2$$

$$v_{t+h} = v_2 + \frac{h}{6}\frac{f_{t+h}}{m}$$

where v_t is the velocity at time t, f_t is the force evaluated at x_t, x_t is the position at time t, x_{t+h} is the position at time $(t+h)$, v_{t+h} is the velocity at time $(t+h)$, f_{t+h} is the force evaluated at x_{t+h} and m is the mass. The rest of the variables are auxiliary variables used to reach the solution at time $(t+h)$.

5.3.4 OMF30 time integration scheme

The recursive formula for this scheme is as follows:

$$v_1 = v_t + \lambda h \frac{f_t}{m} \qquad (5.112)$$

$$x_1 = x_t + \frac{h}{2} v_1$$

$$v_2 = v_1 + (1 - 2\lambda)h \frac{f_1}{m}$$

$$x_{t+h} = x_1 + \frac{h}{2} v_2$$

$$v_{t+h} = v_2 + \lambda h \frac{f_{t+h}}{m}$$

$$\lambda = \frac{1}{2} - \frac{1}{12} \sqrt[3]{36 + 2\sqrt{326}} + \frac{1}{6} \frac{1}{\sqrt[3]{36 + 2\sqrt{326}}}$$

5.3.5 OMF32 time integration scheme

This is an eleven-stage, fourth order scheme. The recursive formula for this scheme is as follows:

$$x_1 = x_t + \rho h\, v_t \qquad\qquad v_1 = v_t + \vartheta h \frac{f_1}{m}$$

$$x_2 = x_1 + \theta h\, v_1 \qquad\qquad v_2 = v_1 + \lambda h \frac{f_2}{m} + \xi h^3 G_2$$

$$x_3 = x_2 + (1 - 2(\theta + \rho))(h/2)v_2 \quad v_3 = v_2 + (1 - 2(\lambda + \vartheta))h \frac{f_3}{m}$$

$$x_4 = x_3 + (1 - 2(\theta + \rho))(h/2)v_3 \quad v_4 = v_3 + \lambda h \frac{f_4}{m} + \xi h^3 G_4$$

$$x_5 = x_4 + \theta h\, v_4$$

$$\qquad (5.113)$$

$$v_{t+\Delta t} = v_4 + \vartheta h \frac{f_5}{m} \qquad \text{where :}$$

$$r_{t+\Delta t} = r_5 + \rho \Delta t\, v_{t+\Delta t}$$

$$\rho = 0.6419108866816235\ E - 01$$

$$\theta = 0.1919807940455741\ E + 00$$

$$\vartheta = 0.1518179640276466\ E + 00$$

$$\lambda = 0.2158369476787619\ E + 00$$

$$\xi = 0.9628905212024874\ E - 03$$

5.3.6 *Forest & Ruth time integration scheme*

Recursive formula for this scheme is as follows

$$
\begin{aligned}
v_1 &= v_t + c_1 h \frac{f_t}{m} \quad & x_1 &= x_t + d_1 h v_1 \\
v_2 &= v_1 + c_2 h \frac{f_1}{m} \quad & x_2 &= x_1 + d_2 h v_2 \\
v_3 &= v_2 + c_3 h \frac{f_2}{m} \\
& & x_{t+h} &= x_2 + d_3 h v_3 \\
& & v_{t+h} &= v_3 + c_4 h \frac{f_{t+h}}{m}
\end{aligned}
\tag{5.114}
$$

where

$$
c_1 = \lambda + 1/2; \quad c_2 = -\lambda; \quad c_3 = -\lambda; \quad c_4 = \lambda + 1/2 \tag{5.115}
$$
$$
d_1 = 2\lambda + 1; \quad d_2 = -4\lambda - 1; \quad d_3 = 2\lambda + 1; \quad \lambda = \tfrac{1}{6}(2^{1/3} + 2^{-1/3} - 1)
$$

Any of these schemes could in principle be applied in a given combined finite-discrete element simulation. To choose the most appropriate scheme, it is necessary to compare the schemes in terms of stability, accuracy and CPU efficiency. The easiest way to accomplish this is to perform comparisons by considering a one degree of freedom mass-spring system

$$
\ddot{x} + \omega^2 x = 0 \tag{5.116}
$$

where x is the position and ω is the natural frequency

$$
\omega = \sqrt{k/m} = 1 \tag{5.117}
$$

and initial conditions at $t = 0$ are given by

$$
x = 0; v = 1 \tag{5.118}
$$

The analytical solution for this system is given by

$$
x = \sin \omega t; v = \cos \omega t \tag{5.119}
$$

The system has been integrated numerically using different time integration schemes. In all cases, the time step varied from 0.01 seconds to 7.00 seconds. For each time step, the total time of the simulation t_e was equal to 40 periods, i.e.

$$
t_e = 40 \cdot 2\pi = 251 \, \text{s} \tag{5.120}
$$

The maximum position (A_F) during the first 15 periods and the maximum position (A_L) during the last 15 periods were recorded, and then they were compared to calculate the so called amplification factor:

$$Amplification\ Factor = \left(\frac{A_L}{A_F}\right)^{\frac{1}{(i_L - i_F)}} \tag{5.121}$$

where i_F and i_L are time steps corresponding to maximum positions A_F and A_L, respectively. The results obtained for the amplification factor as a function of the time step are shown in Figure 5.18.

All the schemes are conditionally stable, i.e. stable for small enough time steps. Some of the schemes exert numerical damping phenomena, while all the schemes exert numerical amplification of amplitude (energy) for large enough time steps. A particular scheme can be considered stable for a given time step if the amplification factor given in Figure 5.18 is less than 1.0. Critical values are given in Table 5.2.

The period error obtained by numerical integration of positions versus time curve using different time integration schemes is shown in Figure 5.19. The total time of the simulation

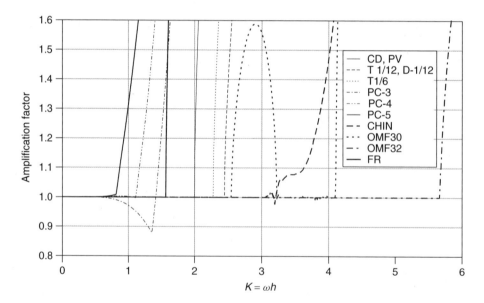

Figure 5.18   Amplification factor as a function of time step for different time integration schemes.

Table 5.2   Values of $K_{critical}$ for each integration scheme

Scheme	$K_{critical}$	Scheme	$K_{critical}$	Scheme	$K_{critical}$
CD	2.000	D-1/12	2.450	PC-5	0.3204
PV	2.000	PC-3	1.417	CHIN	3.050
T-1/12	2.450	PC-4	0.155	OMF30	2.500
T-1/6	2.280	OMF32	3.111	FR	1.572

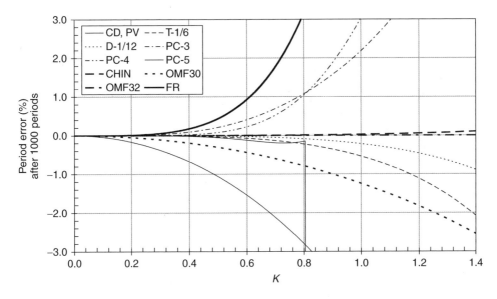

Figure 5.19   Period error for different time integration schemes.

was set to 1000 periods. The number of times that the curve crosses the $x = 0$ line is recorded (n_e), and the time corresponding to the last time (t_e) the position curve crosses the line $x = 0$ is recorded. The approximate period is calculated as shown:

$$T_{apr} = \frac{2t_e}{n_e} \tag{5.122}$$

The period error is then calculated as follows:

$$\varepsilon_{period} = 100\frac{(T_{apr} - 2\pi)}{2\pi} \tag{5.123}$$

The amplitude error is also evaluated from numerical integration of the position-time curve for the one degree freedom by using the root mean square error formula:

$$\varepsilon_{amplitude} = 100\sqrt{\frac{\sum\limits_{i=1}^{i=n}(x - x_{exact})^2}{\sum\limits_{i=1}^{i=n}x_{exact}^2}}\,\% \tag{5.124}$$

To separate the error in period from the error in amplitude, the value of x_{exact} is calculated as

$$x_{exact} = \sin\frac{2\pi}{T_{apr}}t = \sin\omega_{apr}t \tag{5.125}$$

where T_{apr} for a given time step is calculated using equation (5.122).

The amplitude error obtained for different schemes as a function of the size of time step is shown in Figure 5.20. For most of the schemes the error is not a function of t_e. However, those schemes that are unstable or have numerical damping phenomena have the error increasing with t_e. The total CPU as a function of amplitude error for different time integration schemes is shown in Figure 5.21.

None of the schemes considered is unconditionally stable. Both period and amplitude error depend upon the schemes used, and decrease with decreasing time step. Thus it can be assumed that, given a small enough time step, the same accuracy can be achieved regardless of the time integration scheme employed. The only difference is therefore

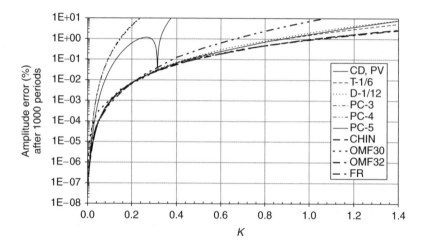

Figure 5.20   Integrated amplitude error after 1000 periods.

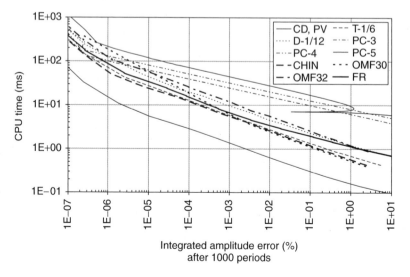

Figure 5.21   CPU time as a function of the integrated amplitude error.

the complexity of the scheme and the CPU time needed to achieve a given accuracy. Higher order schemes for the same accuracy are not necessarily faster (i.e. CPU more efficient) than lower order schemes. The CPU results shown are subject to limitations of a one degree of freedom system, and may be different when applied to large scale combined finite-discrete element systems, where apart from contact force evaluation procedures, contact detection, fracture, fragmentation, fluid coupling, etc. may be involved. The presence of this considerably changes the CPU cost at each time step. However, in this context, higher order schemes requiring multiple force evaluation are much less efficient than lower order schemes.

5.4 THE COMBINED FINITE-DISCRETE ELEMENT SIMULATION OF THE STATE OF REST

Many combined finite-discrete element problems include transient motion that leads to the state of rest. Through energy dissipation mechanisms such as fracture, friction and permanent deformations, the energy of the combined finite-discrete element system is steadily reduced until all the discrete elements are virtually at a state of rest.

There is also a whole class of combined finite-discrete element problems where the state of rest is more important than the transient motion sequence preceding it. These problems are, by nature, static, and require efficient procedures for integration of governing equations.

It has already been mentioned in this chapter that the combined finite-discrete element systems may comprise thousands, even millions, of finite element meshes. Assembling all stiffness matrices and possible problems with inversion of the stiffness matrices has already resulted in the elimination of implicit time integration schemes as possible candidates for solving governing equations. Explicit time integration schemes for both deformable and rigid discrete elements are used instead. In a similar way, for static problems any solver involving the assembly of a stiffness matrix would require unreasonable CPU times, and would be coupled with huge algorithmic difficulties.

Thus in the combined finite-discrete element method, the state of rest and static cases in general are treated as special cases of transient dynamics problems, where energy dissipation mechanisms are such that in a relatively short time a state of rest is achieved or the load is applied at a such a slow rate that no dynamic effects are induced. The method is called *dynamic relaxation*.

In dynamic relaxation the static system is replaced by an equivalent transient dynamic system

$$\mathbf{Kx} + \mathbf{M\ddot{x}} + \mathbf{C\dot{x}} = \mathbf{p} \tag{5.126}$$

or less often with

$$\mathbf{Kx} + \mathbf{C\dot{x}} = \mathbf{p} \tag{5.127}$$

where $\mathbf{K}$ is the stiffness matrix, $\mathbf{M}$ is the mass matrix and $\mathbf{C}$ is the damping matrix. The dynamic relaxation is said to converge if the steady state solution of the equivalent dynamic system is identical to the static solution. By definition, dynamic relaxation assumes an explicit time integration of the equivalent dynamic system. A wide range of explicit schemes is available, however as explained before, the central difference time integration scheme appears to be the optimal one for many problems of practical importance.

The most important advantage of dynamic relaxation in comparison to iterative methods in general is probably the physical meaning, which can be attached to the convergence process itself through gradual motion of the system toward the steady state, which can be expressed in terms of inertia forces. This is very useful in problems with slow monotonic loading and nonlinear problems, in which non-unique solutions may exist. However, the path through which a steady state is reached and the speed at which it is reached are heavily dependent on both the $\mathbf{C}$ and $\mathbf{M}$ matrices.

The recursive formula for dynamic relaxation using the system (5.127) is as follows:

$$
\begin{align}
&0) \quad n = 0; \quad \mathbf{x}_n = 0; \quad \dot{\mathbf{x}}_n = 0; \tag{5.128}\\
&1) \quad n = n + 1;\\
&2) \quad \dot{\mathbf{x}}_n = \mathbf{C}^{-1}(\mathbf{p} - \mathbf{K}\mathbf{x}_{n-1})\\
&3) \quad \mathbf{x}_n = \mathbf{x}_{n-1} + \dot{\mathbf{x}}_n h\\
&4) \quad \text{if the state of rest is not reached go to 1}
\end{align}
$$

The upper limit of the time step for scheme (5.128) to be stable is given by the following inequality:

$$
h < \frac{1}{\omega_n{}^2} \tag{5.129}
$$

where

$$
\omega_i^2 = \frac{\mathbf{v}_i{}^T \mathbf{K} \mathbf{v}_i}{\mathbf{v}_i{}^T \mathbf{C} \mathbf{v}_i} \tag{5.130}
$$

is the frequency of mode i, the shape of which is defined by the eigenvector $\mathbf{v}_i$. Using modal analysis, it can be proven that each mode ω_i converges to the steady state with factor

$$
s = e^{-t\omega_i{}^2} \tag{5.131}
$$

which means that a steady state is practically reached at a time proportional to $1/\omega_i^2$. The longest time will be needed for the lowest frequency ω_1 to reach a steady state, thus the total time needed for the system as a whole to reach a steady state is proportional to $1/\omega_1{}^2$, which leads to the conclusion that the total number of time steps is proportional to $\omega_n{}^2/\omega_1{}^2$.

For problem (5.126), the recursive formula for dynamic relaxation is given by

$$
\begin{align}
&0) \quad n = 0; \quad \mathbf{x}_n = 0; \quad \dot{\mathbf{x}}_n = 0; \tag{5.132}\\
&1) \quad n = n + 1;\\
&2) \quad \ddot{\mathbf{x}}_n = \mathbf{M}^{-1}(\mathbf{p} - \mathbf{K}\mathbf{x}_{n-1} - \mathbf{C}\dot{\mathbf{x}}_{n-1})\\
&3) \quad \dot{\mathbf{x}}_n = \dot{\mathbf{x}}_{n-1} + \ddot{\mathbf{x}}_{n-1} h\\
&4) \quad \mathbf{x}_n = \mathbf{x}_{n-1} + \dot{\mathbf{x}}_n h\\
&5) \quad \text{if the state of rest is not reached go to 1}
\end{align}
$$

For a linear dynamic system, the necessary and sufficient condition for the scheme to be stable is

$$h < \frac{k}{\omega_n} \tag{5.133}$$

where h is the actual time step employed, ω_n is the highest frequency of the system and k is a constant dependent on the damping supplied. For undamped systems $k = 2$, while for an overdamped system, k decreases proportionally with increasing damping.

Several options for choosing **C** matrix of the system (126) are available.

- *Mass proportional damping – underdamped system:* if damping is introduced in the form

$$\mathbf{C} = 2\,\omega_1 \mathbf{M} \tag{5.134}$$

then equation (5.126) can be written as follows:

$$\mathbf{Kx} + \mathbf{M\ddot{x}} + 2\,\omega_1 \mathbf{M\dot{x}} = \mathbf{p} \tag{5.135}$$

By writing the solution in the form

$$\mathbf{x} = \sum_{i=1}^{n} u_i(t)\mathbf{v}_i \tag{5.136}$$

and exploiting the **M**-orthogonality and **K**-orthogonality of eigenvectors $\mathbf{v}_i$

$$\mathbf{v}_i^T \mathbf{Mv}_j > 0 \text{ for } i = j; \text{ and } \mathbf{v}_i^T \mathbf{Mv}_j = 0 \text{ for } i \neq j;$$
$$\mathbf{v}_i^T \mathbf{Kv}_j > 0 \text{ for } i = j; \text{ and } \mathbf{v}_i^T \mathbf{Kv}_j = 0 \text{ for } i \neq j; \tag{5.137}$$

The following equation for mode i is obtained:

$$(\mathbf{v}_i^T \mathbf{Kv}_i)u_i + (\mathbf{v}_i^T \mathbf{Mv}_i)\ddot{u}_i + 2\,\omega_1 (\mathbf{v}_i^T \mathbf{Mv}_i)\dot{u}_i = (\mathbf{v}_i^T \mathbf{pv}_i)u_i \tag{5.138}$$

After substituting

$$\omega_i^2 = \frac{\mathbf{v}_i^T K \mathbf{v}_i}{\mathbf{v}_i^T \mathbf{Mv}_i} \quad \text{and} \quad q_i = \frac{\mathbf{v}_i^T \mathbf{pv}_i}{\mathbf{v}_i^T \mathbf{Mv}_i} \tag{5.139}$$

into (5.144), the equation for mode i can be written as follows:

$$\omega_i{}^2 u_i + \ddot{u}_i + 2\frac{\omega_1}{\omega_i}\omega_i u_i = q_i \quad \text{where} \tag{5.140}$$

$$\xi_i = \frac{\omega_1}{\omega_i} \text{ is the damping ratio}$$

The damping ratio for the lowest frequency is equal to 1. Thus the damping of lowest frequency is equal to the critical damping. The damping ratio for all other frequencies is smaller than 1, which means that all other frequencies are underdamped.

Convergence of the mode i to the zero energy state (state of rest) is given by

$$e^{-\frac{\omega_1}{\omega_i}\omega_i t} \tag{5.141}$$

The time needed to reach a static solution is therefore proportional to

$$\frac{\omega_1}{\omega_i}\omega_i t_s = const \Rightarrow t_s = \frac{const}{\omega_1} \tag{5.142}$$

This means that all frequencies (modes) reach a static solution at the same time. Since the critical time step is proportional to $1/\omega_n$, the total number of time steps required to obtain a static solution is given by

$$n = \frac{t_s}{h} = \frac{const}{\omega_1}\frac{1}{h} = \frac{const}{\omega_1}\omega_n = const\frac{\omega_n}{\omega_1} \tag{5.143}$$

The high frequencies are damped relatively slowly. Their presence in the system converging to the state of rest can, for instance, produces undesirable results. Material that is in a static solution in compression can suddenly, due to the dynamic relaxation, be found to be in tension. In Figure 5.22, $2\omega_1\mathbf{M}$ damping is employed to solve a system comprising 100 different frequencies, with the highest frequency being 1000 rad/s and the smallest frequency being 1 rad/s. It is evident from the results shown in the figure that high frequencies oscillate for a long time. These oscillations can, for instance, produce unphysical behaviour such as brittle fracture in tension, when in fact the equivalent static problem comprises material in compression only. This is due to stress reversal (oscillations), as shown in the figure.

This phenomenon is called overshooting. In this particular scheme, it is present with all modes. In addition, estimation of the lowest frequency can be difficult. The highest frequency is much easier to estimate (from the rigidity of the stiffest finite element) than the lowest frequency. It is also worth mentioning that damping proportional to mass is

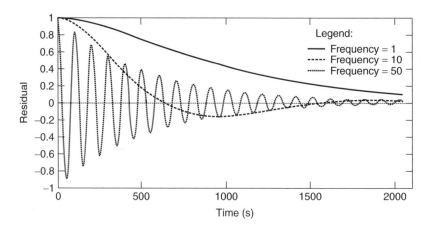

Figure 5.22 Overshooting with mass proportional damping matrix.

an external force which influences considerably the path through which the state of rest is reached.

- *Mass proportional damping – overdamped system:* if damping is introduced in the form

$$\mathbf{C} = 2\,\omega_n\mathbf{M} \tag{5.144}$$

then equation (5.126) can be written as follows:

$$\mathbf{Kx} + \mathbf{M\ddot{x}} + 2\,\omega_n\mathbf{M\dot{x}} = \mathbf{p} \tag{5.145}$$

After applying the modal decomposition as explained above, the governing equation for mode i can be written in the form

$$\omega_i{}^2 u_i + \ddot{u}_i + 2\frac{\omega_n}{\omega_i}\omega_i\dot{u}_i = q_i \ \text{where} \tag{5.146}$$

$$\xi_i = \frac{\omega_n}{\omega_i} \ \text{is the damping ratio}$$

The damping ratio for the highest frequency is equal to 1, which means that the damping of the highest frequency mode is equal to the critical damping. The damping ratio for all other frequencies is greater than 1, which means that all other frequencies are overdamped. The lowest frequency is highly overdamped, and its convergence to the zero energy state is given by

$$e^{-\frac{\omega_1}{2\xi_1}t} = e^{-\frac{\omega_1}{2\omega_n}\omega_1 t} \tag{5.147}$$

The time needed to reach a static solution is therefore proportional to

$$-\frac{\omega_1}{\omega_n}\omega_1 t_s = const \Rightarrow t_s = const\frac{\omega_n}{\omega_1^2} \tag{5.148}$$

As the critical time step for such a highly overdamped system is still proportional to $1/\omega_n$ (because the highest frequency is not overdamped), the total number of time steps required to obtain static solution is given by

$$n = \frac{t_s}{h} = const\frac{\omega_n}{\omega_1^2}\omega_n = const\left(\frac{\omega_n}{\omega}\right)^2 \tag{5.149}$$

The beneficial feature of the scheme is that no overshooting is present, however, it comes at the price of a greatly reduced performance of the scheme in terms of the total CPU time required. On the other hand, only the highest frequency of the system has to be evaluated to calculate the damping matrix, which is usually the easiest to estimate. It is important to note that this very high mass proportional damping comes as external force, and can lead to results which are not physical, i.e. can lead to convergence to a wrong static solution.

- *Stiffness proportional damping:* if damping is introduced in the form

$$\mathbf{C} = \frac{2}{\omega_n}\mathbf{K} \tag{5.150}$$

then equation (5.126) can be written as follows:

$$\mathbf{Kx} + \mathbf{M\ddot{x}} + \frac{2}{\omega_n}\mathbf{K\dot{x}} = \mathbf{p} \tag{5.151}$$

By writing the solution in the form

$$\mathbf{x} = \sum_{i=1}^{n} u_i(t)\mathbf{v}_i \tag{5.152}$$

and exploiting the **M**-orthogonality and **K**-orthogonality of eigenvectors, the following equation for mode i is obtained:

$$(\mathbf{v}_i^T \mathbf{Kv}_i)u_i + (\mathbf{v}_i^T \mathbf{Mv}_i)\ddot{u}_i + \frac{2}{\omega_n}(\mathbf{v}_i^T K\mathbf{v}_i)\dot{u}_i = (\mathbf{v}_i^T \mathbf{pv}_i)u_i \tag{5.153}$$

After substituting

$$\omega_i^2 = \frac{\mathbf{v}_i^T K\mathbf{v}_i}{\mathbf{v}_i^T \mathbf{Mv}_i} \quad \text{and} \quad q_i = \frac{\mathbf{v}_i^T \mathbf{pv}_i}{\mathbf{v}_i^T \mathbf{Mv}_i} \tag{5.154}$$

into (5.144), the equation for mode i can be written as follows:

$$\omega_i{}^2 u_i + \ddot{u}_i + 2\frac{\omega_i}{\omega_n}\omega_i\dot{u}_i = q_i \text{ where} \tag{5.155}$$

$$\xi_i = \frac{\omega_i}{\omega_n} \text{ is damping ratio}$$

The damping ratio for the highest frequency is equal to 1, i.e. the damping of highest frequency is critical damping. The damping ratio for all other frequencies is smaller than 1, which means that all other frequencies are underdamped. Convergence of the mode i to the zero energy state (state of rest) is given by

$$e^{-\frac{\omega_i}{\omega_n}\omega_i t} \tag{5.156}$$

The time needed to reach a static solution is therefore proportional to

$$\frac{\omega_i}{\omega_n}\omega_i t_s = const \Rightarrow t_s = \frac{const \,\omega_n}{\omega_i \quad \omega_i} \tag{5.157}$$

This means that higher frequencies (modes) reach a static solution quicker than lower frequency modes.

Since the critical time step is proportional to $1/\omega_n$, the total number of time steps required to obtain a static solution is given by

$$n = \frac{t_s}{h} = \frac{const \,\omega_n}{\omega_i \quad \omega_i}\omega_n = const\left(\frac{\omega_n}{\omega_i}\right)^2 \tag{5.158}$$

The high frequencies are damped relatively quickly. The lowest frequency reaches the steady state last. The damping does not introduce any artificial external forces to the system. The convergence to the state of rest is natural. However, the total number of steps can be too large. Overshooting associated with highest frequencies is mostly eliminated. However, overshooting associated with small frequencies can still be a problem. This can be easily avoided by simply supplying the load gradually over what is a relatively long time that the system takes to reach the state of rest.

In most combined finite-discrete element simulations, stiffness proportional damping is preferred option. It comes naturally with the deformability of discrete elements, where viscous forces calculated using the rate of deformation tensor act as an energy dissipation mechanism, which is not artificial but is the physical property of the material of discrete elements.

• *Stiffness proportional damping:* if damping is introduced in the form

$$\mathbf{C} = \frac{2}{\omega_1}\mathbf{K} \tag{5.159}$$

then equation (5.126) can be written as follows:

$$\mathbf{Kx} + \mathbf{M\ddot{x}} + \frac{2}{\omega_1}\mathbf{K\dot{x}} = \mathbf{p} \tag{5.160}$$

which, after applying modal analysis, yields the equation for mode i

$$\omega_i{}^2 u_i + \ddot{u}_i + 2\frac{\omega_i}{\omega_1}\omega_i \dot{u}_i = q_i \text{ where} \tag{5.161}$$

$$\xi_i = \frac{\omega_i}{\omega_1} \text{ is damping ratio}$$

The damping ratio for the lowest frequency is equal to 1, i.e. the damping of lowest frequency is critical damping. The damping ratio for all other frequencies is greater than 1, which means that all other frequencies are overdamped.

Convergence of the lowest frequency mode to the zero energy state (state of rest) is given by

$$e^{-\frac{\omega_i}{\omega_1}\omega_i t} = e^{-\frac{\omega_1}{\omega_1}\omega_1 t} = e^{-\omega_1 t} \tag{5.162}$$

The highest frequency mode is highly overdamped, and its convergence to the state of rest is therefore given by

$$e^{-\frac{\omega_n}{2\xi_n}t} = e^{-\frac{\omega_n}{2\omega_n/\omega_1}t} = e^{-\frac{\omega_1}{2}t} \tag{5.163}$$

The time needed to reach a static solution is therefore proportional to

$$\omega_1 t_s = const \Rightarrow t_s = \frac{const}{\omega_1} \tag{5.164}$$

Table 5.3 Comparison of dynamic relaxation schemes

M	C	Overshoot	Time steps	Momentum balance
0	M	NO	$(\omega_n/\omega_1)^2$	NO
M	$2\,\omega_1 M$	YES	(ω_n/ω_1)	NO
M	$2\,\omega_n M$	NO	$(\omega_n/\omega_1)^2$	NO
M	$(1/\,\omega_1)K$	NO	$(\omega_n/\omega_1)^2$	YES
M	$(1/\,\omega_n)K$	YES	$(\omega_n/\omega_1)^2$	YES

Since the critical time step for such an overdamped system is proportional to ω_1/ω^2_n, the total number of time steps required to obtain a static solution is given by

$$n = \frac{t_s}{h} = \frac{const\ \omega_n^2}{\omega_1\ \ \omega_1} = const\left(\frac{\omega_n}{\omega_1}\right)^2 \tag{5.165}$$

This damping does not introduce any artificial external forces to the system, while the convergence to the state of rest is natural.

A summary of the alternative dynamic relaxation schemes listed above is given in Table 5.3: In the first column, the schemes are separated with regard to the mass matrix The second column presents the method of damping. The third column indicates whether overshooting is present or not. The theoretical number of time steps to obtain a steady state solution is given in column 4. An important factor in both dynamic relaxation and transient dynamics is momentum balance. Mass proportional damping introduces external viscous forces into the system. This results in a significant momentum imbalance, which can result in inaccurate stress and strain fields. For nonlinear systems this can lead to a wrong prediction of the response of the system. For instance, some discrete element systems, if subjected to a heavy mass proportional damping, can lead to a very distorted picture of the final static state (state of rest), i.e. the dynamic relaxation scheme may converge to a non-physical static solution.

6

Sensitivity to Initial Conditions in Combined Finite-Discrete Element Simulations

6.1 INTRODUCTION

In the 18th century Pierre Simon de Laplace, in his *Philosophical Essayians on Probabilities*, concluded that for a vast enough intellect, given initial conditions, the future would be just like the past, and that nothing would be uncertain. In the context of the combined finite-discrete element method, this could be interpreted that for a large enough computer, given initial conditions, the motion of each individual particle could be predicted regardless of the size or nature of the physical problem.

However, in systems such as gas, statistical methods were introduced in 1873 by J.C. Maxwell; actually scientists were intuitively aware that a deterministic system can behave in a random way, with randomness occurring in systems with a large number of particles or degrees of freedom. In 1887, H. Poincare discovered that very complicated dynamic behaviour can occur even in a simple system such as an idealised three-body problem. In 1961, Edward Lorenz, while running computer models for weather forecasts, was the first to notice the heavy dependence of the outputs of the weather model on the initial conditions supplied. Rounding errors due to the computer program being interrupted and restarted resulted in a completely different weather prediction. He called this phenomena the 'butterfly effect'.

The common characteristic of all systems mentioned above is that they never find a steady state, i.e. they never repeat themselves. They are sensitive to initial conditions, but they are not periodic. A system that is not periodic and is sensitive to initial conditions is unpredictable.

An irregular unpredictable behaviour that results from nonlinearity in a dynamic system is called 'deterministic chaos', because the model employed is deterministic, but the results are not. It can occur in a system with one degree of freedom, as well as in a system with any number of degrees of freedom.

The Combined Finite-Discrete Element Method A. Munjiza
© 2004 John Wiley & Sons, Ltd ISBN: 0-470-84199-0

6.2 COMBINED FINITE-DISCRETE ELEMENT SYSTEMS

The long-term prediction of chaotic system is impossible. In the case of the combined finite-discrete element method, each body (discrete element) deforms, fractures, fragments and, at the same time, interacts with discrete elements in its vicinity. Thus, the average combined finite-discrete element simulation is highly nonlinear.

This is illustrated in Figure 6.1, where two discrete elements with sharp corners are moving towards each other with velocity **v**.

If digital representation of the coordinates of each of the discrete elements, velocity magnitude and velocity direction were exact, the discrete elements would simply bounce away from each other, as shown in Figure 6.2.

However, a rounding error may result in sharp corners just missing each other, and discrete elements coming in contact with their edges, as shown in Figure 6.3. This time the contact force is normal to the edges that are in contact, and makes discrete elements both move alongside each other and rotate. It is worth noting that with irrational numbers such as the number π, the rounding error is always present regardless of the number of digits that a given CPU can handle. In other words, rounding error is not simply a mater of the number of significant digits. It is much more fundamental than that.

A relatively small-scale combined finite-discrete element problem, where sharp corner type contact situation arises, is shown in Figure 6.4. The problem has a vertical axis of symmetry, as indicated. The initial vertical velocity at point A is supplied. The results obtained using the combined finite-discrete element method show preservation of this initial symmetry (Figure 6.5).

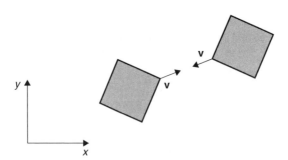

Figure 6.1 Two sharp corners in contact.

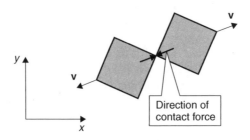

Figure 6.2 Corner to corner contact.

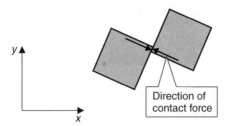

Figure 6.3 Edge to edge contact.

Figure 6.4 Initial conditions.

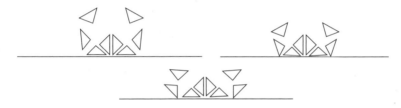

Figure 6.5 Motion sequence before the loss of symmetry.

Figure 6.6 Rounding error influence on resolution of contact impact.

Figure 6.7 Motion sequence after the loss of symmetry.

However, the further motion sequence shown in Figure 6.6 clearly shows corner to corner contact. In the left-hand side of the figure, the bottom triangle has gone under the impacting triangle. In the right-hand side of the figure, the impacting triangle has just missed the bottom triangle.

This slight difference in resolution of contact-impact due to the rounding error has finally resulted in the complete loss of symmetry, as shown in Figure 6.7.

A large multibody system is actually permanently in some sort of bifurcation. Thus, rounding errors or small differences in initial conditions can influence the behaviour of

Figure 6.8 A beam-shaped heap impact problem.

the system or behaviour of its parts. However, it is worth mentioning that a drastic loss of symmetry is only possible with a small system. A large system tends to compensate for this local rounding error type bifurcation by shear numbers of such bifurcations, and much more constrained motion of individual discrete elements.

This is demonstrated by the example shown in Figure 6.8, where a beam shaped heap of triangular discrete elements is impacted by a large triangular discrete element moving at constant velocity. The motion sequence is shown in Figure 6.9. Initially, complete preservation of symmetry is observed despite a large number of discrete elements comprising the system, and a large number of contacts being resolved. This is because the surrounding discrete elements are in a way constraining the motion of discrete elements in contact.

When the density of discrete elements close to the mid-span of the beam decreases, discrete elements in front of the impactor become less constrained. Further resolution of contacts leads to a slight loss of symmetry. The overall behaviour of the system still shows symmetry, although discrete elements close to the tip of the impactor do not match the symmetry requirements. It is worth noting that, given the total number of resolved contacts and possible bifurcations, the fact that the symmetry has been preserved to a great extent demonstrates the robustness of the discretised distributed contact algorithm used to resolve contacts (see Chapter 2). A loss of symmetry and sensitivity to initial conditions or rounding errors is not only due to sharp corner contacts. It is not only limited to 2D problems. Instead, it is almost a second nature of large scale combined finite-discrete element simulations. This is demonstrated by a 3D contact impact problem, shown in Figure 6.10. A rod shaped body has been placed at the centre of a rigid cube shaped box. The rod moves with initial vertical velocity towards the top end of the box.

The simulation is run twice. The first time, the rod is placed exactly in a vertical position. The second time, the rod is slightly inclined relative to the vertical axis of the box. The angle of inclination is very small (10^{-11} radians). In practical terms, the two problems are identical, and one could easily state that there is no difference in initial conditions.

There is almost no difference in the results of simulation either, as shown in Figure 6.11, where both rods have moved towards the top of the box. In Figure 6.12, the rods have bounced off the upper side of the box and are moving towards the bottom side of the box, as shown in Figure 6.13.

In Figure 6.14, the rods have bounced off the bottom side of the box, and are moving towards the upper side of the box.

In Figure 6.15, the rods have bounced again from the upper side of the box, and are moving towards the bottom of the box.

In Figure 6.16, subsequent bouncing from the bottom of the box is shown. This is followed by bouncing from the top of the box, followed by yet another bouncing from the bottom of the box, as shown in Figure 6.17. The rods still seem to be at same position and parallel, i.e. the results of two numerical experiments seem to completely much each other.

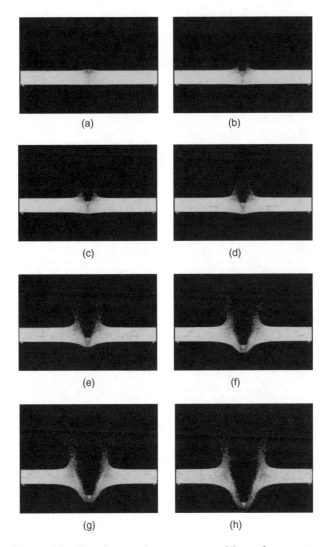

Figure 6.9 Transient motion sequence and loss of symmetry.

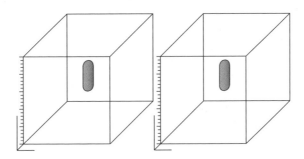

Figure 6.10 Initial conditions: exactly vertical rod (right) and rod inclined at an angle of 10^{-11} radians (left).

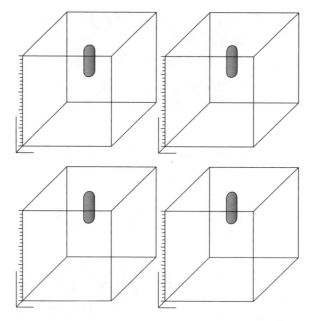

Figure 6.11 Moving towards the top of the box and bouncing back.

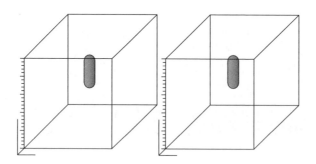

Figure 6.12 Bouncing back.

However, in Figure 6.18 it can be noticed that the left-hand rod is slightly inclined, i.e. slightly different behaviour between the two rods is noticeable.

Further motion shown in Figure 6.19 results in the left-hand rod further rotating, hitting the top of the box at an angle, and thus rotating even further.

In Figure 6.20 the left-hand rod is spinning and moving vertically, while the right-hand rod is still vertical. Further motion results in the left-hand rod being almost horizontal and vertically in a different place from the right-hand rod. Further motion is such that no resemblance between the motion of the left-hand and right-hand rod exists.

It can be observed that an almost negligible difference in initial conditions has resulted in a completely unrelated motion sequence between the two rods. This was a small-scale, frictionless discrete element simulation. Similar sensitivity to initial conditions is

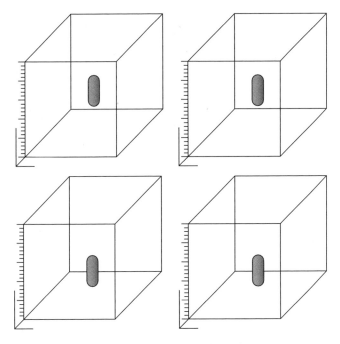

Figure 6.13   Moving towards the bottom.

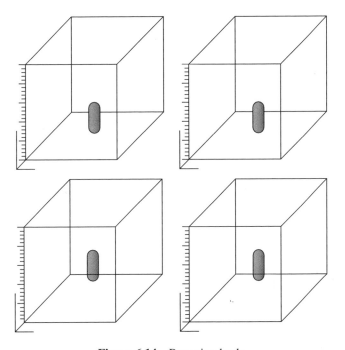

Figure 6.14   Bouncing back.

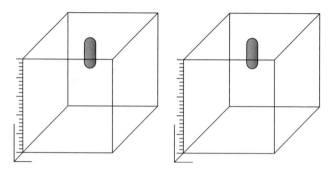

Figure 6.15 Bouncing from the top.

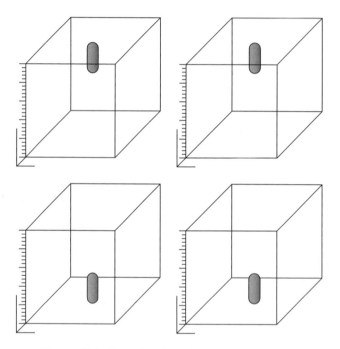

Figure 6.16 Bouncing from the bottom of the box.

obtained for systems comprising a handful of discrete elements. As the system gets larger and discrete elements get packed closer together, each discrete element is constrained by neighbouring discrete elements, and the overall behaviour of the combined finite-discrete element system may get less sensitive to initial conditions.

The motion of a particular discrete element may still be sensitive to initial conditions, but the overall behaviour of the system may be considered not sensitive to initial conditions. For instance, if one makes an X mark on the face of an intact rock mass, it is almost impossible to predict where the rock piece containing this mark will land after the blasting operation. However it is reasonable to expect that the shape of the blast-pile will be predicted with a degree of accuracy.

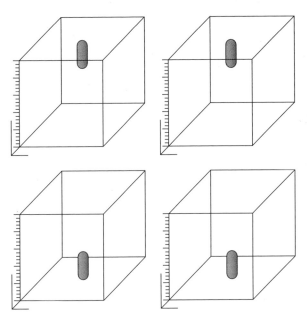

Figure 6.17   Subsequent bouncing from the top side of the box followed by bouncing from the bottom side of the box.

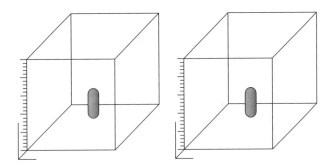

Figure 6.18   Left hand rod is slightly inclined.

Sensitivity to initial conditions is also well established in experimental investigations of combined finite-discrete element systems. In Figure 6.21, the experimental results of the gravitational deposition of wooden cubes into a glass box is shown. The cubes are dropped one-by-one through a regular grid at the top of the box. In total, 648 cubes were dropped. The same experiment was repeated three times, with the same initial and boundary conditions and experimental setup.

One would expect exactly the same results to be obtained in each repeat of the experiment. The final states of rest for all three repeats of the experiment are shown Figure 6.21. It is evident that the results of the three exactly identical experiments are not the same. There is overall similarity of the results, however at a detailed level, the three results differ considerably.

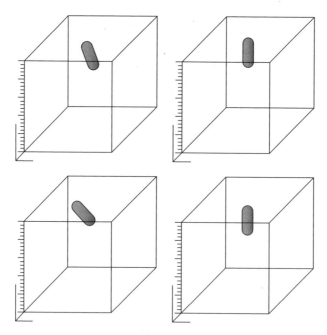

Figure 6.19 Moving up. The left hand rod inclines further, and hits the top at an angle, thus inclining even more.

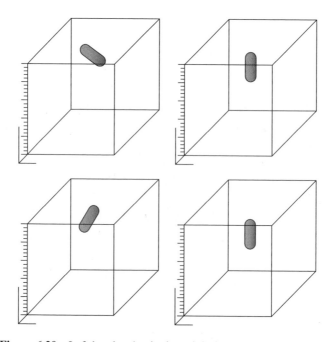

Figure 6.20 Left hand rod spinning, right hand rod is still vertical.

Figure 6.21 Sensitivity to initial conditions in experimental investigations; results of three identical experiments (courtesy of J.P. Latham).

Sensitivity to initial conditions that is observable in the combined finite-discrete element simulations is not purely an academic exercise. It is a physical property of the combined finite-discrete element system, in the same way that energy dissipation is a property of such a system. It has practical implications and needs to be taken seriously. For instance, it can be demonstrated that some 'results', such as the position of an individual discrete element in a big pile of particles, is probably due to chance, i.e. is probably a random variable. A result such as the overall density of the pile at state of rest has no probability associated with it, and is a completely deterministic variable. In short, results of the combined finite-discrete element simulation can be grouped into two categories: probabilistic variables and deterministic variables.

The combined finite-discrete element simulation is based on rational mechanics, and is deterministic in its formulation without inclusion of any theory of probability. However, due to the sensitivity to initial conditions, all results are not deterministic, and some of the results are clearly random variables.

This has important consequences in both verification and validation of combined finite-discrete element models. Verification is in essence a process of checking that the model is implemented properly. One of standard verification tests is to run a problem that contains symmetry. The results should also show the same symmetry. In the combined finite-discrete element method, some loss of symmetry may not indicate that the implementation was wrong, as clearly demonstrated by the 2D example shown in Figure 6.7.

Validation is a process in which the underlying assumptions of any model are tested, usually by comparison to either theoretical or experimental results. In the light of the sensitivity to initial conditions, as explained above, it is important to carefully identify variables to be measured and compared. Probabilistic results may not be of much use in the validation process, and an effort should be made to ensure that the results used for comparison are deterministic results (i.e. results not sensitive to initial conditions). Using, for instance, the positron tracking motion of a single particle may not yield much in terms of results, except confirming that such a result belongs to a group of probabilistic results. Using X-ray imaging to record velocity profiles over the whole domain would probably exhibit no sensitivity to initial conditions, i.e. obtained velocity maps would be very similar, regardless of small perturbations in initial conditions.

In a similar way, measuring total reaction over a large boundary area is likely to produce a deterministic result.

In summary, whenever experimental or numerical investigation into the behaviour of the combined finite-discrete element system is undertaken, due attention has to be given to the phenomena of sensitivity to initial conditions. However, this is not to say that these sensitivities should be an excuse for inconsistencies in the applied model, or errors in the implementation of a particular model.

7

Transition from Continua to Discontinua

7.1 INTRODUCTION

Transition from continua to discontinua in the combined finite-discrete element method occurs through fracture and fragmentation processes. A typical combined finite-discrete element method based simulation, such as rock blasting, may start with a few discrete elements and finish with a very large number of discrete elements.

Fracture in general occurs through alteration, damage, yielding or failure of microstructural elements of the material. To describe this complex, material-dependent phenomenon, the alteration of stress and strain fields due to the presence of microstructural defects and stress concentrations must be taken into account. Several approaches are available, and these include global approaches, local approaches, smeared crack models and single crack models.

Global approaches to fracture are based on the representation of the singularity of the stress field at the crack tip. It was shown by Griffith that the failure of a brittle elastic medium due to such singularity can be characterised by the energy release rate G. The critical value of $G = 2\gamma$ (where γ represents the surface energy) is a material characteristic. The alternative formulation of the Griffith method is achieved through stress intensity factors, which characterise the stress singularity on a semi-local basis in terms of force, while the same singularity is characterised in terms of energy by contour integrals.

Local approaches to crack analysis usually employ a smeared crack approach, with a single crack being replaced by a blunt crack band. This approach has been justified by the fact that engineering materials show a reduction in the load-carrying capacity accompanied by strain localisation after the maximum load-carrying capacity is reached. Beyond the peak load (when the material gradually disintegrates), two types of failure mechanism are observed, namely decohesion and frictional slip. In the first type of failure fracture, zones are observed (cracks), while in the latter failure zones propagate along shear bands (faults). Smeared crack models attempt to describe these processes through constitutive laws, such as a strain softening constitutive law or damage mechanics based formulation. However, standard continuum mechanics formulations incorporating softening fail, as the underlying mathematical problem becomes ill-posed. As a result, the numerical solution

The Combined Finite-Discrete Element Method   A. Munjiza
© 2004 John Wiley & Sons, Ltd   ISBN: 0-470-84199-0

predicts a vanishing energy dissipation upon spatial discretisation refinement. A mathematically well-posed problem is obtained by using an enriched continuum formulation (such as a Micro-polar Cosserat) or higher-order constitutive law (such as a non-local constitutive law, where the higher order gradients of the deformation field are included in the formulation). A relatively straightforward alternative utilising a fracture energy based softening plasticity framework has also been successfully adopted in the past, where a mesh size dependent softening modulus ensures objective energy dissipation.

The local approaches to crack analysis based on a single crack concept are usually based on the Dugdale model or Barenblatt model. The Dugdale model is a relatively simple nonlinear model for a crack with a plastic zone at its tip, where the zone of plastically strained material is replaced by a zone of weakened bonds between the crack walls. As the crack walls separate the bond stress reaches maximum. At the point when the separation reaches a critical value, the bonding stress drops to zero.

The main tasks in describing fracture in the combined finite-discrete element method are to predict crack initiation, predict crack propagation, perform the necessary remeshing, transfer variables from the old to the new mesh and replace the released internal forces with equivalent contact forces. Robustness, accuracy, simplicity and CPU requirements of the fracture algorithms implemented are of major importance, and both single and smeared crack models have been employed in the past. In the rest of this chapter, two most widely employed fracture models are described together with numerical experiments demonstrating the complexity of the combined finite-discrete element simulations involving complex fracture and fragmentation patterns. Fracture and fragmentation is still intensively researched field in the combined finite-discrete element method and Computational Mechanics of Discontinua in general, and currently available simulation techniques are far from optimum.

7.2 STRAIN SOFTENING BASED SMEARED FRACTURE MODEL

In experimental tests of rock and rock-like materials, a gradual load decrease with an increase in displacements is observed. The phenomenon occurs under uniaxial tension as well as under uniaxial pressure and triaxial stress states. Figure 7.1 shows a typical stress-displacement diagram for a rock specimen under uniaxial tension. Due to stress decreasing with increasing strain, pre-failure strains are highly localised in a narrow band, which eventually results in a discontinuity in the form of a crack. The phenomenon of decreasing stress in a localisation band area with increasing strains is called 'strain-softening'.

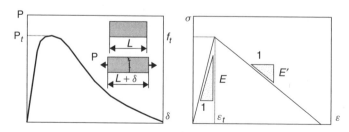

Figure 7.1 A typical stress-displacement diagram for rock under uniaxial tension and idealised stress-strain diagram in the localisation zone.

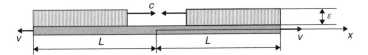

Figure 7.2   Wave propagation in a strain-softening bar in tension.

Actual implementation of strain softening material models into finite element codes has been associated with great difficulties regarding both sensitivity to mesh size and mesh orientation. The problems associated with dynamic analysis are best illustrated by a 1D strain softening bar, shown in Figure 7.2. The bar of length $2L$ with a unit cross-section and mass ρ per unit length has the ends moving simultaneously outward with constant opposite velocities, v, while the initial conditions at time $t = 0$ are given by zero displacements and a zero displacement rate:

$$u(x, t = 0) = \dot{u}(x, t = 0) = 0$$

In the case where the strains never exceed ε_t (strain corresponding to peak stress), the bar remains elastic and the differential equation of motion is hyperbolic:

$$\frac{\partial}{\partial x}(\sigma(\varepsilon)) - \rho\frac{\partial^2 u}{\partial t^2} = 0; \quad \text{or} \quad \frac{\partial \sigma}{\partial \varepsilon}\frac{\partial \varepsilon}{\partial x} - \rho\frac{\partial^2 u}{\partial t^2} = 0; \tag{7.1}$$

After substituting $E_T = \dfrac{\partial \sigma}{\partial \varepsilon}$ and $\varepsilon = \dfrac{\partial u}{\partial x}$, this yields

$$E_T\frac{\partial^2 u}{\partial x^2} - \rho\frac{\partial^2 u}{\partial t^2} = 0 \tag{7.2}$$

The solutions of (7.2) are the two waves travelling in opposite directions (Figure 7.2) at velocity c and meeting at the centre of the bar, at which stage the strain at the midpoint doubles. If this strain exceeds the strain corresponding to the peak stress, the strain-softening region appears at the midpoint of the bar, the tangent modulus of elasticity E_T becomes negative (modulus E' in Figure 7.1), and the differential equation of motion within the strain-softening segment becomes elliptic. This ellipticity means that interaction spreads immediately over finite distances, actually over the entire strain-softening region. The response is in fact the same as if the bar was cut in two at midpoint. Thus, the wave reflects as if the midpoint was a free end. In other words, the strain-softening happens within an infinitesimal segment, and is confined to a single cross-sectional plane of the bar.

A similar problem arises in static analysis. A type of instability can again be investigated on a 1D bar of length $2L$ with prescribed displacements Δ at the bar ends, (Figure 7.3). The governing equation of the problem is given by

$$\frac{\partial}{\partial x}(\sigma(\varepsilon)) = 0; \quad \text{or} \quad \frac{\partial \sigma}{\partial \varepsilon}\frac{\partial \varepsilon}{\partial x} = 0; \quad \text{i.e. } E_T\frac{\partial^2 u}{\partial x^2} = 0; \quad E_T = \frac{\partial \sigma}{\partial \varepsilon} \tag{7.3}$$

where for the softening region, the tangent modulus of elasticity E_T becomes negative. With the assumption that only the segment around the midpoint represents a

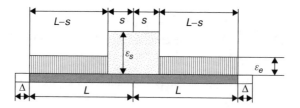

Figure 7.3 Strain softening bar in tension – assumed strain field.

strain-softening segment while the rest of the bar has strains smaller than the strain corresponding to the peak stress, for the equilibrium of stress field it is required that

$$\sigma(\varepsilon_s) - \sigma(\varepsilon_e) = 0 \tag{7.4}$$

and for this equilibrium to be stable it is required that

$$\frac{\partial}{\partial \varepsilon_s}(\sigma(\varepsilon_s) - \sigma(\varepsilon_e)) > 0; \text{ i.e. } \frac{\partial \sigma(\varepsilon_s)}{\partial \varepsilon_s} - \frac{\partial \sigma(\varepsilon_e)}{\partial \varepsilon_e}\frac{\partial \varepsilon_e}{\partial \varepsilon_s} > 0; \tag{7.5}$$

which is equivalent to

$$\frac{\partial \sigma(\varepsilon_s)}{\partial \varepsilon_s} + \frac{\partial \sigma(\varepsilon_e)}{\partial \varepsilon_e}\frac{s}{L-s} > 0$$

It follows that the only stable segment is the segment of length larger than

$$s > -\frac{\dfrac{\partial \sigma(\varepsilon_s)}{\partial \varepsilon_s}}{\dfrac{\partial \sigma(\varepsilon_e)}{\partial \varepsilon_e} - \dfrac{\partial \sigma(\varepsilon_s)}{\partial \varepsilon_s}}L \tag{7.6}$$

Since

$$\frac{\partial \sigma(\varepsilon_s)}{\partial \varepsilon_s} < 0 \text{ and } \frac{\partial \sigma(\varepsilon_e)}{\partial \varepsilon_e} > 0 \tag{7.7}$$

which means that all segments smaller than one given by (7.6) are unstable, which leads to the conclusion that the actual length of the strain-softening segment is zero.

Localisation is the intense straining of a material within thin bands. It is associated with material instability, resulting in bifurcation connected to the loss of ellipticity (in static problems) or to the loss of hyperbolicity (in dynamic problems), and in both dynamic and static problems the strain-softening constitutive law in a classical (local) continuum (i.e. defined in terms of strains), although not mathematically meaningless, dissipates no energy. This is not the case for known strain-softening materials such as rock, which is characterised by finite strain-softening regions. The strains for real engineering softening materials are localised over a relatively small (far smaller than the size of the actual physical or engineering problem) yet finite lengths (characteristic lengths) that reflect the microstructure of the rock, and energy dissipation is therefore well defined.

In finite element based applications, these instabilities are manifested in mesh size and mesh orientation dependency of the solution; the localisation zone width corresponds to the element size, and with a finer mesh the localisation zone width is smaller and, in addition, localisation zones tend to follow preferred directions (along the finite element edges or diagonals) dictated by the mesh. According to how these problems are approached, finite element based discretisations of problems involving localisation can be classified as follows:

- *Sub-h concept* is intended for situations where the localisation bandwidth b is significantly smaller than the element size h. It is realised by implementing a localisation zone into a finite element. The strain field therefore incorporates a localisation band (rather than just a discontinuity), which can be triggered in a state of homogeneous strain. The embedding of localisation zones is usually carried out for low order elements, such as the four node quadrilateral and the three node constant-strain triangle.

- *Super-h* concept is implemented when the localisation band is larger than the element size. The bandwidth is usually uniquely determined by the field equations, such as a non-local strain-softening formulation where the stress at any point is related to the weighted strain within a finite volume about that point, leading to the bandwidth arising from the governing field equations.

- *Iso-h* concept is characterised by the size of finite elements being equal to the width of the localisation zone. The constitutive law is usually modified to include the characteristic length of the discretisation grid, such as the element size or area around the integration point, in order to deal with mesh sensitivity.

Localisation is closely related to smeared crack models, where the localisation zone (crack band) is usually assumed to propagate into the next finite element when the stress in that element reaches a strength limit. In this way, the propagation of the zone is influenced and largely determined by the zone width, because the narrower the zone the larger the stresses ahead. If the crack bandwidth is specified as a material property within a non-local continuum framework, there is nothing wrong with this approach. However, when it is determined by the finite element size as discussed above, additional criteria for band propagation need to be introduced.

The smeared fracture model implemented in the combined finite-discrete element method also uses the concept of localisation band propagation, and in essence belongs to the iso-h group of strain-softening models. The underlying assumptions of the model are:

- the localisation (tensile fracture) occurs on the finite element integration point level;

- the size of the overall model is significantly larger than the size of the finite elements employed (h);

- the strain energy accumulated before the peak stress is reached within an area associated with the integration point undergoing softening can be neglected;

- the plasticity model is assumed to be isotropic, i.e. the accumulated effective plastic strain is monitored in the principal directions only. If, after the strength limit is reached, a full breakage does not occur (stress state on the softening branch), the effective plastic strain is treated as a *scalar state variable* for the next state of deformation, which will be valid for any new rotated principal direction.

To deal with mesh size sensitivity, the local softening material law is formulated in terms of the fracture energy release rate in tension, G_f, and the local control length h:

$$h = \sqrt{\frac{4A}{\pi}} \qquad (7.8)$$

where A is the area associated with the Gauss integration point considered. To avoid limitations to the upper limit of the element size to be used, which arises from the difficulties in numerically capturing the so-called *snap back* in the constitutive law (softening slope return), the fracture energy is assumed to control only the post-peak behaviour, i.e. after the peak stress f_t is reached. The local softening slope for each Gauss-point is then obtained from the energy balance

$$\frac{f_t^2}{2E'}A = 2\gamma h = G_f h; \text{ using } h \text{ from (7.8) yields } E' = \frac{1}{4}\frac{f_t^2}{G_f}\sqrt{\pi A} \qquad (7.9)$$

This modification of the constitutive law resolves the problem of sensitivity of the fracture energy release rate to the mesh size. However, the sensitivity of crack initiation to element size remains. This is because the crack is replaced by a localisation band which is equal to the element size. A further consequence of this is also the sensitivity to mesh orientation. With a deformable discrete element, discretised into finite elements, a critical state of stress (or strain) is reached when an element separates into two or more discrete elements, or a discrete element changes its boundary (if the failure is only partial). At the stage when the strength of material in some Gauss-points is reduced to zero, a crack is assumed to open. The direction of the crack coincides with the direction of the greater principal plastic stretch. A re-meshing of finite elements within every discrete element is therefore performed, and when breakage occurs new boundaries are created.

As mentioned above, the constitutive law taking into account the mesh size and fracture energy of rock is employed to deal with the sensitivity of the model to the mesh size in terms of energy dissipated through the fracture process. The convergence of the solution in terms of energy dissipation is best illustrated by an example (Figure 7.4), showing a strain-softening bar with both ends moving in opposite directions with a constant velocity $v = 0.02$ m/s. The bar is of length $l = 100$ m and rectangular cross-section of $h = 5$ m and $b = 0.2$ m. The material of the bar is a linear softening material with a fracture energy $G_f = 2\gamma = 0.05$ kNm/m^2, modulus of elasticity $E = 900$ kN/m^2, tensile strength $f_t = 1$ kNm/m^2 and density $\rho = 1000$ kg/m^3. The combined finite-discrete element simulation of the problem using coarser and finer meshes both resulted in the bar being broken in two parts that move further away. The total energy of the bar (i.e. kinetic energy plus

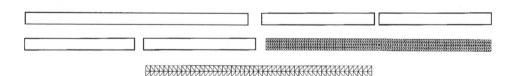

Figure 7.4 Failure sequence of a strain softening bar together with finite element meshes employed.

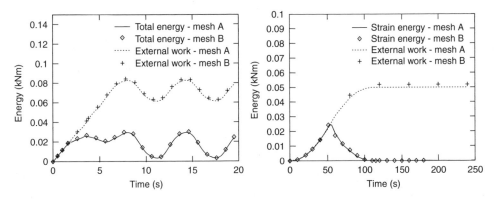

Figure 7.5   Energy balance for the strain softening bar: dynamic case (left) and static case (right).

strain energy), together with the external work that is needed to keep the ends of the bar moving at constant velocity, is shown in Figure 7.5. The difference between the external work and total energy of the bar is equal to the energy dissipated due to fracture, i.e.

$$bhG_f = 0.05 \text{ kNm/m}^2 \tag{7.10}$$

It is worth noting that after the bar has been broken, the two pieces still oscillate, which makes the total energy of the bar and external work oscillate as well, although the difference remains constant. Static analysis of the bar (both ends moving very slowly) results in a similar convergence in terms of energy. These results indicate convergence of the solution only in terms of the energy dissipation, while the sensitivity to mesh alignment remains despite the constitutive law taking into account mesh size; limited remeshing around the crack tip due to the creation of new boundaries may constrain this sensitivity.

The smeared crack approach described in this section is able to produce complicated fracture and fragmentation patterns, as shown in Figure 7.6, where a 2D combined finite-discrete element approximation of a bench blasting problem is shown. The geometry parameters of the problem are described by a bench of 13 m, burden distance of 7 m and spacing of 7 m. The explosive charge consisted of a 9 m long *ANFO* charge in a borehole of 380 mm diameter. The initial density of the explosive charge was $\rho = 80 \text{ kg/m}^3$, while the detonation energy and detonation velocity were $Q_e = 3700 \text{ kJ/kg}$ and $VOD = 1725 \text{ m/s}$, respectively. The properties of rock are described by modulus of elasticity $E = 28 \text{ GPa}$, Poisson ratio $\nu = 0.1$, density $\rho = 4.2 \text{ t/m}^3$ and fracture energy release rate $G_f = 0.25 \text{ kNm/m}^2$. Any problem involving rock blasting is by nature a 3D problem, with rock involving natural 2.5D type discontinuities and being inhomogeneous. The combined finite-discrete element simulation shown is only 2D simulation, with rock being assumed as a homogeneous and isotropic material without any initial discontinuities.

Initially, there is only one discrete element comprising four boreholes. Borehole pressure results in the formation of radial cracks that propagate towards the surface. As the stress wave reflects from the free surface, it is followed by cracks that propagate toward the boreholes. Eventually, these two crack systems meet and the fragmentation process begins, resulting in a network of interconnected cracks together with a large number of separate rock fragments in transient motion, eventually resulting in a muck-pile. It

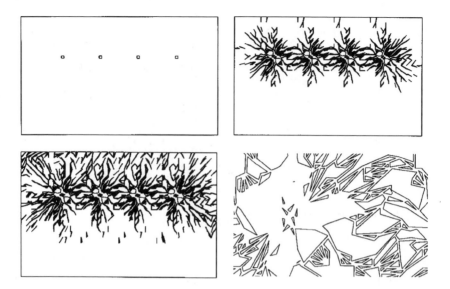

Figure 7.6 Fracture sequence in 2D combined finite-discrete element simulation of bench blasting; the last picture shows an enlarged fracture pattern.

is worth noting that the initial finite element mesh employed is a very coarse one. As the fracture pattern develops, the finite element mesh is refined to accommodate newly created boundaries (as previously described in this section), while the energy dissipation mechanisms due to strain softening are controlled in respect to the mesh size.

A typical sequence for a muck-pile formation is shown in Figure 7.7. Initially, fractured rock forms an unstable column with individual rock blocks moving horizontally and accelerating downwards under gravity (self-weight). The bottom of the column collapses first, and this collapse is propagated toward the upper layers of rock. In the next stage the lower layers of rock decelerate due to confinement induced through boundaries, which results in contact interaction between rock fragments and intensive energy dissipation.

Figure 7.7 Initial collapse of fractured rock.

The problem of sensitivity to mesh orientation remains, and it is not easy to provide an answer to the question of the extent to which the obtained fracture pattern is a function of the initial mesh. This is because the final fracture pattern does not depend only on the initial mesh, but also on all subsequent remeshings, which are in turn governed by the creation of new boundaries. In other words, the mesh pattern is influenced by mesh orientation, and in turn the transient meshes are a result of boundaries created by the fracture pattern. The problem of sensitivity of the fracture pattern to mesh orientation in the context of the smeared fracture and fragmentation model is also coupled with algorithmic complexities involving permanent remeshing due to the creation of new boundaries and the problems associated with it (transfer of variables, tracing of new contacts, permanently changing size, topology and CPU and RAM requirements during execution of the computer problem). Thus, in recent developments of the combined finite-discrete element method, the emphasis is being placed on discrete crack based approaches.

7.3 DISCRETE CRACK MODEL

As explained above, the smeared crack model for fracture and fragmentation is coupled with numerical difficulties and algorithmic complexities. Bearing in mind the other complexities involved in combined finite-discrete element simulations, such as contact detection and contact interaction, the transition from continua to discontinua algorithms must be optimised both in terms of CPU time and RAM requirements. Recent research efforts regarding fracture modelling in the context of the combined finite-discrete element method have therefore also included the single crack model. The model presented in this section is actually a combination of the smeared and single crack approaches. It was designed with the aim of modelling multiple-crack situations, progressive fracture and failure, including fragmentation and the creation of a large number of rock fragments of general shape and size.

The model presented in this section is aimed at mode I loaded cracks only. It is based on the approximation of stress-strain curves for rock in direct tension, (Figure 7.8). A typical stress-strain curve for rock consists of the hardening branch (before the peak stress is reached) and strain-softening part, which represents decreasing stress with increasing strain.

The strain-hardening part of the stress-strain curve presents no difficulties when implemented in the combined finite-discrete element method, and is therefore implemented in a standard way through the constitutive law. The strain-softening part of the stress-strain

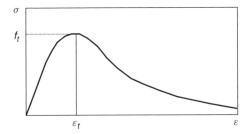

Figure 7.8 Typical strain softening curve defined in terms of strains.

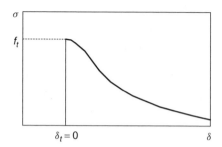

Figure 7.9   Strain softening defined in terms of displacements.

curve is connected with localisation of strains, loss of ellipticity (hyperbolicity) of the governing equation, ill-posed problems and general sensitivity to mesh size and mesh orientation. To deal with these problems, formulation of the strain softening by means of stress and displacements is adopted, as shown in Figure 7.9.

The area under the stress-displacement curve represents the energy release rate, $G_f = 2\gamma$, where γ is the surface energy, i.e. the energy needed to extend the crack surface by unit area. The softening stress-displacement relationship is implemented in the combined finite-discrete element method through the single crack model, i.e. using bonding stress, as shown in Figure 7.10.

In theory, the separation $\delta = \delta_t = 0$ coincides with the bonding stress being equal to the tensile strength f_t, i.e. no separation occurs before the tensile strength is reached. With increasing separation $\delta > \delta_t$ the bonding stress decreases, and at separation $\delta = \delta_c$ the bonding stress drops to zero. Bonding stress for separation $\delta_t < \delta < \delta_c$ is given by

$$\sigma = z f_t \tag{7.11}$$

i.e. a scaled tensile strength, with the scaling (softening) function z being defined in such a way that it represents a close approximation of the stress-displacement curve. Thus, a heuristic formula for z is adopted:

$$z = \left[1 - \frac{a+b-1}{a+b} \exp\left(D \frac{a+cb}{(a+b)(1-a-b)} \right) \right] [a(1-D) + b(1-D)^c] \tag{7.12}$$

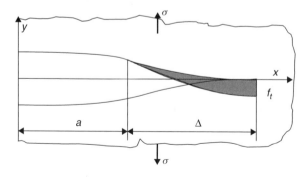

Figure 7.10   Single crack model with bonding stress.

where the variable D is given by

$$D = \begin{cases} 0, & \text{if } \delta \le \delta_t \\ 1, & \text{if } \delta > \delta_c \\ \dfrac{\delta - \delta_t}{\delta_c - \delta_t}, & \text{otherwise} \end{cases} \tag{7.13}$$

while the parameters a, b and c are obtained from experimental stress displacement curves by curve fitting. Note that for any value of these parameters, the above heuristic formula results in a bonding stress of f_t for $D = 0$ and a bonding stress equal to zero for $D = 1$. The tangent at the stress displacement curve at $D = 0$ is horizontal. Thus, parameters a, b and c control the slope of the curve at $D = 1$ and the shape of the curve (curvature) at $D = 0$ together, with the inflection point.

In the discrete crack model, it is assumed that the crack walls coincide with the finite element edges. Thus initially the total number of nodes for each of the finite element meshes (every single discrete element is associated with its separate finite element mesh) is doubled, and nodes are held together through a penalty function method. Thus the separation δ_t is a function of the penalty term p employed. In the limit no separation of adjacent edges takes place before stress f_t is reached, i.e.

$$\lim_{p \to \infty} \delta_t = 0 \tag{7.14}$$

With increasing separation $\delta > \delta_t$ the bonding stress decreases, and at separation $\delta > \delta_c$ it is zero and the crack is assumed to propagate.

In finite element discretisation of the governing equations, only approximate stress and strain fields close to the crack tip are obtained. With the bonding stress model as described above, the stress and strain fields close to the crack tip are influenced by the magnitude and distribution of the bonding stress close to the crack tip. For the bonding stress to have a significant effect on stress distribution results, it is necessary that the size of finite elements close to the crack tip be smaller than the actual size of the plastic zone. The coarser mesh results in bonding stress in all elements close to the crack walls being reduced to zero, except for the few elements adjoining the crack tip. The propagation of the crack is therefore influenced by the orientation of those elements close to the crack tip. The coarse finite element mesh does not accurately represent the stress field in the proximity of the crack tip, and as a result, the stress field obtained is influenced by the mesh topology close to the crack tip. i.e. the de-bonding and separation of crack walls occurs on an element-by-element basis.

One way to avoid this problem is to have an element size close to the crack tip much smaller than the size of the plastic zone. The approximate length of the plastic zone Δ for a plane stress mode I loaded crack can be approximated from Muskhelishvili's exact solution for a crack loaded in mode I. For an infinite body under plane stress conditions, Muskhelishvili's solution gives at points $y = 0$ normal stress σ_y and crack opening displacement δ as a function of the coordinate x:

$$\sigma_y = \sigma \frac{1}{\sqrt{1 - (a/x)^2}} \quad \text{and} \quad \delta = \sigma \frac{4a}{E} \sqrt{1 - (x/a)^2} \tag{7.15}$$

where the maximum crack opening displacement at $x = 0$ is given by

$$\delta_0 = \sigma \frac{4a}{E} \tag{7.16}$$

The lower value estimate of the plastic zone length for a short crack can be obtained by substituting $\Delta = a$, $\delta_o = \delta_t$ and $\sigma = f_t$ into (7.16).

$$\Delta = \frac{E}{4 f_t} \delta_t \tag{7.17}$$

In a similar way, for a long crack, Westergaard's asymptotic solution for the crack opening near the crack tip $(x = a)$ is obtained from equation (7.15) by substituting

$$1 - \left(\frac{x}{a}\right)^2 \doteq 2 \left(1 - \frac{x}{a}\right) \tag{7.18}$$

i.e.

$$\delta_{x \doteq a} = \sigma \frac{4a}{E} \sqrt{2 \left(1 - \frac{x}{a}\right)} \tag{7.19}$$

For a plane stress mode I crack, the strain energy release rate is given in terms of stress intensity factor K_1 (Irwin's formula):

$$G_f = 2\gamma = \frac{K_1^2}{E} \text{ and } K_1 = \sigma \sqrt{\pi a} \tag{7.20}$$

from which it follows that

$$\sigma = \sqrt{\frac{G_f E}{\pi a}} \tag{7.21}$$

which, when substituted into (7.19), gives

$$\delta_{x \doteq a} = \sqrt{32 \frac{G_f (a - x)}{\pi E}} \tag{7.22}$$

The lower value estimate of the plastic zone length for a long crack can therefore be estimated by substituting

$$\Delta \doteq (a - x); G_f \doteq \delta_t f_t \text{ and } \delta \doteq \delta_t \tag{7.23}$$

into (7.22), which then yields

$$\Delta \doteq \frac{\pi E \delta_t}{32 f_t} \tag{7.24}$$

The meaning of the plastic zone length is best illustrated by the problem shown in Figure 7.11. For a very coarse mesh, the stress field close to the crack tip is almost

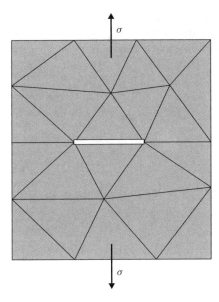

Figure 7.11 A crack tip approximation using a very coarse mesh – thick line indicates the failure pattern.

ignored by a weak formulation of the boundary value problem, and the failure load appears to be approximately

$$\frac{\sigma}{f_t} = 1 \tag{7.25}$$

where σ and f_t are the critical load and maximum bonding stress, respectively. The critical load predicted is wrong by an order of magnitude, for the simple reason that no plastic zone or stress concentration has been taken into account due to the element size being greater than the size of the plastic zone. The direction of the crack propagation is mostly influenced by the global elastic stress field, and does not depend much upon either the crack presence or element orientation.

For a very fine mesh, the element size is in general only a fraction of the size of the plastic zone. Thus the plastic zone itself spreads through a relatively large number of finite elements, and stress and strain fields within the plastic zone are well represented by the finite element approximation. In addition, by the Sant-Venant principle the stress field towards the end of the plastic zone is not influenced much by the small change in the shape of the crack walls corresponding to the finite element size (which is assumed to be only a fraction of the plastic zone size). In other words, for a very fine mesh the critical load, together with the general direction of crack propagation, should be accurately predicted due to the insignificant effect of element orientation on the stress field within the plastic zone.

For intermediate element sizes, the critical load is overestimated. The stress field in the vicinity of the crack tip is, to a large extent, influenced by the size, shape and orientation of the finite elements in the vicinity of the crack tip. Because crack propagation is locally driven (i.e. the direction of the crack propagation is determined on the basis of the stress and strain fields in the vicinity of the crack tip), it follows that the direction of crack propagation is to a large extent influenced by the topology of the finite element mesh in the vicinity of the crack tip. In other words, the result of analysis will be sensitive to the topology of the finite element mesh employed. The fracture pattern is therefore a function of both problem parameters (overall geometry, applied load, material properties) and model parameters (mesh orientation, mesh size).

Sensitivity to mesh size and orientation is due to the singularity of the stress field at the crack tip. The influence of such singularity can be illustrated on the example of the following function:

$$\sigma = \frac{1}{\sqrt{x}} \tag{7.26}$$

being approximated by constant value finite elements, as shown in Figure 7.12.

The relative error of such an approximation can be estimated by

$$\varepsilon = \frac{\dfrac{1}{\sqrt{(n-1)h}} - \dfrac{1}{\sqrt{nh}}}{\dfrac{1}{\sqrt{(n-1)h}}} 100\% = \left(1 - \sqrt{\frac{n-1}{n}}\right) 100\% \tag{7.27}$$

It is evident that the relative error of approximation does not decrease with the decreasing element size. This means that, for a zero length plastic zone, no mesh refinement would increase the ability of the model to predict the fracture pattern. In other words, fracture models (often found in the literature) based on the sudden release of stress (acoustic release) cannot be objective, regardless of the finite element mesh or other type of grid employed. In contrast, the combined single and smeared crack model is based on the assumption of a finite plastic zone. Thus, very fine meshes in conjunction with the combined single and smeared crack model should result in accurate prediction of both the critical load and fracture pattern.

The problem is that such fine meshes are in many practical applications simply not affordable. Small scale problems such as rock crushers, the fracture of smaller size blocks, and so on, are already affordable on modern day computers. Coarser meshes would also make large scale problems affordable in terms of CPU time (for instance, blasting

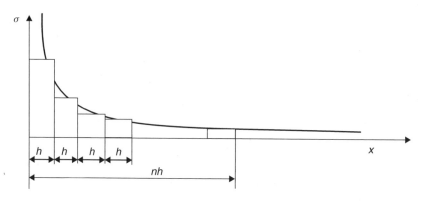

Figure 7.12   A singular stress field approximated by constant stress finite elements.

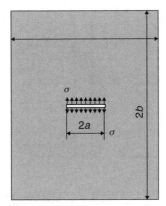

Figure 7.13   A thin plate with a crack parallel to the edges.

operations). However, such meshes, when coupled with the combined single and smeared crack model, result in stress and strain fields in the vicinity of the crack tip being inaccurate by an order of magnitude. Thus, the resulting fracture patterns are extremely sensitive to the local element size and element orientation. With over one billion elements, it is feasible that some real scale bench blasting problems can be accurately modelled using the combined smeared and discrete crack model as described in this chapter.

The theoretical conclusions about mesh sensitivity are best demonstrated through a set of numerical experiments designed to look at the influence of the element size on fracture results. A finite size thin plate with a crack parallel to the edges subjected to uniform inner pressure employed for this purpose is given in Figure 7.13.

The geometry of the plate is defined by:

$$a = 10\,\text{mm}$$
$$b = 30\,\text{mm}$$

(7.28)

while the following material properties are assumed:

$$\text{Modulus of Elasticity } E = 26.6\,\text{GPa}$$
$$\text{Poisson's ratio } \nu = 0.205$$
$$\text{Tensile strength } f_t = 5\,\text{MPa}$$
$$\text{Density } \rho = 2340\,\text{kg/m}^3$$

(7.29)

The load is increasing with time:

$$\sigma = 20{,}000t \text{ MPa}$$

(7.30)

where t is time in seconds and σ is the uniform pressure on crack walls.

The problem is solved for two different energy release rates ($G_f = 2\gamma = 3\,\text{N/m}$ and $G_f = 2\gamma = 30\,\text{N/m}$), each time using four different element sizes, meshes A, B, C and D, as shown in Figure 7.14. Mesh A is characterised by an element size of $h = 10\,\text{mm}$.

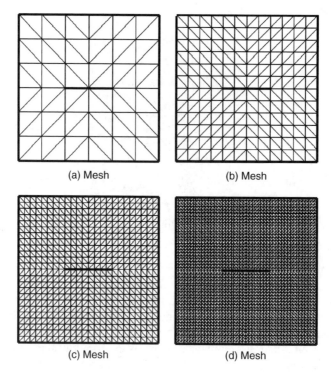

(a) Mesh   (b) Mesh

(c) Mesh   (d) Mesh

Figure 7.14   Finite element meshes used in the analysis of the thin plate.

Mesh size B is comprised of finite elements of size $h = 5$ mm, while meshes C and D are comprised of finite elements of size $h = 2.5$ mm and $h = 1.25$ mm, respectively. In all cases, plane stress loading conditions are assumed with corresponding Lamé constants

$$\lambda' = \frac{\nu E}{1 - \nu^2} = \frac{0.205\, 26.6}{1 - 0.205^2} = 5.59 \text{ GPa} \tag{7.31}$$

$$\mu = \frac{E}{2(1 + \nu)} = \frac{26.6}{2(1 + 0.205)} = 11.04 \text{ GPa}$$

For a very low energy release rate of $2\gamma = 3$ N/m, the fracture sequence obtained using a very coarse mesh (mesh A) is shown in Figure 7.15. The grey scale indicates spatial distribution of the σ_{yy} stress component, while a thick line indicates the crack walls and a thin line indicates the plastic zone. It is evident that the size of the plastic zone is equal to the size of the finite elements employed.

The results obtained using a finer mesh (mesh B) are shown in Figure 7.16. The size of plastic zone is shorter this time, however it is still equal to the finite element size.

A further decrease in element size (mesh C) results in the fracture sequence shown in Figure 7.17. The size of the plastic zone is still equal to the size of the finite elements employed. Even the greatest reduction in element size investigated (mesh D), as shown in Figure 7.18, results in the plastic zone stretching over only two elements.

It can be concluded that the size of the plastic zone in all the examples shown is governed by the size of the finite elements employed. This is because the theoretical size

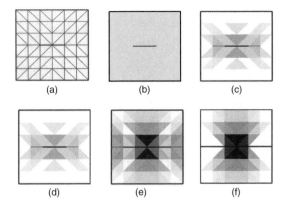

Figure 7.15 Finite element mesh (mesh A) employed and fracture sequence obtained for $2\gamma = 3$ N/m. The frames shown correspond to (b) $t = 0$ ms, (c) $t = 0.09$ ms, (d) $t = 0.15$ ms, (e) $t = 0.21$ ms, (f) $t = 0.25$ ms; i.e. transient loads (b) $\sigma = 0$ MPa, (c) $\sigma = 1.8$ MPa, (d) $\sigma = 3.0$ MPa, (e) $\sigma = 4.2$ MPa, (f) $\sigma = 5.0$ MPa.

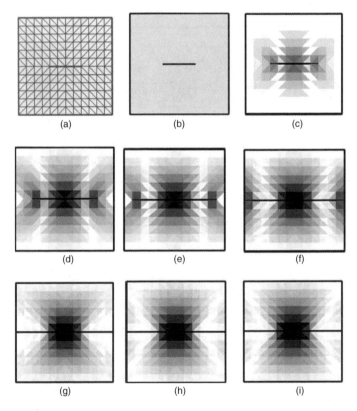

Figure 7.16 Finite element mesh (mesh B) employed and fracture sequence obtained for $2\gamma = 3$ N/m. The frames shown correspond to (b) $t = 0$ ms, (c) $t = 0.09$ ms, (d) $t = 0.15$ ms, (e) $t = 0.16$ ms, (f) $t = 0.17$ ms, (g) $t = 0.18$ ms, (h) $t = 0.19$ ms, (i) $t = 0.20$ ms; i.e. transient loads (b) $\sigma = 0$ MPa, (c) $\sigma = 1.8$ MPa, (d) $\sigma = 3.0$ MPa, (e) $\sigma = 3.2$ MPa, (f) $\sigma = 3.4$ MPa, (g) $\sigma = 3.6$ MPa, (h) $\sigma = 3.8$ MPa, (i) $\sigma = 4.0$ MPa.

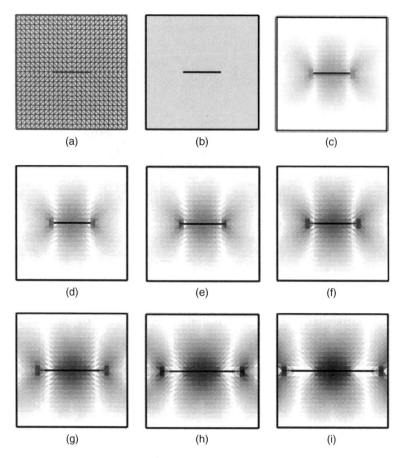

Figure 7.17 Finite element mesh (mesh C) employed and fracture sequence obtained for $2\gamma = 3$ N/m. The frames shown correspond to (b) $t = 0$ ms, (c) $t = 0.05$ ms, (d) $t = 0.08$ ms, (e) $t = 0.09$ ms, (f) $t = 0.11$ ms, (g) $t = 0.13$ ms, (h) $t = 0.14$ ms, (i) $t = 0.15$ ms; i.e. transient loads (b) $\sigma = 0$ MPa, (c) $\sigma = 1.0$ MPa, (d) $\sigma = 1.6$ MPa, (e) $\sigma = 1.8$ MPa, (f) $\sigma = 2.1$ MPa, (g) $\sigma = 2.6$ MPa, (h) $\sigma = 2.8$ MPa, (i) $\sigma = 3.0$ MPa.

of the plastic zone as explained earlier is smaller than the size of the finite elements employed. Thus, none of the meshes employed is able to model the plastic zone, and a further reduction in element size would be necessary to get an accurate representation of the stress and strain fields close to the crack tip.

The size of the plastic zone is a function of the fracture energy release rate, as shown earlier. This is demonstrated through the same thin plate with a crack parallel to the edges. The material properties, loading and geometry are all the same as in the example described above. The only difference is that this time a much larger fracture energy release rate $G_f = 2\gamma = 30$ N/m is assumed, resulting in a much larger plastic zone.

The problem is solved using four different meshes. The fracture sequences obtained using meshes C and D are shown in Figures 7.19 and 7.20, respectively. It is worth noting that in both cases the size of the plastic zone (indicated by a thin line) is much larger than the size of the individual finite elements employed. The size of plastic zone obtained

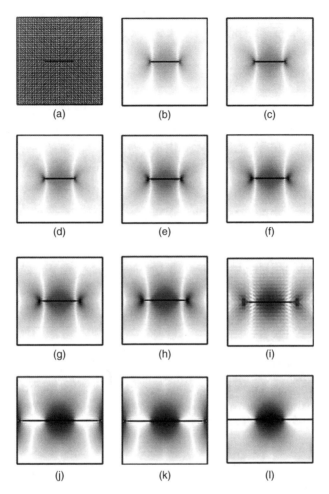

Figure 7.18 Finite element mesh (mesh D) employed and fracture sequence obtained for $2\gamma = 3$ N/m. The frames shown correspond to (b) $t = 0.05$ ms, (c) $t = 0.07$ ms, (d) $t = 0.08$ ms, (e) $t = 0.09$ ms, (f) $t = 0.10$ ms, (g) $t = 0.11$ ms, (h) $t = 0.12$ ms, (i) $t = 0.13$ ms, (j) $t = 0.14$ ms, (k) $t = 0.15$ ms, (l) $t = 0.16$ ms; i.e. transient loads (b) $\sigma = 1.0$ MPa, (c) $\sigma = 1.4$ MPa, (d) $\sigma = 1.6$ MPa, (e) $\sigma = 1.8$ MPa, (f) $\sigma = 2.0$ MPa, (g) $\sigma = 2.2$ MPa, (h) $\sigma = 2.4$ MPa, (i) $\sigma = 2.6$ MPa, (j) $\sigma = 2.8$ MPa, (k) $\sigma = 3.0$ MPa, (l) $\sigma = 3.2$ MPa.

using mesh C is almost the same as the size of the plastic zone obtained using mesh D. In other words, the solution has converged in terms of the size of the plastic zone and stress and strain fields close to the crack tip. The result is that the fracture patterns obtained using both meshes are the same. This is also the case with load levels at the instance of crack propagation.

As mentioned earlier, the combined finite-discrete element algorithms also include transient dynamics procedures. Thus the load is applied as a time dependent variable, and the results obtained include inertia effects, while strain rate effects are ignored. The load is increasing with time, and the load level at the instance that the crack commences propagating is registered in each case. In further text this load is termed the *fracture load*. The

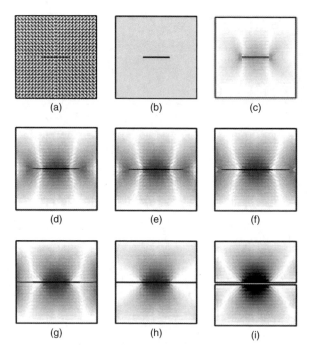

Figure 7.19 Finite element mesh (mesh C) employed and fracture sequence obtained for $2\gamma = 30\,\text{N/m}$. The frames shown correspond to (b) $t = 0\,\text{ms}$, (c) $t = 0.17\,\text{ms}$, (d) $t = 0.27\,\text{ms}$, (e) $t = 0.28\,\text{ms}$, (f) $t = 0.30\,\text{ms}$, (g) $t = 0.31\,\text{ms}$, (h) $t = 0.32\,\text{ms}$, (i) $t = 0.50\,\text{ms}$; i.e. transient loads (b) $\sigma = 0\,\text{MPa}$, (c) $\sigma = 3.4\,\text{MPa}$, (d) $\sigma = 5.4\,\text{MPa}$, (e) $\sigma = 5.6\,\text{MPa}$, (f) $\sigma = 6.0\,\text{MPa}$, (g) $\sigma = 6.2\,\text{MPa}$, (h) $\sigma = 6.4\,\text{MPa}$, (i) $\sigma = 10.0\,\text{MPa}$.

results of the above numerical experiments are summarised in terms of fracture load in Figure 7.21, where the obtained fracture load as a function of the size of finite elements employed is shown. Element size has been normalised in such a way that mesh A is characterised by an element size of $h = 1$, mesh B by an element size of $h = 0.5$, mesh of C by an element size of $h = 0.25$ and mesh D by an element size of $h = 0.125$.

For different fracture energy release rates, different fracture loads are obtained. In general, the greater the fracture energy release rate, the greater the fracture load for the same mesh. However, for the same fracture energy release rate it can be seen that different fracture loads are obtained depending on the mesh employed, i.e. the element size. The smallest fracture load is in general obtained for mesh D (the finest mesh), while the largest fracture load is obtained for mesh A (the coarsest mesh). In addition, it is observed that the fracture loads obtained using different element sizes correspond to a straight line. The slope of this straight line depends upon the fracture energy release rate. For a small fracture energy release rate, the slope of the line is considerable. However, for a fracture energy release rate of 30 N/m this straight line is almost horizontal. A horizontal straight line would indicate that the same fracture load has been obtained irrespective of element size – in other words, the solution has converged to the exact solution. The large slope of the straight line obtained for the small energy release rate of 3 N/m indicates that the solution is far from the exact solution.

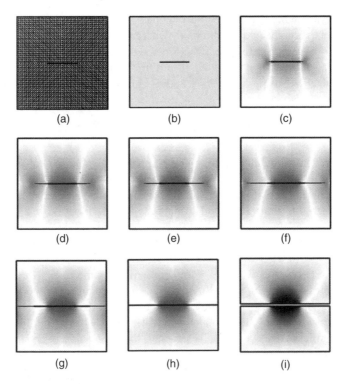

Figure 7.20 Finite element mesh (mesh D) employed and fracture sequence obtained for $2\gamma = 30\,\text{N/m}$. The frames shown correspond to (b) $t = 0\,\text{ms}$, (c) $t = 0.17\,\text{ms}$, (d) $t = 0.27\,\text{ms}$, (e) $t = 0.28\,\text{ms}$, (f) $t = 0.30\,\text{ms}$, (g) $t = 0.31\,\text{ms}$, (h) $t = 0.32\,\text{ms}$, (i) $t = 0.50\,\text{ms}$; i.e. transient loads (b) $\sigma = 0\,\text{MPa}$, (c) $\sigma = 3.4\,\text{MPa}$, (d) $\sigma = 5.4\,\text{MPa}$, (e) $\sigma = 5.6\,\text{MPa}$, (f) $\sigma = 6.0\,\text{MPa}$, (g) $\sigma = 6.2\,\text{MPa}$, (h) $\sigma = 6.4\,\text{MPa}$, (i) $\sigma = 10.0\,\text{MPa}$.

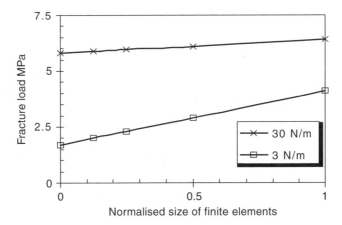

Figure 7.21 Fracture load as a function of the energy release rate and element size.

An exact solution would in theory be obtained by a finite element mesh comprising zero size elements, as indicated by the point where the straight line intersects the y-axis. The difference between the fracture loads obtained using meshes C and D is one half of that from using meshes B and C, which in turn is one half of the difference between fracture loads obtained using meshes A and B. In other words, if the element size is halved, the error is also halved and the order of error as a function of element size h is thus given by

$$\varepsilon = O(h) \tag{7.32}$$

That the combined smeared and discrete crack approach is able to deal with complicated fracture patterns is demonstrated by a combined finite-discrete element simulation of explosive induced fragmentation of a square concrete block, shown in Figure 7.22. The block is 2 m in length, while the depth (thickness) of the block is 0.1 m. A continuous cut from the centre of the block to the top edge is assumed (thick vertical line). Thus, the pressure of detonation gases is assumed to act on the walls of the cut. The ignition of explosive charge (0.8 kg/1 m *depth* = 0.08 kg) takes place at the bottom of the cut, and spreads with detonation velocity of 7580 m/s towards the top of the cut. Due to the detonation gas pressure on cut walls, a stress wave is produced in the rock. Initially, it has an elongated shape due to the ignition taking place at the bottom of the cut.

As the stress wave travels toward the boundaries of the block, radial cracks appear behind the wave. The wave eventually reaches the free surface and reflects from it, causing cracks on the boundary. The final fracture pattern is a result of inner cracks propagating outwards and outer cracks propagating inwards, together with a system of secondary cracks appearing.

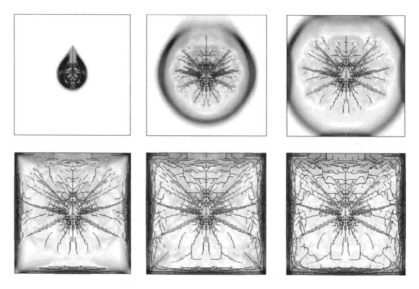

Figure 7.22   Fracture sequence for 2 m square block at 0.08 m, 0.24 ms 0.32 ms, 0.40 ms, 0.48 ms and 0.72 ms after initiation of explosive charge (finer mesh).

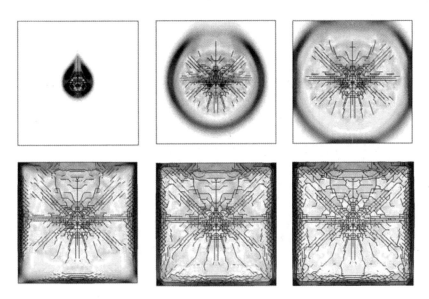

Figure 7.23 Initial stress wave: 0.08 ms, 0.24 ms, 0.3 ms2, 0.40 ms, 0.48 ms, 0.72 ms after ignition) after ignition (coarser mesh).

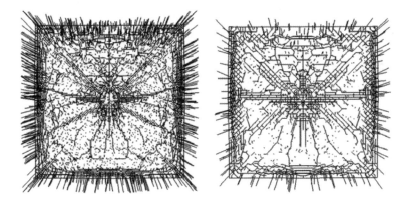

Figure 7.24 Final velocity fields at 0.72 ms after ignition: finer mesh (left) and coarser mesh (right).

The same sequence of events for exactly the same problem modelled with a three times coarser mesh (three times smaller number of finite elements) is shown in Figure 7.23.

The final velocity fields for both meshes are shown in Figure 7.24, which indicates convergence of the solution in terms of mesh size both in terms of the fracture pattern and velocity fields obtained.

The convergence in terms of kinetic energy for different meshes and amounts of explosive is shown in Figure 7.25.

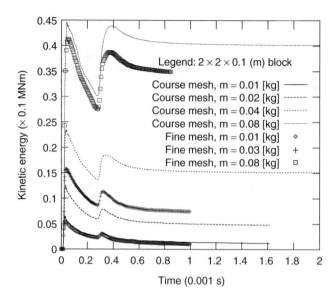

Figure 7.25   Kinetic energy as a function of time for different meshes and amounts of explosive.

7.4   A NEED FOR MORE ROBUST FRACTURE SOLUTIONS

The fracture and fragmentation algorithms proposed in the context of the combined finite-discrete element method are in general sensitive to both element size and element orientation. This applies to both smeared strain softening localisation based fracture algorithms (proposed in the early days of the combined finite-discrete element development), and single crack based models proposed in recent years, including the most recent combined single and smeared fracture algorithm.

Only for extremely fine meshes can one expect accurate representation of the stress and strain fields close to the crack tip. In such cases, these fields are not a function of either relative size of individual elements or relative orientation of individual elements. The undulations and errors in the local crack wall geometry are a function of the size of the finite elements employed, and diminish with decreasing size of finite elements, while the length of plastic zone stays constant. Thus the influence of the undulations in the crack walls on the stress field within the plastic zone also diminishes with the decreasing size of finite elements employed. This leads to the logical conclusion that with very fine meshes both critical load and fracture pattern are not sensitive to either mesh size or mesh orientation.

However, extremely fine meshes for complex fracture patterns are difficult to realise due to extensive CPU requirements. Thus a problem of this type is and remains a so-called 'grand challenge problem,' that is likely to be addressed by hardware of the future. Alternatively, an algorithmic break-through may occur.

8

Fluid Coupling in the Combined Finite-Discrete Element Method

8.1 INTRODUCTION

Many engineering problems of discontinuous media involving fracture and fragmentation also involve interaction between solid discrete elements and fluid. There are two types of such problems:

- The Computational Fluid Dynamics (CFD) problem with combined finite-discrete element coupling. In this type of problem, the discrete elements are submerged inside a fluid.

- The combined finite-discrete element problem with Computational Fluid Dynamics coupling. In this type of problem, most of the volume is filled by solid discrete elements, while the fluid fills the gaps and voids between discrete elements.

8.1.1 CFD with solid coupling

A typical problem of the first type is, for instance, deposition of solid blocks into water in coastal defence construction operations. In this problems, interaction with fluid influences the way the solid particle move through fluid, and therefore the transient behaviour of the solid system and the final state of rest is greatly influenced by interaction with water. Problems of this type are best considered as problems of Computational Fluid Dynamics with deformability, kinematics and dynamics of individual particles being resolved using the combined finite-discrete element method. Standard CFD models for both compressible and incompressible flow, together with various turbulence models, are readily available in the form of either finite difference based techniques or finite volume based techniques. Both finite volume and finite difference techniques use discretisation of the spatial domain to approximately satisfy the governing equations in integral form. Finite difference and finite volume based CFD techniques are based on Eulerian formulation. The essence of this formulation is that the grid is fixed in space and does not move with particles of fluid.

The Combined Finite-Discrete Element Method   A. Munjiza
© 2004 John Wiley & Sons, Ltd   ISBN: 0-470-84199-0

The fluid moves over the grid. Inertia terms are taken into account through convective terms of the type

$$\frac{dv_x}{dt} = \frac{\partial v_x}{\partial t} + \frac{\partial v_x}{\partial x}\frac{\partial x}{\partial t} + \frac{\partial v_x}{\partial y}\frac{\partial y}{\partial t} + \frac{\partial v_x}{\partial z}\frac{\partial z}{\partial t} \tag{8.1}$$

$$= \frac{\partial v_x}{\partial t} + \frac{\partial v_x}{\partial x}v_x + \frac{\partial v_x}{\partial y}v_y + \frac{\partial v_x}{\partial z}v_z$$

where v_x, v_y and v_z are velocity components in the x, y and z directions of the inertial reference frame fixed to the Eulerian grid. These velocity components do not correspond to fluid particles, but to points fixed in space. The acceleration field

$$a_x(x, y, z, t) = \frac{dv_x}{dt} \tag{8.2}$$

is an instantaneous acceleration field, which for any spatial point (x, y, z) represents acceleration of the fluid particle that at a given time instance, occupies a spatial position defined by specific spatial coordinates x, y and z. At every time instance, t it is a different particle. This is achieved by defining velocity at grid points which are not fixed to fluid particles and do not move in space, i.e. the coordinates of these grid points do not change in time. If the grid is regular, there is even no need to remember the coordinates of the grid points. However, the formulation is nonlinear in terms of velocity.

The combined finite-discrete element method is based on Lagrangian grid, where nodes of the finite element are fixed to the moving solid particles and move together with solid particles in space. Thus, inertia terms are simply expressed as

$$\frac{dv_x}{dt} = \frac{\partial v_x}{\partial t} \tag{8.3}$$

where, for a specific solid particle defined by its fixed initial coordinates at time $t = 0$, i.e. coordinates x_i, y_i and z_i

$$v_x = v_x(x_i, y_i, z_i, t) \tag{8.4}$$

is the velocity component in the x-direction of the inertial reference frame. This velocity component represents the velocity of the same specific solid particle at all time instances. This solid particle can be recognised by the fact that at time $t = 0$ (initial configuration) it occupied the spatial point

$$(x_i, y_i, z_i) \tag{8.5}$$

where x_i, y_i and z_i are clearly not functions of time. In practical terms, this is achieved through defining velocity at the nodes of the finite element mesh. These nodes are fixed to solid particles and move together with solid particles. Thus, the coordinates of the nodes change in time. The formulation is linear in terms of velocity. However, the coordinates of the nodal points have to be updated at every time instance considered (usually every time step).

In short, the CFD grid comprises grid points fixed in space and fluid particles moving relative to these grid points, while the combined finite-discrete element method comprises grid points moving together with solid particles.

The coupling of the two requires a Lagrangian grid to be superimposed over the Eulerian grid. As the solid particles are by definition of the problem loosely packed, and most of the domain is filled by the fluid, the primary grid is the Eulerian CFD grid. As a consequence, the primary solver is the Eulerian CFD solver. Thus, these are in essence CFD problems. The influence of fluid on the motion of solid particles is obtained through transferring the fluid pressure and drag forces onto the discrete elements.

There are two approaches available to take into account the influence of solid particles on the CFD model:

- Resolving the interaction of each individual particle with the fluid.

- Averaging the interaction between solid particles and the fluid through introduction of solid 'density'. Introducing solid density goes against the spirit of the combined finite-discrete element method, because it is in essence a continuum formulation. For it to be valid, the continuum assumption must be valid, in which case the combined finite-discrete element method is not necessary. If the continuum assumption is not valid, by introducing a continuum based formulation for what is a discontinuum problem, all discontinua-based phenomena may be automatically neglected, and the main purpose of employing the combined finite discrete element method defeated.

8.1.2 Combined finite-discrete element method with CFD coupling

Problems of the second type are more common. A typical example of this type of problem is fracture and fragmentation of solid using explosives. Fracture and fragmentation is the result of the interaction of a detonation gas at high pressure with a fracturing solid. After the initiation of the explosive charge, the detonation propagates through the charge at the velocity of detonation. The detonation process results in a phase change of the explosive material so that, as a result, hot detonation gas at high pressure is produced. The energy released in the detonation process depends upon the type of explosive used, and is usually expressed as specific explosive energy (i.e. energy per unit mass of explosive charge). Through expansion of the detonation gas, part of this energy is transferred onto the solid, causing it to break, fracture and/or fragment, depending on the amount of explosive employed.

Detonation gas induced fracture and fragmentation involves interaction between the detonation gas and fracturing solid. The gas exerts pressure onto the free surface of the solid, causing the solid to accelerate, deform, fracture and displace at a velocity field that changes with time. This in turn results in the expansion of the detonation gas, its partial penetration into the cracks and voids created and its flow through the cracks and voids and between fracturing solid blocks. The expansion of the detonation gas, flow of the detonation gas and mechanical work done by the gas results in a decrease in the pressure of the gas.

For this reason, the problem of detonation gas induced fracture and fragmentation is employed in this chapter as a typical example of the combined finite-discrete element method with fluid coupling. There are two aspects to this coupled problem:

- The evaluation of gas pressure as the surface load for a solid.
- The deformability, fracture and fragmentation of a solid under gas pressure.

The models for gas pressure evaluation vary from a fixed, user-supplied pressure-time diagram for the pressure of detonation gas, which is then applied to the user-defined surfaces of the solid (for instance, the borehole walls) to the full scale gas flow models through the fracturing solid. It is noted that these models are not concerned with the detonation process as such. They concentrate on the expansion of the detonation gas and its penetration into cracks.

The problem with user predefined gas pressure is that the resulting fracture and fragmentation process (although modelled by a computer using complex algorithms and large CPU times) is in essence supplied by the user, in a sense that the extent of fracture and fragmentation depends upon the magnitude and duration of the gas pressure applied to the borehole walls.

Full scale gas flow models can be classified as follows:

- Models based on tracing gas flow through individual cracks. This is the most realistic approach. However, it requires robust crack detection algorithms, which also incorporate procedures for the detection of connectivity between individual cracks and voids. These procedures have proven much more difficult than the detection of contacts between solid particles. This is because the geometry of solid particles is readily available, while the geometry of voids is only implicitly defined and has to be deduced from the geometry of solid particles.

- Models based on porous media-based idealisation of fractured solid. The fracturing solid is approximated by a porous medium requiring estimation of the porosity. The errors arising from this process and the extent to which the final results are influenced by transient geometry of the fracturing solid are not easy to estimate. Full scale gas flow models are also coupled with considerable algorithmic complexities and extensive CPU requirements.

In real problems, the detonation gas pressure drops rapidly due to a relatively small change in volume, and most of the work done by the gas is done while the pressure is high. Only a small portion of energy remains at lower pressures of detonation gas. Thus, although it is an experimental fact that the detonation gas penetrates cracks and voids, the extent of gas flow through the larger section of the solid domain remains an open question. For instance, in some blasting operations, small explosive charges in closely spaced boreholes are used to induce only partial breaking of the rock without significant fracture and fragmentation. In these, the gas flow effects are clearly small. On the other hand, in operations of explosive induced loosening of soil, a significant gas flow may be present.

In short, successful modelling of detonation gas induced fracture and fragmentation may not always require a complex, full scale gas flow model. However, proper evaluation of gas pressure with gas expansion should include the preservation of energy balance. This is recognised in a set of models that take into account the equation of state of the detonation gas and spatial distribution of gas pressure due to gas flow without considering a full scale model of gas flow. In these models, detonation gas penetration into the cracks and the spatial gradient of detonation gas pressure are taken into account through a set of problem parameters such as gas penetration depth. In this chapter, such a detonation gas expansion model, together with a model for spatial pressure distribution in the context of the combined finite-discrete element modelling of detonation gas induced fragmentation, is described in detail.

8.2 EXPANSION OF THE DETONATION GAS

Once the explosive charge is initiated through the detonation process, hot detonation gas at high pressure is created. The pressure of the detonation gas can exceed 10 GPa. The detonation process is characterised by the Velocity Of Detonation (VOD). The velocity of detonation of modern explosives can exceed 7000 m/s. This means that the detonation process is in general very fast in comparison to the fracture and fragmentation processes. An acceptable approximation in such cases would be that the detonation process is completed instantaneously. With this assumption, the combined finite-discrete element simulation starts with a borehole filled with detonation gas at high pressure. An alternative approach is to consider the detonation process while at the same time considering coupled combined finite-discrete element simulation. In such a case, the mass of detonation gas and the spatial distribution of the gas within the borehole changes as the detonation progresses.

In both cases, the coupled combined finite-discrete element simulation of interaction of the detonation gas with a fracturing and fragmenting solid must consider:

- Expansion of the detonation gas, coupled with cooling of the detonation gas and rapid pressure drop. The gas pushes against the free surfaces of the solid, thus transferring part of its energy onto the solid in the form of mechanical work.

- Penetration of detonation gas into cracks and voids coupled with escape of the gas through these cracks and voids. This is in essence a fluid flow problem of the second type, as defined above.

8.2.1 Equation of state

The gas side of the coupled combined finite-discrete element simulation is subject to the same CPU and RAM requirements to which the rest of the combined finite-discrete element procedures are subject. From this point of view, it is ideal if expansion of the detonation gas is described in analytical form. To facilitate this, the following equation of state for the detonation gas is assumed:

$$p = \left[\frac{1}{v} + a \left(\frac{1}{v} \right)^b \right] RT \tag{8.6}$$

where p is the pressure of the detonation gas, v is the specific volume of the detonation gas, T is the temperature of the detonation gas, while a, b and R are constants.

The same equation can be written in terms of the density of detonation gas $\rho = 1/v$:

$$p = \left[\rho + a\rho^b \right] RT \tag{8.7}$$

8.2.2 Rigid chamber

The simplest case of detonation gas expansion is the hypothetical case of an explosive charge being placed in a rigid thermally insulated chamber, and filling only a part of the chamber; the detonation gas expands until the rigid chamber is completely filled with the detonation gas. Because the chamber is rigid, no mechanical work is done by the expanding gas. The chamber walls being impermeable to the flow of heat (adiabatic walls) means that no heat transfer between the chamber walls and gas takes place. The detonation gas expands into the chamber, and when all the transient motion has ceased the gas reaches equilibrium. In this process the gas has done no mechanical work, and no energy in the form of heat has been exchanged. As a result, no change in the internal energy of detonation gas occurs.

In a state of thermodynamic equilibrium, the internal energy of the unit mass of detonation gas, u, depends only upon the thermodynamic state of the gas, which in this case is completely described by the specific volume v and temperature T. Consequently, any change in u is given by

$$du = \left(\frac{\partial u}{\partial T}\right)_v dT + \left(\frac{\partial u}{\partial v}\right)_T dv \tag{8.8}$$

where any changes in u depend only upon the initial and final states, and not on the path (for instance, changing v at constant T and then changing T at constant v yields the same u as changing T at constant v and then changing v at constant T).

With an additional assumption that the internal energy per unit mass of the detonation gas at a given temperature does not depend upon the specific volume, any change in internal energy can be given in terms of the temperature only:

$$du = \left(\frac{\partial u}{\partial T}\right)_v dT = c_v\, dT \tag{8.9}$$

where c_v is the heat capacity per unit mass of the detonation gas at constant volume.

The heat capacity per unit mass of the detonation gas is in general a function of the temperature of the detonation gas. However, for the sake of simplicity it is assumed that c_v is constant and calculated in such a way that the internal energy of the detonation gas at maximum temperature is accurately represented. In other words, c_v is such that the specific explosive energy of the explosive is accurately represented.

No change in internal energy implies no change in temperature, and consequently the temperature of the detonation gas after the expansion is completed and equilibrium state is reached is equal to the temperature of the detonation gas before the expansion, i.e. expansion is also isothermal and

$$u_c = u_0; \quad T_c = T_0 \tag{8.10}$$

where subscripts $_o$ and $_c$ indicate equilibrium states before and after the expansion.

From the equation of state, it follows that

$$T_0 = \frac{p_o}{R(\rho_o + a\rho_o{}^b)} = T_c = \frac{p_c}{R(\rho_c + a\rho_c{}^b)} \tag{8.11}$$

which yields the expression for the pressure of the detonation gas after expansion, p_c, as function of the density of detonation gas after expansion:

$$\frac{p_c}{p_o} = \frac{\rho_c + a\rho_c{}^b}{\rho_o + a\rho_o{}^b} = \frac{\rho_c}{\rho_o}\frac{1 + a\rho_c{}^{b-1}}{1 + a\rho_o{}^{b-1}} = \frac{\rho_c}{\rho_o}\frac{1 + a\rho_o{}^{b-1}(\rho_c/\rho_o)^{b-1}}{1 + a\rho_o{}^{b-1}} \tag{8.12}$$

By substituting the expression for density ρ in terms of specific volume v

$$\rho = \frac{1}{v} \qquad (8.13)$$

into equation (8.12), the gas pressure in terms of specific volume is obtained:

$$\frac{p_c}{p_o} = \frac{v_o}{v_c} \frac{1 + (a/v_o^{\,b-1})(v_o/v_c)^{b-1}}{1 + a/v_o^{\,b-1}} \qquad (8.14)$$

As mentioned before, the solution to the expansion of the detonation gas has been obtained in a closed analytical form.

For a particular explosive, parameters a and b are obtained by an inverse method. For instance, in Figure 8.1, the irreversible adiabatic expansion of nitroglycerine detonation gas is shown. By curve fitting, parameters a and b are estimated $a = 10.99e{-}10\,\mathrm{m^9 kg^3}$ and $b = 4.0$.

The results for pressure as a function of specific volume obtained using equation (8.14) compare well with the results given by Johansson. As the specific volume of the gas increases, the slope of the *log p log v* curve decreases. This slope asymptotically approaches -1 as the specific volume becomes large. In other words, at very low densities the detonation gas approaches the behaviour of an ideal gas. At high density the slope of the curve is given by

$$\ln\left(\frac{p_c}{p_o}\right) = \ln\left(\frac{v_o}{v_c} \frac{1 + (a/v_o^{\,b-1})e^{(b-1)\ln(v_o/v_c)}}{1 + a/v_o^{\,b-1}}\right) \qquad (8.15)$$

$$\ln(p_c) = -\ln(v_c) + \ln\left(1 + \frac{a}{v_o^{\,b-1}}e^{(b-1)\ln(v_o/v_c)}\right)$$

$$+ \ln(p_o) + \ln(v_o) - \ln(1 + a/v_o^{\,b-1}) \qquad (8.16)$$

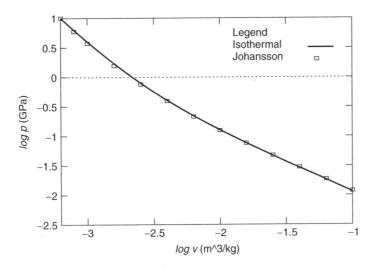

Figure 8.1 Irreversible adiabatic expansion of nitroglycerine detonation gas.

By substituting temporarily $\ln(p_c) = y$ and $\ln(v_c) = x$, equation (8.16) becomes

$$y = -x + \ln\left(1 + \frac{a}{v_o^{b-1}}e^{(b-1)(\ln(v_o)-x)}\right) + \ln(p_o) + \ln(v_o) - \ln(1 + a/v_o^{b-1}) \quad (8.17)$$

and by differentiation, the slope of the $\ln p - \ln v$ curve is obtained:

$$\frac{d(\ln p)}{d(\ln v)} = \frac{dy}{dx} = -1 - \frac{\dfrac{a(b-1)}{v_o^{b-1}}e^{(b-1)(\ln(v_o)-x)}}{1 + \dfrac{a}{v_o^{b-1}}e^{(b-1)(\ln(v_o)-x)}} \quad (8.18)$$

In the limit when the specific volume becomes large

$$\lim_{x \to \infty} \frac{dy}{dx} = -1 \quad (8.19)$$

and when the density becomes large

$$\lim_{x \to -\ln v_o} \frac{dy}{dx} = -1 - \frac{\dfrac{a(b-1)}{v_o^{b-1}}}{1 + \dfrac{a}{v_o^{b-1}}} \quad (8.20)$$

For nitroglycerine expansion given in Figure 8.1, the initial density was $1600\,\text{kg/m}^3$, thus initial specific volume is $v = 1/1600\,\text{m}^3/\text{kg}$, which yields the initial slope of the *log p* versus *log v* expansion curve:

$$\lim_{x \to \ln v_o} \frac{dy}{dx} = -1 - \frac{\dfrac{10.99e - 10(4-1)}{(1/1600)^{4-1}}}{1 + \dfrac{10.99e - 10}{(1/1600)^{4-1}}} = -3.45 \quad (8.21)$$

Thus, the slope of the curve changes from -1 at zero density of the detonation gas to -3.45 at large densities of the detonation gas.

8.2.3 Isentropic adiabatic expansion of detonation gas

If the detonation gas is placed in a thermally insulating chamber, the walls of which are able to move, and initially the chamber is completely filled with detonation gas, (i.e. at initial density the volume of the gas equals the volume of the chamber), adiabatic reversible expansion of the detonation gas takes place.

Because no heat exchange between the chamber walls and gas takes place, the change in internal energy of the gas is due to the mechanical work only:

$$du = -p\,dv \quad (8.22)$$

From the assumption that at constant temperature T the internal energy u does not depend upon the specific volume v, it follows that any changes in internal energy are due to the temperature change:

$$du = c_v dT = -p dv \qquad (8.23)$$

From equation of state, it follows that the pressure of detonation gas is given by

$$p = (\rho + a\rho^b)RT \qquad (8.24)$$

which, when substituted into (8.23), yields

$$c_v dT = -(\rho + a\rho^b)RT dv \qquad (8.25)$$

Substituting

$$v = \frac{1}{\rho} \text{ and thus } dv = -\frac{1}{\rho^2}d\rho \qquad (8.26)$$

into (8.25) gives

$$\frac{c_v}{R}\frac{dT}{T} = \left(\frac{1}{\rho} + a\rho^{b-2}\right)d\rho \qquad (8.27)$$

and after integration

$$\int_{T_c}^{T}\frac{1}{T}dT = \int_{v_c}^{v_v}\frac{R}{c_v}\left(\frac{1}{\rho} + a\rho^{b-2}\right)d\rho \qquad (8.28)$$

with assumption that R and c_v are constants and that $R/c_v = k - 1$, one obtains

$$\ln(T/T_c) = \ln(\rho/\rho_c)^{k-1} + (k-1)\frac{a}{b-1}(\rho^{b-1} - \rho_c^{b-1}) \qquad (8.29)$$

$$= \ln(\rho/\rho_c)^{k-1} - (k-1)\frac{a\rho_c^{b-1}}{b-1}[1 - (\rho/\rho_c)^{b-1}]$$

$$= \ln\left[(\rho/\rho_c)^{k-1}e^{-(k-1)\frac{a\rho_c^{b-1}}{b-1}[1-(\rho/\rho_c)^{b-1}]}\right]$$

or

$$T/T_c = (\rho/\rho_c)^{k-1}e^{-(k-1)\frac{a\rho_c^{b-1}}{b-1}[1-(\rho/\rho_c)^{b-1}]} \qquad (8.30)$$

From equation of state, it follows that

$$p/p_c = \frac{\rho}{\rho_c}\frac{1 + a\rho_c^{b-1}(\rho/\rho_c)^{b-1}}{1 + a\rho_c^{b-1}}\frac{T}{T_c} \qquad (8.31)$$

and after substitution into (8.30)

$$p/p_c = (\rho/\rho_c)^k \frac{1 + a\rho_c^{b-1}(\rho/\rho_c)^{b-1}}{1 + a\rho_c^{b-1}} e^{-(k-1)\frac{a\rho_c^{b-1}}{b-1}[1-(\rho/\rho_c)^{b-1}]} \tag{8.32}$$

By substituting $\rho = 1/v$

$$p/p_c = (v_c/v)^k \frac{1 + (a/v_c^{b-1})(v_c/v)^{b-1}}{1 + a/v_c^{b-1}} e^{-(k-1)\frac{a/v_c^{b-1}}{b-1}[1-(v_c/v)^{b-1}]} \tag{8.33}$$

It also follows that

$$\ln\frac{p}{p_c} = k\ln\frac{v_c}{v} - (k-1)\frac{a/v_c^{b-1}}{b-1}[1 - e^{(b-1)\ln(v_c/v)}] + \ln\frac{1 + a/v_c^{b-1}e^{(b-1)\ln(v_c/v)}}{1 + a/v_c^{b-1}} \tag{8.34}$$

Again, as a result of a conveniently assumed equation of state, an analytical solution to the gas expansion has been obtained.

8.2.4 Detonation gas expansion in a partially filled non-rigid chamber

If the detonation gas is placed in a thermally insulating chamber, the walls of which are able to move, and initially the chamber is only partially filled with detonation gas (i.e. at initial density the volume of the gas is smaller than the volume of the chamber), the expansion of the gas is coupled with the motion of the chamber walls. The initial expansion of the detonation gas until the chamber is filled also involves pressure being exerted onto the chamber walls and mechanical work being done by the expanding gas. Thus, in contrast to the rigid chamber, the process would not be irreversible adiabatic.

This situation occurs in rock blasting when the borehole is only partially filled with explosive. After initiation of the explosive charge, the detonation gas first fills the borehole, exerting the pressure onto the borehole walls and gradually transferring energy onto the rock in the form of mechanical work.

In the initial stages, even if the chamber walls are free to move, their motion is much restricted by inertia. In real applications the mass of the fracturing solid is much larger than the mass of explosive charge used to break the solid. Thus, the acceleration and motion of the solid is relatively slow in comparison to the rate at which any transient motion of the detonation gas yields thermodynamic equilibrium.

Consequently, at every stage of the motion of the chamber walls, it is assumed that the detonation gas is in a state of thermodynamic equilibrium, and that its expansion in a non-rigid chamber takes place in two stages:

- The first stage involves no mechanical work, and is completed when the chamber is completely filled with detonation gas.

- The second stage involves expansion of the detonation gas in which the chamber walls move together with the detonation gas and the internal energy of the detonation gas is transferred into mechanical energy of the chamber walls.

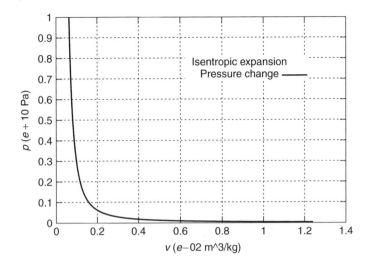

Figure 8.2 Pressure versus specific volume.

This means that for the first stage the gas pressure is given by (8.14), while the gas pressure for the second stage of expansion is calculated using equation (8.33). Because no heat exchange between the chamber walls and gas takes place, both stages of expansion involve adiabatic expansion, although pressure decrease in the first stage is slower, while the pressure decrease in the second stage is more rapid due to the gas expansion being coupled with mechanical work.

The assumption that no heat transfer taking place between detonation gas and chamber walls is due to the very short time in which expansion takes place. Given the law of thermal conductivity of rock, no significant change in the internal energy of the detonation gas due to heat transfer takes place.

In Figure 8.2 pressure versus specific volume is shown. The figure shows a rapid reduction in pressure due to the increase of the specific volume. The pressure drops from 10 GPa to almost atmospheric pressure with proportionally much smaller change in the specific volume. Different pressure-specific volume curves applied to the same explosive induced fracture and fragmentation process in general produce different transient stress and strain fields, and consequently different fracture patterns. Pressure as a function of time can be expressed as the summation of harmonics of different frequencies. The natural frequency of a dynamic load is one of the most important parameters in numerical simulation of dynamic solid systems. In this light, it is evident that user-supplied pressure-time curves lead to non-objective results. However, it is paradoxical that even very complicated models taking into account gas expansion can yield non-objective pressure-time curves.

In Figure 8.3 the internal energy change versus pressure of the detonation gas is shown. Again, the reduction in internal energy with the reduction in pressure is very rapid. This means that most of the energy of the detonation gas is released at relatively high pressure. Most of the mechanical work is done by the expanding gas at relatively high pressure. The portion of energy remaining after the initial pressure drop is relatively small. For instance, in rock blasting operations, the specific energy of the explosive is usually shown as a sum of kinetic energy, strain energy, fracture energy, etc. Results shown in Figure 8.3 indicate

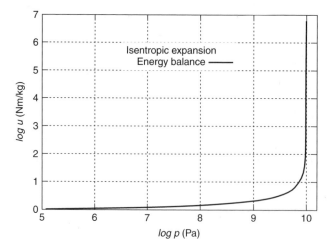

Figure 8.3 Internal energy versus pressure.

that none of these energy components changes considerably after the significant pressure drop. Different curves are likely to result in different energy partition, different stress and strain fields and different fracture and fragmentation patterns, and finally, different sized distribution curves characterising the pile of blasted rock.

In the context of the implementation of gas expansion curves into the combined finite-discrete element simulation of explosive induced fracture and fragmentation, these results imply that for the early stages of expansion, a very accurate integration of governing equation in the temporal domain must be performed. Thus, very small time steps must be used for the explicit integration scheme employed, at least before a significant reduction in pressure takes place.

8.3 GAS FLOW THROUGH FRACTURING SOLID

Expansion of the detonation gas and interaction with the fracturing solid also involves penetration of gas into the cracks and gas flow through cracks and voids. The initial gas flow is linked to existing cracks and voids. After this initial stage, the secondary flow roots form due to the cracking process, fracturing and eventual fragmentation of the solid. Thus, the gas exerts pressure on the crack walls and, at the same time, it expands into the cracks formed by the fracture and fragmentation process.

In this section, a relatively simple model for gas flow through the voids created by the fracturing and fragmenting solid is described. The basic assumptions for the model are as follows:

- The gas is present only within a predefined zone (in further text referred to as the 'gas zone') around the explosive charge, say the borehole. This zone is supplied to the model as input.

- Gas flow through the cracks and voids can be approximated by the steady state flow of ideal gas through ducts of a constant cross-section area. It is also assumed that all the ducts are of the same cross-section area and length. The length is characterised by the size of gas zone, while the area of individual ducts and the number of ducts is derived from the current size and distribution of individual solid fragments within the gas zone.

- The spatial distribution of pressure within the gas zone is described by the pressure drop within individual ducts.

- Gas flow at the entry to the duct is described by the steady state flow of ideal gas in a converging no-friction duct.

8.3.1   Constant area duct

To obtain an analytical expression for gas flow through cracks, a constant area duct compressible flow of ideal gas is employed. Standard formulae describing such flow are employed (Figure 8.4).

The detonation gas flow close to the entrance of the duct is described by the converging-diverging duct flow, and is assumed to be isentropic. The algorithm employed can be summarised as follows:

Step 1: Assume the Mach number for the flow at the exit of the duct, Ma.
Step 2: Using equation (8.35), calculate the Mach number at the entrance of the duct, Ma_1:

$$\frac{1}{2k}\left(\frac{1}{Ma_2^2} - \frac{1}{Ma_1^2}\right) \tag{8.35}$$

$$+ \frac{k+1}{4k}\ln\left\{\left(\frac{Ma_2}{Ma_1}\right)^2 \frac{2+(k-1)Ma_1^2}{2+(k-1)Ma_2^2}\right\} + \frac{fL}{8A/P} = 0$$

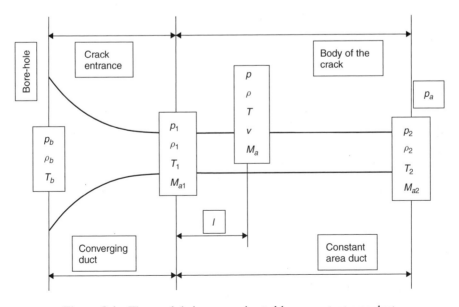

Figure 8.4   The crack being approximated by a constant area duct.

where $k = c_p/c_v$ is the specific heat ratio, f is an averaged friction factor along the length of the duct, A is the area of the duct and P is the perimeter of the duct.

Step 3: Assuming isentropic flow close to the entrance of the constant area duct, calculate:

(a) Pressure at the entrance of the duct, p_1:

$$p_1 = p_b \left\{ \frac{1}{1 + [(k-1)/2]Ma_1^2} \right\}^{k/(k-1)} \tag{8.36}$$

where p_b is the pressure of the detonation gas inside the borehole. This formula comes from the assumption that the gas flow at the entrance to the crack can be described by the flow of ideal gas through a converging no-friction duct.

(b) Temperature at the entrance of the duct, T_1:

$$T_1 = T_b \frac{1}{1 + [(k-1)/2]Ma_1^2} \tag{8.37}$$

·where T_b is the temperature of the detonation gas inside the borehole. This formula also comes from the assumption that the gas flow at the entrance to the crack can be described by the flow of ideal gas through a converging no-friction duct.

(c) Density at the entrance of the duct, ρ_1 is obtained

$$\rho_1 = \rho_b \left\{ \frac{1}{1 + [(k-1)/2]Ma_1^2} \right\}^{1/(k-1)} \tag{8.38}$$

where ρ_b is the density of the detonation gas inside the borehole. This formula also comes from the assumption that the gas flow at the entrance to the crack can be described by the flow of ideal gas through a converging no-friction duct.

Step 4: Assuming nonisentropic flow with friction for the rest of the constant area duct, calculate pressure at the exit of the duct, p_2:

$$p_2 = p_1 \frac{Ma_1}{Ma_2} \left\{ \frac{1 + [(k-1)/2]Ma_1^2}{1 + [(k-1)/2]Ma_2^2} \right\}^{1/2} \tag{8.39}$$

and if the pressure at the exit of the duct is different from atmospheric pressure p_a, and the assumed Mach number at the exit of the duct is smaller than 1, go to the step 1.

Step 5: At the end of iteration, calculate the velocity at the entrance of the constant area duct:

$$v_1 = Ma_1 \sqrt{kRT_1} \tag{8.40}$$

and gas flow rate

$$q = \frac{dm}{dt} = v_1 A \rho_1 \tag{8.41}$$

where m is the total remaining mass of the detonation gas.

For a cross-section at distance l from the entrance to the constant area duct, the corresponding Mach number Ma is calculated using the formula

$$\frac{1}{2k}\left(\frac{1}{Ma^2}-\frac{1}{Ma_1^2}\right) \tag{8.42}$$

$$+\frac{k+1}{4k}\ln\left\{\left(\frac{Ma}{Ma_1}\right)^2\frac{2+(k-1)Ma_1^2}{2+(k-1)Ma^2}\right\}+\frac{fl}{8A/P}=0$$

while the pressure is evaluated from the Mach number as follows:

$$p=p_1\frac{Ma_1}{Ma}\left\{\frac{1+[(k-1)/2]Ma_1^2}{1+[(k-1)/2]Ma^2}\right\}^{1/2} \tag{8.43}$$

The above procedure yields both spatial pressure distribution within the gas zone and the rate of gas loss due to gas flow through the cracks and voids, while all calculations involve only inexpensive (in terms of CPU time) 1D flow simulation. The density of the remaining detonation gas decreases due to both gas expansion (i.e. change of the volume of the gas $w=dV/dt$, where V is the total volume) and due to the flow of gas, i.e.

$$\rho=\frac{m}{V};\quad \frac{d\rho}{dt}=\frac{1}{V}\frac{dm}{dt}-\frac{m}{V^2}\frac{dV}{dt} \tag{8.44}$$

The specific volume is given as $v=1/\rho$, thus the change in specific volume is given by

$$\frac{dv}{dt}=-\frac{1}{\rho^2}\frac{d\rho}{dt}=-\frac{V^2}{m^2}\left(\frac{1}{V}\frac{dm}{dt}-\frac{m}{V^2}\frac{dV}{dt}\right)=-\frac{V}{m^2}\frac{dm}{dt}+\frac{1}{m}\frac{dV}{dt} \tag{8.45}$$

$$=-\frac{1}{m}v\frac{dm}{dt}+\frac{1}{m}\frac{dV}{dt}=-\frac{1}{m}vq+\frac{1}{m}w$$

where the change of volume of the gas zone $w=dV/dt$ is in actual implementation calculated from the rate of mechanical work done by the expanding gas. In temporal discretisation and actual implementation of the algorithm, at every time step, nodal forces due to the pressure of the detonation gas are calculated together with incremental displacements and nodal velocities. Thus, the work rate is merely calculated as a product of nodal forces due to gas pressure and nodal velocities.

The gas zone for each borehole is best defined as a polygon around the borehole, (Figure 8.5). The area of individual ducts A is defined as the average crack opening displacement within the gas zone:

$$A=\frac{\int_\Gamma \delta d\Gamma}{\int_\Gamma d\Gamma} \tag{8.46}$$

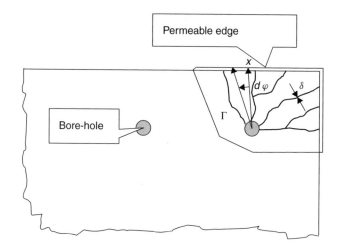

Figure 8.5 Gas zone with individual cracks of variable opening.

The length of individual ducts l is defined as the average length from the borehole to the permeable wall of the gas zone, i.e.

$$l = \frac{\int_S x \, d\varphi}{\int_S d\varphi} \tag{8.47}$$

where x is the distance from the borehole to the permeable edge and S is the permeable edge.

The total number of ducts is thus given as

$$A = \frac{\int_\Gamma d\Gamma}{l} \tag{8.48}$$

8.4 COUPLED COMBINED FINITE-DISCRETE ELEMENT SIMULATION OF EXPLOSIVE INDUCED FRACTURE AND FRAGMENTATION

The equation of state for the detonation gas together with the gas flow model complement other numerical algorithms comprising the combined finite-discrete element simulation. To demonstrate the combined finite-discrete element simulation coupled with gas flow through cracks, simulation of explosive induced fragmentation of a 2×2 m square block shown in Figure 8.6 is performed. The thickness (depth) of the block is 0.1 m. Initially, the block has a continuous cut from the centre to the top end. The ignition of the explosive charge takes place at the bottom of the cut, and propagates toward the top of the cut at a VOD of 7580 m/s.

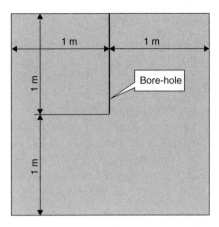

Figure 8.6   2 m square block of thickness 0.1 m with explosive charge placed in the borehole.

Figure 8.7   Fracture sequence for a 2 m square block at 0.08 ms, 0.24 ms, 0.32 ms, 0.40 ms, 0.48 ms and 0.72 ms after ignition of a 0.03 kg explosive charge.

In Figure 8.7 the stress wave propagation due to the ignition of an explosive charge consisting of 0.3 kg of explosive per 1 m depth of the block, i.e. 0.03 kg of explosive for a block of 0.1 m in thickness. The following model parameters are assumed:

$$a = 10.99e - 10 \text{ m}^9/\text{kg}^3$$
$$b = 4$$
$$k = 1.15$$
$$VOD = 7580 \text{ m/s}$$

(8.49)

initial density $\rho_o = 1600 \text{ kg/m}^3 = 1/v_o$

$p_0 = 1e + 10 \text{ Pa}$

$\dfrac{\text{initial volume of the cut}}{\text{initial volume of charge}}$, i.e. $\dfrac{V_c}{V_o} = \dfrac{v_c}{v_o} = 1.90$

Due to the ignition taking place at the bottom of the cut, the initial stress wave is elongated towards the top edge of the block. At a later stage, the front of the stress wave takes an almost circular form, with the top end of the wave front reaching the top edge of the block just before 0.24 ms after ignition. Behind the wave-front radial cracks appear, with the crack front moving much slower than the wave front. Thus, by the time the wave front has reached the free edges of the block inner cracks still propagate toward the edges of the block.

Further propagation of the wave front, leads to the reflection of the stress wave from the free surface of the block causing outer cracks on the boundary of the block. The final fracture pattern is a result of the inner cracks propagating outwards and outer cracks propagating inwards, together with the opening of secondary cracks.

It can be observed that 0.03 kg of explosive, although sufficient to break the block, is insufficient to cause a substantial fragmentation of the block. Thus, the same problem was solved with explosive charge being increased to 0.08 kg. Again, initial stress wave propagation is followed by propagation of inner cracks, and as the stress wave reflects from the boundary, outer cracks appear. The combined propagation of inner and outer cracks yields the final fracture pattern. The increased mass of the explosive charge has resulted in a considerable fragmentation of the block, (Figure 8.8).

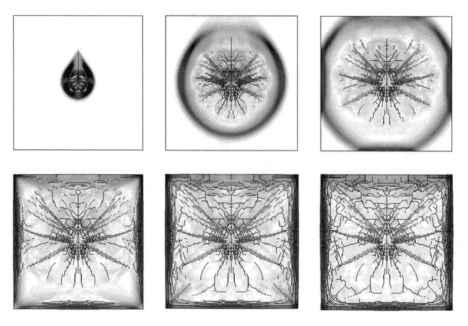

Figure 8.8 Stress wave and fracture sequence in a 2 m block at 0.08 ms, 0.24 ms, 0.32 ms, 0.40 ms, 0.48 ms and 0.72 ms after ignition of 0.08 kg explosive charge.

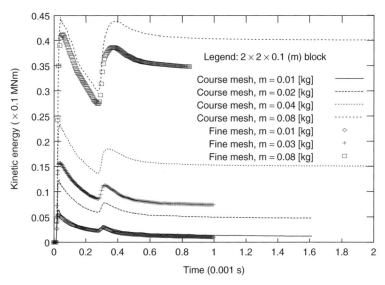

Figure 8.9 Total kinetic energy of all solid fragments (versus time after ignition) for different amounts of explosive.

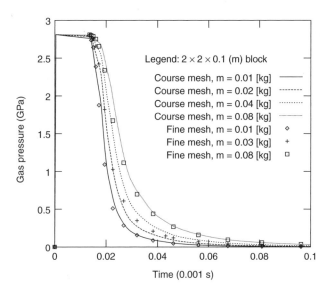

Figure 8.10 Pressure of detonation gas versus time after ignition for a 2 m square block.

The kinetic energy versus time diagrams for a series of explosive charges and different discretisation grids are shown in Figure 8.9. The kinetic energy increases with the increase in the amount of explosive used. In all the results shown, the specific explosive energy of 6280 kNm/kg is employed.

The coupled simulation of explosive induced fragmentation also yields pressure, temperature, density and the internal energy of the detonation gas. The detonation gas pressure versus time for different amounts of explosive is shown in Figure 8.10. The pressure is in general greater for a larger amount of explosive, and this is also true for the duration of the pressure. It can be observed that the pressure-time curve obtained shows at first instance a relatively slow decrease in pressure (up to 0.02 ms). This is due to the inertia of the rock. Once the solid domain of the rock acquires an initial velocity field, the pressure drop is rapid. This rapid pressure drop continues for approximately 0.02 ms. After that the pressure becomes so small that it can almost be neglected. However, it continues decreasing, although at much slower rate.

When this time-scale is compared to the time-scale of fracture and fragmentation processes, it is evident that a significant pressure drop occurs much before the completion of the fracture and fragmentation processes.

8.4.1 Scaling of coupled combined finite-discrete element problems

Linear problems such as linear elasticity problems are relatively easy to scale. This means that if a solution for, say, a 2 m block is available, a solution for a 0.2 m block can be obtained analytically using the scaling laws. Coupled combined finite-discrete element analysis is highly nonlinear, and simple scaling laws do not apply. This is demonstrated by the coupled problem shown in Figure 8.11.

The problem comprises a $0.2 \times 0.2 \times 0.01$ m square block. The amount of explosive supplied is 5.5 g per 0.1 m of thickness, i.e. $(5.5/0.1)0.01 = 0.55$ g. All the other parameters are the same as those given for the 2×2 m block. The stress wave propagation and transient fracture patterns are shown in Figure 8.12.

The kinetic energy of the rock fragments as a function of time is shown in Figure 8.13. The figure also shows the pressure of the detonation gas as calculated by the model. The initial pressure drop is relatively rapid. The kinetic energy peaks twice, and the second peak coincides with an intensive fragmentation taking place, together with a significant

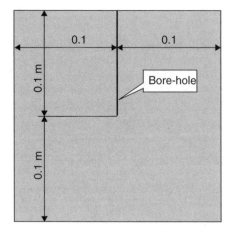

Figure 8.11 A 0.2 m square block of thickness 0.01 m and 0.55 g explosive charge: wave fronts at 0.018 ms and 0.032 ms.

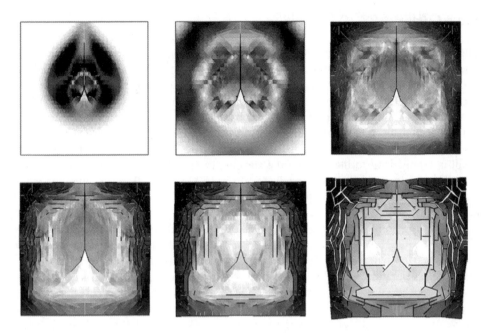

Figure 8.12   A 0.2 m square block and 0.55 g explosive charge: wave fronts and fracture patterns at: 0.018 ms, 0.032 ms, 0.048 ms, 0.064 ms, 0.080 ms and 0.24 ms.

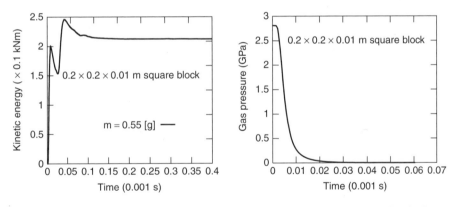

Figure 8.13   A 0.2 m square block with 0.55 g explosive charge: kinetic energy of solid fragments and pressure of detonation gas.

drop in the pressure of detonation gas. This again reinforces the importance of the initial stages of the gas expansion process.

This problem is in essence a 2 m block reduced in size by 10 times. However, the fracture patterns obtained and pressure duration do not reflect any of the scaling of the initial geometry.

8.5   OTHER APPLICATIONS

The same principles employed with coupled simulations involving explosive induced fracture and fragmentation apply to problems where most of the domain is filled by solid particles; the fluid fills most or at least some of the cracks and voids between the solid particles, and fluid flow occurs through the network of interconnected cracks and voids.

All of these problems are characterised by relatively small domains associated with fluid flow and the CFD part of the coupled model. These domains (voids, cracks) are discontinuous, and are distributed in between solid particles. The mass of fluid is therefore small in comparison to the mass of the surrounding solid. For instance, in the explosive induced solid fragmentation the mass of solid is 1000 to 30,000 greater than the mass of the detonation gas. Consequently, the inertia of the fluid is small in comparison to the inertia of the solid.

The moving solid takes with it the cracks and voids acting as a non-inertial frame of reference for the fluid locked inside these cracks and voids. It can be said that each crack or void has its own moving Eulerian grid. Inertia effects due to the grid moving, and therefore accelerating, are so small that their effects on, for instance, pressure distribution inside a crack or void are negligible. The CFD formulation can in most cases ignore inertia effects due to the acceleration of Eulerian grids. In some cases (heavy fluid, small pressure differentials, larger size of voids), these inertia effects have to be taken into account.

The common characteristic of this type of coupled problem is that the Eulerian CFD grids are superimposed on Lagrangian finite element grids. With the increase in the size of voids and space between solid particles becoming larger, there comes a point when solid particles become completely submerged in fluid. These then become coupled problems of the first type, with a Lagrangian grid being superimposed on the Eulerian grid that is now fixed in space. These types of problems are in essence CFD problems, and are outside the scope of this book.

9

Computational Aspects of Combined Finite-Discrete Element Simulations

Combined finite-discrete element simulations comprising a few hundred to a few thousand discrete elements are classified as small scale combined finite-discrete element simulations. Small scale combined finite-discrete element simulations usually require minimum hardware resources in terms of CPU and RAM space, and can be accomplished on almost any machine. In this light, no special software requirements are necessary, and the simplest software configurations can do the job.

Combined finite-discrete element simulations comprising tens of thousands of discrete elements are classified as medium scale combined finite-discrete element simulations. Medium-scale combined finite-discrete element simulations require all the sophisticated algorithmic details, such as efficient contact detection procedures, as detailed in the rest of this book. Neither special hardware requirements nor special software designs are necessary. Medium scale simulations easily fit well within the constraints of modern day hardware in terms of both CPU and RAM requirements, and run times are usually very short.

9.1 LARGE SCALE COMBINED FINITE-DISCRETE ELEMENT SIMULATIONS

Combined finite-discrete element simulations comprising hundreds of thousands or a few millions of discrete elements are classified as large scale combined finite-discrete element simulations. RAM requirements for such simulations are measured in gigabytes. Run times are measured in hours, days or even weeks. Present day desktop workstations can easily handle jobs of this type. However, special care must be taken to:

- Minimise RAM requirements.
- Minimise CPU requirements.
- Minimise data storage requirements.
- Minimise RISK (computer crash, power failure).
- Maximise transparency, portability and robustness.

The Combined Finite-Discrete Element Method   A. Munjiza
© 2004 John Wiley & Sons, Ltd   ISBN: 0-470-84199-0

9.1.1 Minimising RAM requirements

Minimising RAM requirements usually involves the special design of an in-core database. A combined finite-discrete element simulation is in essence a process where certain sets of data are repeatedly modified to arrive at an intermediary or final set of data, which is then given a physical interpretation such as position, orientation, velocity, etc. of discrete elements. Such sets of data are conveniently organised in what is termed a 'database'.

The purpose of this database is to store all the information about the combined finite-discrete element system. All the data in the database is very frequently accessed by the CPU. Thus, access to such a database must be very fast. To achieve this, it is an imperative to place the whole database within the available RAM space for the given hardware. This is the reason why such a database is called the 'in-core database'.

Modern hardware and operating systems do give a possibility of RAM requirements of a specific computer job exceeding the available RAM space. For instance, if a given computer has 1 gigabyte of RAM space, with a UNIX operating system one could theoretically have a computer job of any size of in-core database. This is termed 'virtual memory'. Thus, one can have an in-core database of, say, 1.5 gigabytes. However, at any given time while the job is running only up to a maximum of 1 gigabyte will reside within the RAM space. The rest of the in-core database will have to sit somewhere else. This is usually specially allocated space on the hard disk.

Virtual memory is accessed through what is termed 'paging'. The access speed is governed by the access speed of the physical storage device. As access to a hard disk is extremely slow in comparison to access to the RAM type storage, access to the in-core database will in general be very slow if the whole or even a small part of the in-core database has to be stored on the hard-disk.

The concept of virtual memory is now available across a wide range of operating systems. Detailed procedures explaining the paging process are outside the scope of this book. However, for the sake of understanding at least the basic idea behind virtual memory, one can imagine the paging being a cut and paste operation. One page of data is cut from RAM and pasted to the hard disk, while another page is cut from the hard disk and pasted to the RAM.

This pasting operation is very expensive. In a typical combined finite-discrete element simulation at least 99% of the in-core database is accessed at least once every time step. Most often the access is required several times every time step. This can translate to more data than the in-core database being copied onto the hard disk every time step. This is generally time consuming, even in cases when only a small proportion of the in-core database is residing on the hard-disk.

In addition to the virtual memory, present day workstations enable what is termed 'multitasking'. Multitasking is described as two or more jobs running simultaneously on the same processor. In reality this is not the case; it appears that the jobs are running simultaneously, when in fact the processor is juggling the jobs. What happens is that the processor works for very short time on job A. It then suspends job A and works for a very short time on job B. It then comes back to the job A, etc. In a similar fashion, if there are more than two jobs at any given time only one job is running, while the rest of the jobs are waiting. Waiting times and run times are very short, so it appears to the user that all the jobs are running at the same time. Very often in combined finite-discrete element simulations one is in a dilemma – should jobs be run simultaneously or one after the other?

Say that job A takes one day running on its own, and job B takes one day running on its own. Thus if only one machine is available, after day one the results of the job A will be ready for post-processing, and after day two the results of job B will be available for post-processing. Should the jobs be run simultaneously, the best case scenario is that after two days the results of both jobs are available. Very often, the situation is much worse. Jobs A and B, when run simultaneously, compete not only for CPU time but also for RAM space. If both jobs involve small scale combined finite-discrete element simulations, it is most likely that they will both fit nicely within the available RAM space, and no paging will be necessary. However, if each job requires more than 50% of the available RAM space, both of them will not be able to fit within the available RAM at the same time, and some data from the in-core database will have to sit on the hard disk. This will bring a familiar situation of the frequent need to access the hard disk. The result will be a significant, sometimes tenfold, increase in run time.

It is evident that the RAM requirements of combined finite-discrete element simulations must be taken very seriously. There are several things that can be done:

- Avoid multitasking, i.e. running more than one combined finite-discrete element simulation on the same machine at the same time.
- Run only combined finite-discrete element simulations that require less space than the available RAM space on the given machine.
- Design the in-core database in such a way that the size of the database is reduced to a minimum. Two approaches are available, namely an object orientated in-core database and a relational in-core database.

9.1.2   Minimising CPU requirements

Minimisation of CPU requirements can be achieved by:

- more efficient algorithms,
- more efficient implementations, and
- the use of faster hardware including parallel, distributed and grid computing options.

9.1.3   Minimising storage requirements

Combined finite-discrete element simulations require a massive number of results to be stored during the simulation. Hard disk requirements can easily run into hundreds and thousands of gigabytes. These can reduce the CPU efficiency, but can also make certain combined finite-discrete element simulations impossible without a large disk storage space being readily available. The solution is data compression before any results are stored onto the hard disk.

9.1.4   Minimising risk

Large scale combined finite-discrete element simulations can run for days, even weeks. Restart procedures are necessary to reduce the risk of loosing valuable CPU times due to, say, the computer being accidentally turned off. In such cases, it is customary to save whole in-core database say 10 or 20 times during the run time. In this way, the maximum loss cannot exceed 10% or 5% of the total required run time. The job that has crushed once will therefore require only 10% or 5% longer run time than it would take if the crush did not occur.

9.1.5 Maximising transparency

Large scale combined finite-discrete element simulations are so diverse in applications that two different applications may not even look alike. The algorithms employed in application A may completely differ from algorithms employed in application B. Both applications may share the basic common concepts, such as contact detection, contact interaction, solver and discretisation. A general purpose combined finite-discrete element computer program should in theory handle a wide range of applications. This can be done in two ways:

- Implement common algorithms in such a way that they fit all applications. However, large scale combined finite-discrete element simulations are very demanding on computer resources, and generalised implementation of common algorithmic procedures will neither minimise RAM nor CPU requirements. In fact, such an approach is more likely to maximise both. Thus the classic approach of 'all tools in one computer program' is not an option in large scale combined finite-discrete element simulations.
- Produce a virtual workbench-like or library-like software environment, which would enable a relatively fast and automated process of building a specialised computer program for a given application, or even for a given problem.

The end goals of reusability, transparency and CPU and RAM efficiency seem to contradict each other. Indeed, to achieve both in languages such as FORTRAN is impossible. A bold move towards one of the modern day languages is necessary. In terms of speed, transparency and speed of implementation, it is now widely recognised that modern languages such as C or C++ offer unique advantages in comparison to languages such as FORTRAN (in terms of transparency) and in comparison to languages such as Java in terms of speed.

9.2   VERY LARGE SCALE COMBINED FINITE-DISCRETE ELEMENT SIMULATIONS

Very often practical application of both discrete elements and the combined finite-discrete element method require systems comprising hundreds of millions of discrete elements. These simulations are termed very large scale combined finite-discrete element simulations. RAM requirements for such systems are at the upper limit of the most present day affordable single processor machines. The problems with RAM requirement are best illustrated by the fact that these systems may require tens, even hundreds of gigabytes of RAM space. With the virtual memory concept, this space must be addresses. Virtual memory can be viewed as an array with the first variable being at address 1, second variable at address 2, third variable at address 3, nth variable at address n. The address of an individual variable is in essence a number pointing to where this variable is in the array. If this number is, say, a 32-bit number, the total number of different variables is given by

$$n = 2^{32} = 4.29 \text{ billion} \qquad (9.1)$$

Exact implementation of the virtual memory concept varies from operating system to operating system, and can be different on different workstations. Irrespective of these different implementations, it is evident from equation (9.1) that there is a limit to the size of problem that can be handled with a 4 byte address. The problem of RAM space can be resolved in two ways:

- Implementation of parallel, distributed or grid computing options, with both RAM and CPU requirements being distributed to different processors or different computers. With a distributed option comprising a 1000 workstation cluster, the above limit would increase by 1000 times. The same applies to parallel or grid computing options.
- The 64 bit address increases the above RAM limit by 4.29 billion times, making virtual RAM space almost unlimited for all practical purposes of very large scale combined finite-discrete element simulations.

9.3 GRAND CHALLENGE COMBINED FINITE-DISCRETE ELEMENT SIMULATIONS

Some combined finite-discrete element problems may require over one hundred billion discrete elements. Some may even require over one trillion or even a quadrillion of discrete elements. Combined finite-discrete element simulations of this type are beyond the power of present day affordable desktop workstations. Thus they are termed 'grand challenge combined finite-discrete element simulations'.

Grand challenge combined finite-discrete element simulations are grand in terms of both CPU and RAM requirements. If a one million discrete element combined finite-discrete element problem takes 24 hours of CPU on present day workstations, a typical one trillion combined finite-discrete element simulation would probably take 1,000,000 workstation-days, which translates into

$$\frac{1,000,000}{365} = 2740 \text{ years} \qquad (9.2)$$

It is evident that either massive parallelisation or massive distributed computing are at present the only options. The distributed computing option would probably require a cluster of 30,000 present day affordable workstations for a year. At present day costs, assuming that a workstation is outdated after two years, the cost of such a job would be

$$\frac{30,000 \cdot 2000}{2} = \pounds 30,000,000 \qquad (9.3)$$

This is a very large sum indeed to pay to run a single combined finite-discrete element job. The problem that such a job could solve would, for instance, involve rock blasting of a total volume of

$$10 \cdot 10 \cdot 10 = 1000 \, \text{m}^3 \tag{9.4}$$

discretised to the resolution of 1 mm (which would give a reasonable accuracy). The market value of all the rock would probably not exceed £10,000. Coarser discretisation to 1 cm resolution (with a greatly reduced accuracy and micromechanical processes associated with fines mostly neglected) would result in the total amount of rock being simulated equal to, say

$$40 \cdot 50 \cdot 500 = 1,000,000 \, \text{m}^3 \tag{9.5}$$

and a probable market value of the rock £10,000,000, which is still much less than the cost of a computer simulation of such a problem.

Grand challenge combined finite-discrete element simulations are clearly beyond the CPU and RAM limits of present day computers, and at present day RAM and CPU prices are not affordable for most practical engineering, industrial and scientific applications. However, it is worth mentioning that even a one billion system seemed impossible five years ago. A one million system was beyond the limits of most available hardware in the early 1990s, while even a one hundred thousand system was beyond limits of the most powerful and most expensive computers in the early 1980s. It is only 25 years since those early days. If Moore's law is to continue in the future, the next 25 years may even bring systems comprising a quadrillion of discrete elements. This is illustrated by an imaginary scenario of the market price of a single processor of performance of a modern workstation being reduced to 1 penny. An array of

$$1000 \cdot 1000 \cdot 1000 = 1,000,000,000 \, \text{processors} \tag{9.6}$$

would cost £10,000,000. Say that each processor can handle 1,000,000 discrete element system in one day. A grid of such processors, if communication between processors is neglected, would handle a one thousand trillion (one quadrillion) particle system in 24 hours. This is equivalent to a cube of rock 1000 by 1000 by 1000 m comprising discrete elements of an average size of 1 cm. The probable value of such a cube of rock would be £10 billion. Assuming that the computer is outdated after two years, and that computer capacity is used 100%, the cost of computer simulations would be

$$\frac{10,000,000}{2 \cdot 365} = £1370 \tag{9.7}$$

This is negligible in comparison to the probable market price of the rock. This clearly demonstrates that there is no need to be pessimistic about grand challenge combined finite-discrete element simulations. On the contrary, one has to look forward to the day when these will become small-scale combined finite-discrete element simulations because of the massive CPU power available. A pile of rock is a relatively low technology problem. However, there is a whole class of similar discontinua problems that will require grand challenge combined finite-discrete element simulations. That such simulations will become possible is beyond doubt; the only question is whether the suitable hardware will become available in ten or twenty years time. In the meantime, the role of computational mechanics of discontinua is to develop suitable accurate and robust computational solutions.

9.4   WHY THE C PROGRAMMING LANGUAGE?

It is most often the case that combined finite-discrete element systems are large. This makes the reduction of CPU and RAM requirements very important. This can be done in several ways:

- design fast algorithmic solutions,
- use robust hardware, or
- implement the algorithms in such a way that the program is made to run faster and is more efficient in terms of RAM.

There are several ways in which a particular algorithm can be implemented in a more efficient way. Most of them are connected to the computer language used. For instance, applications implemented in assembler are usually faster than those implemented in FOR-TRAN. The problem is that assembler is a low level language, where transparency becomes difficult as soon as the code exceeds few hundred lines. At the other end of the spectrum are high level, object-oriented languages such as C++ or Java. The problem with those is that speed is not always easy to achieve, especially with Java.

The C computer language seems to cover both ends of the spectrum. Features such as pointers, pointers to functions, dynamic memory allocation, shift operators, etc. make this language very suitable for writing very fast codes that are gentle in terms of RAM requirements. It is worth mentioning, for instance, that the UNIX operating system is written in C. Thus, a computer program written in C becomes almost an extension of the operating system itself.

Features such as structures, private and public data and methods (functions) make this language suitable for writing very large transparent codes. In fact C++ is an extension of C, where additional features such as inheritance, classes and virtual objects are added. Some C++ compilers compile the C++ code by first translating it into C and then using a C compiler to compile the C translation of the code.

9.5   ALTERNATIVE HARDWARE ARCHITECTURES

Combined finite-discrete element simulations are computationally much more intensive than the more familiar finite element methods. Thus, many CPU critical simulations are almost impossible on present day sequential architectures. A solution to at least some of these CPU and/or RAM intensive combined finite-discrete element simulations is sought in the form of parallel and distributed computing.

9.5.1   Parallel computing

Implementation of parallel computing in combined finite-discrete element simulations requires access to parallel machines. These machines are characterised by a number of

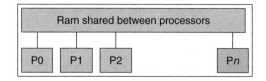

Figure 9.1 Parallel architecture with processors sharing memory.

processors working together. For instance, in Figure 9.1 a shared memory architecture is shown. Processors communicate through sharing RAM. Software libraries give shared memory parallel programmers a flexible interface for developing parallel applications for shared memory platforms, ranging from the desktop to the supercomputer. Using such libraries, one can either parallelise parts of the program such as CPU demanding loops, or complete regions of the program.

A typical performance of the shared memory approach is shown in Figure 9.2. It is evident from the figure that the speed increase is not linear, i.e. it lags behind the increase in number of processors.

A distributed parallel architecture model is shown in Figure 9.3. Each processor has a separate RAM space associated with it. The processes executing in parallel have a separate address space. The processors communicate with each other through message passing. Communication occurs when a portion of one process's address space is copied into another process's address space.

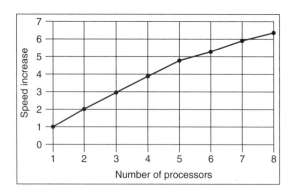

Figure 9.2 Speedup using the shared memory approach. Total number of particles-60 million.

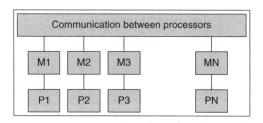

Figure 9.3 Parallel architecture with separate RAM space associated with each processor.

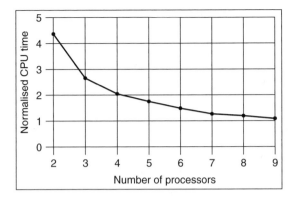

Figure 9.4   Time per time step using a distributed parallel architecture model and domain decomposition. Total number of particles-0.6 billion.

A typical performance of a distributed parallel architecture model is shown in Figure 9.4. The speed increase is not linear with the number of processors. However, it is evident that a much larger problem can be addressed by this model than by the shared memory approach.

9.5.2 Distributed computing

The biggest problem associated with parallel hardware architectures is that such architectures are not required by the majority of applications in which computers are used. In other words, applications which require parallel architectures are few and far between. Thus a particular parallel architecture is unlikely to be manufactured in large quantities. This makes parallel architectures disproportionately much more expensive than sequential architectures. For instance, a fixed number of float point operations executed on a parallel machine can be over 1000 times more expensive than the same number of float point multiplications executed on a sequential machine.

The situation is made even worse by the very large difference between applications requiring parallel computing. For instance, parallelisation of a chess game would ideally require a different parallel hardware architecture from the parallelisation of language translation or vision processing. Finite element applications would ideally require a different parallel hardware configuration from discrete element applications. 2D simulations would require a significantly different parallel architecture from 3D applications if optimum performance is to be achieved.

There have been attempts in the past to design a general purpose parallel hardware. The problem with these attempts is that they seem to be optimal architectures for none of the applications they were intended for. In addition, the parallelisation libraries end up being extremely hardware-dependent, not only in terms of performance, but very often in terms of detail and even syntax. This is not to say that parallelisation is impossible. Parallelisation is certainly a viable solution for many CPU and RAM critical applications. The problem is that very often it is not affordable. Parallelisation is almost like a custom-designed machine, because no parallel computer has ever been produced in quantities measured in millions of machines.

This is in sharp contrast with, for instance, the PC home computer market. The price of the latest generation of PC processors is reducing dramatically, fast approaching a critical £10 level, and has the potential to go even lower. This is obviously because of the massive quantities made, reducing development and capital costs to negligible levels. In this environment £100,000 could soon be buying 10,000 processors, with the potential of being enough money to buy 100,000 processors. £100,000 is a relatively small amount of money with which at present day prices not much parallel hardware could be bought. It is also a small amount in absolute terms, and many enterprises could theoretically afford to invest this much in high performance computing every year. If only 10 CPU critical jobs are run a year on such a machine, the capital cost would be £10,000 per job. In many fields that require high performance computing this is not a high price. These cheap, powerful, fast von Neumann architectures come with fast mathematical libraries and incorporate fast inter-computer communications. World Wide Web developments during the last decade have resulted in the speed of network cards changing from 10 megabits per second to 100 megabits per second, and recently to over 3 gigabits per second. This has been fast recognised as a cheap opportunity to build massive custom made distributed high performance computing clusters. The basic idea is not to build a parallel machine, but to distribute the CPU and/or RAM intensive computing job over different computers connected to a network. So, for instance, the visual part of an application sits on one machine, the sound part is on another, frequently used data is distributed over another couple of machines, and so on. There are a number of emerging concepts that are related to distributed computing. These include fabric computing, on-demand computing, organic IT, utility computing, grid computing, etc. In the context of the combined finite-discrete element method, one could think of different aspects of the combined finite-discrete element simulation residing on different machines connected to a network.

However, a better approach is to divide a domain into sub-domains. For instance, if discrete elements are all packed inside a box, as shown in Figure 9.5, the box can be subdivided into five parts. Each part with all the discrete elements in it is then loaded as a separate problem on its own PC. These five sub-domains share the boundaries of the sub-domains. These are not real boundaries, and discrete elements are free to move

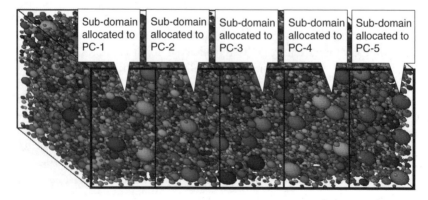

Figure 9.5   A typical combined finite-discrete element domain divided into sub-domains for distributed computing purposes.

from one sub-domain to the other. Discrete elements can also interact over the sub-domain boundaries.

Thus, a PC running the problem comprising discrete elements from one sub-domain must communicate to the PC running the problems comprising the discrete elements from the neighbouring domains. To solve the problem shown in Figure 9.5, a cluster of five PCs is needed. The simplest case is when all five PCs are identical. These PCs need to communicate to each other as shown in Figure 9.6.

Very often 2D domains are subdivided in a chess-pattern type subdivision as shown in Figure 9.7. It is evident that each PC is connected to the four neighbouring PCs. Similar domain subdivision and PC assignment in 3D would require each PC communicating to six neighbouring PCs. A cluster of PCs dedicated to distributed computing represents in essence a parallel computing architecture, except that each element of such an architecture

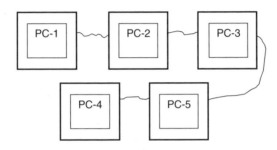

Figure 9.6 An optimal configuration of PCs with network connectivity between PCs for problem shown in Figure 9.5.

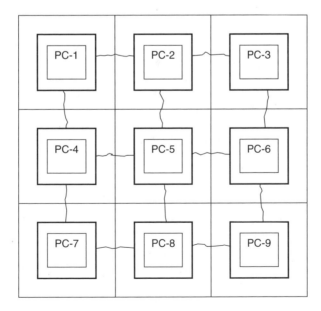

Figure 9.7 A 2D domain divided into sub-domains with a PC assigned to each sub-domain, together with connectivity between PCs.

is a separate PC with its own CPU and RAM, and most often hard disk. In some distributed computing applications, different computer types can be used, even different operating system. For combined finite-discrete element simulations this is not desirable, and the best solution is probably a cluster of identical PCs configured in such a way that mapping from sub-domains onto the PC cluster is relatively simple. In this way, communications between PCs are kept to a minimum, and the amount of time each CPU spends actually processing the job assigned to it is maximised.

9.5.3 Grid computing

Grid computing can be defined as a massive integration of computer systems available through a network to offer performance unattainable by any single machine. Grid computing enables the virtualisation of computing resources distributed over a grid. These include processing, network bandwidth and storage capacity, which are used to create a single virtual image of the system, For grid-based combined finite-discrete element simulations, domain subdivision is obviously one of the ways of exploiting the grid. An alternative and more feasible way is to use the grid as a virtual experimentation lab. In that way, instead of using the grid as an alternative to massively parallel or cluster-based simulations, it is used to complement them. Researchers located at various locations around the globe can in essence simultaneously work together on the same problem. An example of such a class of problems is parametric studies, where it may not be necessary to repeat large scale computations –the results can be parameterised instead, and made available to both research and industry. Such combined finite-discrete element projects would rely on individual users harnessing the unused processing power and coordinating the work towards a common goal. Thus, both money and resources are saved, the project is speeded up and cooperation among individual researchers is brought to a new level. A particular combined finite-discrete element simulation is nothing more but a numerical experiment. Much like a given physical experiment in the lab, the results of such a numerical experiment should be independently verified and/or validated. Verification in essence is about confirming that everything was done properly and that the results are not a consequence of a coding error in the program. Validation is the next step, matching the results of a numerical experiment to the results of a physical experiment. To grid-enable such processes, in the era when numerical experimentation is to some extent replacing or complementing physical experimentation, it is not enough to communicate results through journals. The space available for journal publications is too small to record all the details of a numerical experiment. While a physical experiment can very often be recorded on a couple of pages of written text, a numerical experiment may involve a large amount of input data, problem parameters, methods used, algorithms employed, implementation details, etc. The only way to solve the problem is to have both hardware and software tools in virtual form, enabling easy abstraction of such problems and also easy access to problem data through a virtualised distributed database.

In summary, various options are available to address very large scale and grand challenge discontinua problems. However, it appears that, due to a need for communication between processing units, there is a limit to what speedup can be achieved using any of the above listed options. Existing parallel, distributed and grid computing options are able to achieve better CPU and RAM performance, thus increasing the size of the problem that can be addressed by one to two orders of magnitude. However, this appears to be an upper limit. A further increase in the number of processing units may not result in significant improvements in performance. There exist many problems similar in nature to combined finite-discrete element problems. There will always be problems characterised as grand challenge, irrespective of the CPU power available. However, in the long run, the solutions similar in nature to distributed computing are likely to prevail. Cheap standard components produced at massive scale will become very affordable. The design of such components is by nature modular. A combination of such cheap components will enable problem specific architectures to be built at affordable prices. For this to succeed, the following conditions need to be met:

- Massively produced modules comprising CPU and RAM are cheap and used in PC and other computer types.
- Massively produced modules comprising CPU and RAM enable at least six simultaneous ultra-fast communication channels to at least six neighbouring CPUs connected to it.
- Massively produced modules comprising CPU and RAM are physically such that they can be attached to each other, forming different physical patterns; or cheap devices enabling the same are available.

In short, the design of parallel machines will eventually be left to the users of such machines. Much like the hardware part of a sequential machine was separated from the software part, so will the assembly, design and functionality of a parallel machine be separated from the hardware manufacturing process. For this to happen, what is needed is that manufactures of sequential modular components such as the CPU take into account the need for communications between these modules, and the need to link them physically together to build a parallel machine. In this light there will be no reason for such disproportionality in the price of parallel and sequential machines.

10

Implementation of some of the Core Combined Finite-Discrete Element Algorithms

Combined finite-discrete element simulations are computationally much more complex than the more familiar finite element methods, and this must be taken into account both in terms of software design and implementation. In this chapter, some problems associated with the implementation of the combined finite-discrete element method are addressed. This is done through listing and explaining the implementation of the core combined finite-discrete element algorithms such as contact detection, contact interaction and deformability.

Originally the combined finite-discrete element method was implemented in an object oriented fashion using C++. However, because many readers may not be familiar with C++, a much simpler C-based implementation of these core algorithms is described in this chapter.

10.1 PORTABILITY, SPEED, TRANSPARENCY AND REUSABILITY

10.1.1 Use of new data types

C enables the definition of new data types. This is used, for instance, for object-orientated programming. However, very often one may wish to use the combined finite-element program in a single precision mode for one problem (or machine), and in double precision mode for another problem (or machine). This flexibility can save both programming and CPU time and reduce memory requirements.

10.1.2 Use of MACROS

Combined finite-discrete element simulations are CPU intensive, thus it is very important that the features of the C language that speed up C programs are used when implementing

The Combined Finite-Discrete Element Method   A. Munjiza
© 2004 John Wiley & Sons, Ltd   ISBN: 0-470-84199-0

```
#define YMATINV2(m,minv,det)\
  { det=m[0][0]*m[1][1]-m[1][0]*m[0][1]; \
      minv[0][0]= m[1][1]/det; minv[1][0]=-m[1][0]/det;\
      minv[0][1]=-m[0][1]/det; minv[1][1]= m[0][0]/det; \
  }
```

Listing 10.1   The MACRO for small matrices.

```
#define V3DTranToGl(x1,y1,z1,a,b,c,x2,y2,z2,x3,y3,z3)\
  {   (x1)=(a)*(x2)+(b)*(x3)+(c)*((y2)*(z3)-(z2)*(y3)); \
      (y1)=(a)*(y2)+(b)*(y3)+(c)*((z2)*(x3)-(x2)*(z3)); \
      (z1)=(a)*(z2)+(b)*(z3)+(c)*((x2)*(y3)-(y2)*(x3));\
  }
```

Listing 10.2   Transformation of vector components from the global triad of orthonormal vectors to a local triad of orthonormal vectors.

combined finite-discrete element algorithms. One of the most important such features is MACROS. In the C language, a MACRO is almost like a function (subroutine, for readers familiar with FORTRAN). The difference is that wherever a MACRO is called, the compiler actually includes the body of the MACRO in the place where the call to the MACRO is made. In other words, no actual function call occurs. At compilation time, the body of the MACRO is itself copied into the code together with all the local variables being passed directly to the MACRO.

For instance, a macro for 2D matrix inversion is shown in Listing 10.1. The macro takes as input matrix m, calculates the determinant, calculates the inverse matrix and returns the determinant det and the inverse matrix minv. A very frequent call to this macro saves the CPU overheads of calling a function, making that part of the program in which this macro is called run faster.

Transformation of vector components from one triad of orthogonal unit vectors to another triad of unit vectors can be elegantly implemented using macros, thus making the code shorter and more transparent without decreasing the CPU efficiency. The macro used for the transformation of vector components from a local triad into a global triad is shown in Listing 10.2.

The macro takes vector components in the local triad (a,b,c) and returns vector components in the global triad (x1,y1,z1). Global components of the two unit vectors of the local triad (components x2,y2,z2 and x3,y3,z3) are used in this transformation. The third unit vector of the local triad is calculated using the cross product of the first two vectors, i.e. the third vector is equal to the cross product of vectors (x2,y2,z2) and (x3,y3,z3).

10.2   DYNAMIC MEMORY ALLOCATION

The dynamic memory allocation works in such a way that at the time of compilation of the code, no memory is allocated for dynamic variables, arrays, etc. Memory is instead allocated as and when required at run time. As soon as this memory is not required it is returned, and can be used again. It would appear that allocating the memory during the run time can only slow down the program. This is not necessarily the case. For instance,

```
DBL **TalDBL2(m2,m1)
   INT m2; INT m1;
{  INT isize,i2;
   DBL   *p1;
   DBL   **p2;
   void  *v1;

   isize=sizeof(DBL*)*(m2+3)+
      sizeof(DBL )*(m2*m1+3);
   if(isize==0)return DBL2NULL;
   v1=MALLOC(isize);
   p2=(DBL**)v1;
   p1=(DBL*)v1;
   p1=p1+((m2+1)*sizeof(DBL**))/sizeof(DBL)+2;
   for(i2=0;i2<m2;i2++)
   { p2[i2]=p1+i2*m1;
   }
   return p2;
}
```

Listing 10.3   Dynamic allocation of the two-dimensional array – file frame.c.

most of the combined finite-discrete element in-core database takes the form of arrays. Fixed size 2D or 3D arrays are very difficult to handle, while access to individual numbers is expensive. For readers familiar with FORTRAN, it is worth mentioning that the same happens when multidimensional arrays are used in FORTRAN codes.

Well designed dynamic arrays are usually much faster and allow better flexibility without any change in the syntax. Dynamic allocation of the two dimensional array is shown in Listing 10.3. The function TalDBL2 calculates the size of the array and allocates space for it.

This is followed by breaking the array into m2 one-dimensional arrays. Each of these one-dimensional arrays is of size m1 elements. Each of the m2 one-dimensional arrays is allocated a pointer. All these pointers are stored as a one-dimensional array of pointers. A pointer p2 to the first element of this one-dimensional array of pointers is returned. This pointer is therefore a pointer to a pointer. The advantage of such an array is the way

```
DBL FunctionUsingDynamicArray(p2)
   DBL **p2
{  DBL x;
   DBL y;
   DBL *p1
   p1=*(p2+5);
   y=*(p1+7);
   /* equivalent syntax */
   x=p2[5][7];
   return (x-y);
}
```

Listing 10.4   An example of accessing an element of a two-dimensional dynamic array.

in which it is passed to a function, as shown in Listing 10.4. Access of a single element of the array p2 involves first the reading of pointer p1 stored at address (p2+5). This is followed by reading the array element from address (p1+7). Thus, any element of a two-dimensional array is accessed by two subsequent dereferencing of a pointer. Subsequent dereferencing of a pointer to a pointer can also be written as shown in the line before the last line of Listing 10.4. FunctionUsingDynamicArray will obviously return zero, because both variables x and y are assigned the same element of the dynamic array p2.

10.3   DATA COMPRESSION

Combined finite-discrete element simulations usually require repeated outputs. During a single combined finite-discrete element simulation, it is not unusual to produce over 1000 output instances. Each output instance requires that almost the entire in-core database be written on the hard disk. To minimise the disk space required, these data are compressed before any output is created. The following requirements are met in this process:

- To facilitate portability of the output files, all output files are created in ASCII format.

- Data is compressed just before being written on the hard disk, thus there is no need to have an uncompressed file on the hard disk at any time.

- When the output data is read from the hard disk, say by a postprocessor, it is read in a compressed format and uncompressed by the postprocessor itself. In this way, a need for an uncompressed file sitting on the hard disk waiting to be postprocessed is avoided. The reason for this is because, for instance for animation purpose, the postprocessor needs quick access to all output files at the same time. Uncompressing them all and leaving them on the hard disk would often make postprocessing impossible because of the large storage requirements, which may not be met by the available space on the hard disk.

Thus, in the C implementation of the combined finite-discrete element method, all output data is written onto the hard disk in a special compressed ASCII format. Output is written in an object-oriented format, with each line of output comprising all data about a single object, say a single finite element or a single discrete element.

10.4   POTENTIAL CONTACT FORCE IN 3D

10.4.1   Interaction between two tetrahedrons

The theoretical details of the potential contact force algorithm in 3D are explained in Chapter 2. The easiest way to implement the potential contact force is to use the discretised contact solution approach, i.e. to discretise the domain of each discrete element into finite elements, and process potential-contact-force-based interaction between contacting finite elements.

It is most likely that the simpler finite element geometries would result in a more efficient contact solution than the alternative, more complex finite element geometries.

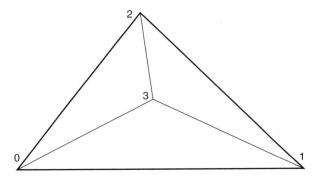

Figure 10.1 Numbering of nodes of target or contactor tetrahedron.

```
/*******************************************************/
/*          loop over target surfaces           */
/*******************************************************/
for(itars=0;itars<4;itars++)
{ ipt[0]=ieltop[itars][itarth];
  ipt[1]=ieltop[1][itarth];
  ipt[2]=ieltop[2][itarth];
  ipt[3]=ieltop[3][itarth];
  if(itars>0) ipt[3]=ieltop[itars-1][itarth];
  if((itars==1)||(itars==2)) ipt[1]=ieltop[3][itarth]];
  if(itars>1) ipt[2]=ieltop[0][itarth];
  /*  nested  loop over contactor surfaces  */
  for(icons=0;icons<4;icons++)
  { ipc[0]=ieltop[icons][iconth];
  ...........continued in the next listing...............
```

Listing 10.5 A loop over surfaces (itars) of the target tetrahedron.

Thus in this C implementation of the combined finite-discrete element method, a four-noded tetrahedron is employed. (Figure 10.1). Note that the numbering of nodes starts with zero. This is because in the C language, the first element of a one-dimensional array is the 0th element.

The source code of the function processing interaction between two contacting tetrahedra is described in detail in this section. The code shown here is compatible with the description of the potential contact force algorithm described in Chapter 2. First, a loop over surfaces of the target tetrahedron itars is opened (Listing 10.5). The target surfaces are as follows:

- Target surface 0: 012-3.
- Target surface 1: 132-0.
- Target surface 2: 230-1.
- Target surface 3: 310-2.

The first three numbers indicate the target surface itself, while the last number indicates the remaining node of the tetrahedron that has a particular target surface as the base.

```
.........continued from previous listing.............
ipc[1]=ieltop[1][iconth];
ipc[2]=ieltop[2][iconth];
ipc[3]=ieltop[3][iconth];
if(icons>0) ipc[3]=ieltop[icons][iconth]];
if((icons==1)||(icons==2)) ipc[1]=ieltop[3][iconth];
if(icons>1)                 ipc[2]=ieltop[0][iconth];
/* set nodal coordinates */
..........continued in the next listing...............
```

Listing 10.6   A loop over surfaces (icons) of the contactor tetrahedron.

```
.........continued from previous listing...............
for(i=0;i<3;i++)
{ xt[i]=d1nccx[ipt[i]];
  yt[i]=d1nccy[ipt[i]];
  zt[i]=d1nccz[(ipt[i]];
  xc[i]=d1nccx[(ipc[i]];
  yc[i]=d1nccy[(ipc[i]];
  zc[i]=d1nccz[(ipc[i]];
}
xt[3]=xcent; yt[3]=ycent; zt[3]=zcent;
xc[3]=xcenc; yc[3]=ycenc; zc[3]=zcenc;
xorig=xc[0]; yorig=yc[0]; zorig=zc[0];
for(i=0;i<4;i++)
{  xt[i]=xt[i]-xorig; yt[i]=yt[i]-yorig; zt[i]=zt[i]-zorig;
   xc[i]=xc[i]-xorig; yc[i]=yc[i]-yorig; zc[i]=zc[i]-zorig;
}
.........continued in the next listing.................
```

Listing 10.7   Extracting coordinates from the in-core database.

Inside the loop of target surfaces a loop over surfaces icons of the contactor tetrahedron is nested, as shown in Listing 10.6. The contactor surfaces are as follows:

- Contactor surface 0: 012-3.
- Contactor surface 1: 132-0.
- Contactor surface 2: 230-1.
- Contactor surface 3: 310-2.

where the first three numbers indicate the contactor surface itself, while the last number indicates the remaining node of the tetrahedron, for which the contactor surface is the base.

In Listing 10.7, the coordinates of both the target and contactor tetrahedra are extracted. The following variables are used: xcent, ycent and zcent are the coordinates of the centre of the target tetrahedron; xcenc, ycenc and zcenc are the coordinates of the centre of the contactor tetrahedron; xc[0] is used as the origin of the local coordinate system.

In Listing 10.8, the components of the unit vector normal to the contactor surface are calculated as a cross product of the two edges of the surface. A local triad of orthogonal unit vectors is then defined, with only the first two vectors xe and ye being stored.

```
.......continued from previous listing.............
/* contactor normal, e-base and target points in e-base */
V3DCro(xnc,ync,znc,xc[1],yc[1],zc[1],xc[2],yc[2],zc[2]);
V3DNor(xe[0],xnc,ync,znc);
xe[0]=xc[1]; ye[0]=yc[1]; ze[0]=zc[1];
V3DNor(xe[1],xe[0],ye[0],ze[0]);
V3DCro(xe[1],ye[1],ze[1],xnc,ync,znc,xe[0],ye[0],ze[0]);
........continued in the next listing..............
```

Listing 10.8   Unit normal to the current contactor surface.

```
........continued from previous listing.............
/* contactor normal, e-base and target points in e-base */
for(i=0;i<4;i++)
{ V3DDot(dct[i],xnc,ync,znc,xt[i],yt[i],zt[i]);
  V3DDot(ut[i],xt[i],yt[i],zt[i],xe[0],ye[0],ze[0]);
  V3DDot(vt[i],xt[i],yt[i],zt[i],xe[1],ye[1],ze[1]);
}
if((dct[0]<=R0)&&(dct[1]<=R0)&&(dct[2]<=R0))continue;
.........continued in the next listing.............
```

Listing 10.9   Distances and local coordinates of the nodes of the target sub-tetrahedron.

In Listing 10.9, the distance dct[i] of each of the nodes of the target sub-tetrahedron from the current contactor surface is calculated, together with ut[i] and vt[i] coordinates of the nodes of the current sub-tetrahedron. If the distances are such that no contact is possible, the program moves onto a new contactor surface, thus continue is used, which obviously refers to the loop over contactor surfaces.

The next step is the calculation of S-points, i.e. intersection points between the current contactor surface and the edges of the current target sub-tetrahedron, as explained in Chapter 2. This is done using two nested for loops, as shown in Listing 10.10. If the number of S-points is such that no contact is possible, the program moves onto a new contactor surface, thus continue is used, which obviously refers to the loop over contactor surfaces.

In Listing 10.11, S-points are rearranged and C-points are calculated. C-points are in essence the nodes of the contactor tetrahedron that belong to the current contactor surface. The local coordinates of these nodes are calculated.

Now the problem has become a 2D problem. There are nspoin S-points all inside the current contactor surface. The C-points are also all inside the current contactor surface. There are in total three C-points forming a triangle, i.e. one of the bases of the contactor tetrahedron. This triangle is in further text referred to as the C-triangle. S-points form a convex polygon. In further text this polygon is referred as an S-polygon. Processing of this 2D problem involves resolving intersection between the C-triangle and S-polygon. First, the distances of the C-points and S-points from the edges of the S-polygon and edges of the C-triangle, respectively, are calculated, as shown in Listing 10.12.

In this listing, the total number of C-points inside the S-polygon is also counted (inners), together with the total number of S-points inside the C-triangle (innerc).

```
.............continued from previous listing.............
/* u,v coordinates of S-points and C-points   */
nspoin=0;
for(i=0;i<3;i++)
{ for(j=0;j<2;j++)
  { inext=i+1;
    if(inext>2)inext=0;
    if(j==0)inext=3;

    if(((dct[i]>R0)&&(dct[inext]<R0))||
      ((dct[i]<R0)&&(dct[inext]>R0)))
    { factor=ABS(dct[i]-dct[inext]);

    if(factor>EPSILON)
    { factor=ABS(dct[i]/factor);
      us[nspoin]=factor*ut[inext]+(R1-factor)*ut[i];
      vs[nspoin]=factor*vt[inext]+(R1-factor)*vt[i];
      inners[nspoin]=0;
      nspoin=nspoin+1;
} }}}

if((nspoin<3)||(nspoin>4))continue;
.............. continued in the next listing...........
```

Listing 10.10 Distances and local coordinates of the nodes of the target sub-tetrahedron.

```
.............continued from previous listing.............

/* u,v coordinates of S-points and C-points   */

/* check ordering of S-points */
if(((us[1]-us[0])*(vs[2]-vs[0])-(vs[1]-vs[0])*(us[2]-us[0]))<R0)
{ i=0;
  j=nspoin-1;

  while(i<j)
  { k=inners[i]; inners[i]=inners[j]; inners[j]=k;
    tmp=us[i];  us[i]=us[j];      us[j]=tmp;
    tmp=vs[i];  vs[i]=vs[j];      vs[j]=tmp;
    i++; j--;
} }

for(i=0;i<3;i++)
{ V3DDot(uc[i],xc[i],yc[i],zc[i],xe[0],ye[0],ze[0]);

  V3DDot(vc[i],xc[i],yc[i],zc[i],xe[1],ye[1],ze[1]);

  innerc[i]=0;
}
.............continued in the next listing...........
```

Listing 10.11 Ordering of S-points and calculation of C-points.

```
.............continued from previous listing.............
      /* distances of C-points from S edges */
niners=0; ninerc=0;
for(i=0;i<nspoin;i++)
{ inext=i+1;
  if(inext>=nspoin)inext=0;
  for(j=0;j<3;j++)
  { jnext=j+1;
    if(jnext>2)jnext=0;
    dcs[j][i]=(uc[jnext]-uc[j])*(vs[i]-vc[j])-
          (vc[jnext]-vc[j])*(us[i]-uc[j]);
    dsc[i][j]=(us[inext]-us[i])*(vc[j]-vs[i])-
          (vs[inext]-vs[i])*(uc[j]-us[i]);
    if(dsc[i][j]>=R0)
    { innerc[j]=innerc[j]+1;
      if(innerc[j]==nspoin)ninerc=ninerc+1;
    }
    if(dcs[j][i]>=R0)
    { inners[i]=inners[i]+1;
      if(inners[i]==3)niners=niners+1;
} } }
      .............continued in the next listing...........
```

Listing 10.12 Distance dcs[j][i] from the edges of a C-triangle to S-points (nodes of S-polygon); and distance dsc[i][j] from the edges of an S-polygon to C-points (edges of C-triangle).

The intersection between the C-triangle and S-polygon is called a B-polygon. A B-polygon is the area through which the current target sub-tetrahedron is interacting with the current surface of the current contactor tetrahedron. The nodes of the B-polygon polygon are called B-points.

Three sceanarios are possible:

- C-triangle is inside the S-polygon, as shown in Listing 10.13.
- S-polygon is inside the C-triangle (Listing 10.14).
- General case of polygon-triangle intersection (Listing 10.15).

A general case of polygon-triangle intersection is illustrated in Figure 10.2, where both S and B points are shown. In the example shown in the figure, S-points form a triangle.

```
.............continued from previous listing.............
      /* distances of C-points from S edges */
  /* B-points */
  if(ninerc==3) /* triangle inside polygon      */
  { nbpoin=3;
    for(i=0;i<nbpoin;i++)
    { ub[i]=uc[i]; vb[i]=vc[i];
  } }
      ...........continued in the next listing...........
```

Listing 10.13 Calculation of B-points when the C-triangle is inside an S-polygon.

```
............continued from previous listing..............
else if(niners==nspoin) /* polygon inside triangle      */
{ nbpoin=nspoin;
  for(i=0;i<nbpoin;i++)
  { ub[i]=us[i]; vb[i]=vs[i];
  }}
else        /* intersection points polygon triangle */
{ nbpoin=0;
  for(i=0;i<nspoin;i++)
  { if(inners[i]==3)
    { ub[nbpoin]=us[i]; vb[nbpoin]=vs[i]; nbpoin++;
    }}
  for(i=0;i<3;i++) /* grab inner C-points */
  { if(innerc[i]==nspoin)
    { ub[nbpoin]=uc[i]; vb[nbpoin]=vc[i]; nbpoin++;
    }}
.............. continued in the next listing...........
```

Listing 10.14   Calculation of B-points when an S-polygon is inside a C-triangle.

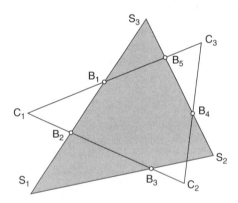

Figure 10.2   Intersection of the target sub-tetrahedron with the base of the contactor sub-tetrahedron.

C-points also form a triangle. The intersection between these two triangles is a set of points called B-points. B-points form a convex polygon called a B-polygon. As shown in Listing 10.15, first the intersection between the edges of the B-polygon and C-triangle are found, i.e. B-points. These points do not necessarily form a polygon, thus they are put into a sequence (sorted) as shown in Figure 10.2.

Once the coordinates of B-points are known, penetration (i.e. contact overlap at B-points) is calculated as shown in Listing 10.16. From penetration, the contact force potential at the B-points is evaluated.

The result so far is the interaction polygon (B-polygon) between the target sub-tetrahedron and one of the surfaces of the target tetrahedron. Contact potential at the B-points is known. As the contact potential over target sub-tetrahedron is a linear function

```
............continued from previous listing...........
for(i=0;i<nspoin;i++)        /* intersection points   */
{ inext=i+1; if(inext>=nspoin)inext=0;
  for(j=0;j<3;j++)
  { jnext=j+1; if(jnext>2)jnext=0;
    if((((dsc[i][j]>R0)&&(dsc[i][jnext]<R0))||
       ((dsc[i][j]<R0)&&(dsc[i][jnext]>R0)))&&
      (((dcs[j][i]>R0)&&(dcs[j][inext]<R0))||
       ((dcs[j][i]<R0)&&(dcs[j][inext]>R0))))
    { factor=ABS(dsc[i][j]-dsc[i][jnext]);
      if(factor<EPSILON){ factor=RP5;              }
      else          { factor=ABS(dsc[i][j]/factor); }
      ub[nbpoin]=(R1-factor)*uc[j]+factor*uc[jnext];
      vb[nbpoin]=(R1-factor)*vc[j]+factor*vc[jnext];
      nbpoin++;
} } }
for(i=1;i<nbpoin;i++)
{ if(vb[i]<vb[0])
  { tmp=vb[i]; vb[i]=vb[0]; vb[0]=tmp;
    tmp=ub[i]; ub[i]=ub[0]; ub[0]=tmp;
} }
for(i=1;i<nbpoin;i++)
{ tmp=ub[i]-ub[0]; if(ABS(tmp)<EPSILON)tmp=EPSILON;
  anb[i]=(vb[i]-vb[0])/tmp;
}
for(i=1;i<nbpoin;i++) /* sort B-points */
{ for(j=i+1;j<nbpoin;j++)
  { if(((anb[i]>=R0)&&(anb[j]>=R0)&&(anb[j]<anb[i]))||
      ((anb[i]<R0)&&((anb[j]>=R0)||(anb[j]<anb[i]))))
    { tmp=vb[i]; vb[i]=vb[j]; vb[j]=tmp;
      tmp=ub[i]; ub[i]=ub[j]; ub[j]=tmp;
} } } }
if(nbpoin<3)continue;
/* Target-plain normal and penetration at B-points */
```

Listing 10.15 Calculation of B-points for the general case of intersection between a C-triangle and S-polygon.

of the spatial coordinates, this implies that the contact potential distribution over the B-polygon is also a linear function of the spatial coordinates. This function, when multiplied with the penalty parameter, represents the distributed normal contact force. To integrate this force, the B-polygon is subdivided into simplex shapes and integration in analytical form over each of the shapes is performed (Figure 10.3). This is implemented in Listing 10.17.

Equivalent nodal representation of this force for both the contactor and target tetrahedra is then obtained. Nodal forces for the contactor tetrahedron are shown in Listing 10.18. The distributed contact force acting on the contactor tetrahedron is the surface traction force acting on the current surface of the contactor tetrahedron. Thus, only the tree nodes defining this surface share the contact force, while the fourth node of the tetrahedron has

```
.........continued from previous listing.............
V3DCro(xnt,ynt,znt,xt[1]-xt[0],yt[1]-yt[0],zt[1]-zt[0],
        xt[2]-xt[0],yt[2]-yt[0],zt[2]-zt[0]);
V3DDot(theigh,xt[3]-xt[0],yt[3]-yt[0],zt[3]-zt[0],xnt,ynt,znt);
/* penetration at origin of the e-base and dp/du dp/dv; */
V3DDot(peneto,xc[0]-xt[0],yc[0]-yt[0],zc[0]-zt[0],xnt,ynt,znt);
V3DDot(penetu,xe[0],ye[0],ze[0],xnt,ynt,znt);
V3DDot(penetv,xe[1],ye[1],ze[1],xnt,ynt,znt);
peneto=peneto/theigh;
penetu=penetu/theigh;
penetv=penetv/theigh;
for(i=0;i<nbpoin;i++)
{ penetb[i]=peneto+ub[i]*penetu+vb[i]*penetv;
}
```

Listing 10.16 Calculation of penetration (contact overlap) at B-points.

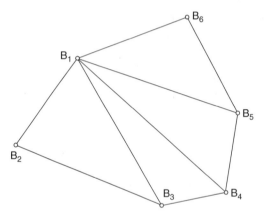

Figure 10.3 Integration of contact force over the intersection polygon.

no contact force assigned to it. The nodes that define the contactor surface are called C-points. There are three such nodes, thus the loop goes from zero to three.

Nodal forces for the target tetrahedron are shown in Listing 10.19. The contact force on the target tetrahedron is a volumetric potential force. This force is due to the contact force potential defined over the target tetrahedron. Thus, this force is first divided over the four nodes of the current sub-tetrahedron. Three of these nodes are also nodes of the target tetrahedron, while the fourth node is the central node of the target tetrahedron. The contact force assigned to this central node is therefore divided to all four nodes of the target tetrahedron–one quarter to each node. That is where the constant RP25=0.25=1/4 comes from.

Adding forces to the global vector is shown in Listing 10.20. The topology of both the contactor and target tetrahedra is used to mach nodes of the tetrahedra with global force vectors d1ncfx, d1ncfy, d1ncfz. It is worth noting that these vectors are one-

```
.........continued from previous listing.............
forco=R0; uforc=R0; vforc=R0; /* force and center of force */
for(i=1;i<(nbpoin-1);i++)
{ penetr=penetb[0]+penetb[i]+penetb[i+1];
  if(penetr>EPSILON)
  { force=((ub[i]-ub[0])*(vb[i+1]-vb[0])-
         (vb[i]-vb[0])*(ub[i+1]-ub[0]))*penetr*penalty;
    fact0=(RP5*penetb[0]+
            RP25*(penetb[i]+penetb[i+1])) / penetr;
    facti=(RP5*penetb[i]+
           RP25*(penetb[0]+penetb[i+1])) / penetr;
    fact1=R1-fact0-facti;
    if(ABS(force+forco)>EPSILON)
    { uforc=(forco*uforc+force*
        (fact0*ub[0]+facti*ub[i]+fact1*ub[i+1])) / (forco+force);
      vforc=(forco*vforc+force*
        (fact0*vb[0]+facti*vb[i]+fact1*vb[i+1])) / (forco+force);
      forco=forco+force;
    } } }
```

Listing 10.17   Integration of contact force over the B-polygon.

```
.........continued from previous listing.............
/*          resultant at C-points */
for(i=0;i<4;i++)
{ fc[i]=R0; ft[i]=R0;
}

tmp=((uc[1]-uc[0])*(vc[2]-vc[0])-(vc[1]-vc[0])*(uc[2]-uc[0]));
for(i=0;i<3;i++)
{ j=i+1; if(j>2)j=0; k=j+1; if(k>2)k=0;
  fc[k]=forco*
 (((uc[j]-uc[i])*(vforc-vc[i])-(vc[j]-vc[i])*(uforc-uc[i])) / tmp);
}
```

Listing 10.18   Calculating resultant force at C-points.

dimensional arrays forming a part of a single two-dimensional array and using dynamic memory allocation (see Section 10.2).

10.5   SORTING CONTACT DETECTION ALGORITHM

As explained in Chapter 3, the sorting contact detection algorithm is based on sorting one-dimensional arrays. A general function for sorting such arrays is given in Listings 10.21 and 10.22. In Listing 10.21, local variables are given, while in Listing 10.22, the code performing actual sorting is given. The sorting follows procedure, which is described

.........continued from previous listing.............

```
/*         resultant at T-points */
tmp=((ut[1]-ut[0])*(vt[2]-vt[0])-(vt[1]-vt[0])*(ut[2]-ut[0]));

inext=-1;
if(ABS(tmp)<RP1*theigh)
{ inext=0; tmp=ABS(ut[1]-ut[0])+ABS(vt[1]-vt[0]);
  for(i=0;i<3;i++)
{ j=i+1;
   if(j>2)j=0;
   if(tmp>(ABS(ut[j]-ut[i])+ABS(vt[j]-vt[i])))
   { tmp=ABS(ut[j]-ut[i])+ABS(vt[j]-vt[i]);  inext=i;
}}

j=inext+1;
if(j>2)j=0;
if(ABS(zt[j])>ABS(zt[inext]))inext=j;
j=inext+1; if(j>2)j=0;  k=j+1;  if(k>2)k=0;
tmp=(ut[k]-ut[j])*(vt[3]-vt[j])-(vt[k]-vt[j])*(ut[3]-ut[j]);
}

for(jnext=0;jnext<3;jnext++)
{ i=jnext;
  j=i+1;
  if(j>2)j=0;
  k=j+1;
  if(k>2)k=0;
  if(i==inext)i=3; if(j==inext)j=3; if(k==inext)k=3;
  ft[k]=forco*
  ((((ut[j]-ut[i])*(vforc-vt[i])-(vt[j]-vt[i])*(uforc-ut[i]))/tmp);
}

ft[3]=RP25*ft[3];
for(i=0;i<3;i++)
{ ft[i]=ft[i]+ft[3];
}
```

Listing 10.19 Calculating resultant force at T-points, i.e. nodes of the target tetrahedron.

in detail in Chapter 3. Arrays are sorted by repeated swapping of couples of numbers that match the swapping criteria.

10.6 NBS CONTACT DETECTION ALGORITHM IN 3D

The list of local variables used for the NBS contact detection algorithm in 3D is shown in Listing 10.23. The input for the NBS contact detection algorithm is comprised of arrays i1eccx, i1eccy and i1eccz. These arrays contain integerised coordinates x,y and z for each discrete element.

.........continued from previous listing.............

```
/* add forces into global vector */
for(i=0;i<4;i++)
{ d1ncfx[ipc[i]]=d1ncfx[ipc[i]]+fc[i]*xnc;
  d1ncfy[ipc[i]]=d1ncfy[ipc[i]]+fc[i]*ync;
  d1ncfz[ipc[i]]=d1ncfz[ipc[i]]+fc[i]*znc;
  d1ncfx[ipt[i]]=d1ncfx[ipt[i]]-ft[i]*xnc;
  d1ncfy[ipt[i]]=d1ncfy[ipt[i]]-ft[i]*ync;
  d1ncfz[ipt[i]]=d1ncfz[ipt[i]]-ft[i]*znc;
}}}}
```

Listing 10.20   Adding forces to the global force vector.

```
        void TsortINT(n,nsort,nrear,i2)
/* Sorts nsort arrays   - smallest...largest */
/* and rearrange nrear-nsort arrays same way  */
  INT n; INT nsort; INT nrear; INT **i2;
{ INT iblock,i,j,k,isort,tmp,m;
  INT *il, *ilr;
  INT ibig[50];
  INT iend[50];
  ibig[0]=0;
  iend[0]=n-1;
  iblock=0;
```

Listing 10.21   Sorting a set of one-dimensional integer arrays– local variables.

First the space boundaries are calculated using a loop over all discrete elements, as shown in Listing 10.24. The second step is a calculation of total number of cells in the x, y and z direction (ncelx, ncely, ncelz) (Listing 10.25).

Once the required number of cells in each direction is known, dynamic memory allocation is performed and dynamic arrays are created (Listing 10.26).

A one-dimensional integer array i1cfz[ncelz] is allocated to store the first element (head) of each of the lists containing discrete elements mapped onto the same layer of cells (z direction).

Two one-dimensional arrays, i2cfy[0] and i2cfy[1], are allocated for storing the heads of y-lists for layers (iz-1) and iz, respectively (where iz is the current central layer). Two one-dimensional arrays, i2cfx[0] and i2cfx[1], are allocated to store the heads of x-lists. Array i2cfx[0] stores the heads of all x-lists for the central row iy; array i2cfx[1] stores the heads of all x-lists containing all the discrete elements from all y-rows that are neighbouring rows of the central row iy according to the contact map. Arrays i1cnz, i1cny and i1cnx are one-dimensional arrays used to store, for each discrete element, the next discrete element that is in the same z-list, y-list or x-list, respectively. Listing 10.27 represents initialisation, where all lists are initialised in such a way that they represent empty lists. The first loop in the listing makes all z-lists empty lists. The second loop makes all lists of both sets of y-lists empty. The third loop makes all lists of both sets of x-lists empty.

In Listing 10.28, all discrete elements are placed onto the corresponding z-list according to the z-coordinates of each discrete element. This is done using the loop over all discrete

```
while(iblock>=0)
{ i=ibig[iblock];
  j=iend[iblock];
  iblock=iblock-1;
  if(j>i)
  { tmp=0; m=0; isort=0;
    while((isort<nsort)&&(m==tmp))
    { i1=i2[isort];
      isort=isort+1;
      tmp=i1[j];
      m=tmp;
      for(k=i;k<j;k++)
      { tmp=MINIM(tmp,i1[k]);
        m=MAXIM(m,i1[k]);
    }}
    if(tmp!=m)
    { m=(tmp+m)/2;
      while(i<=j)
      { while((i<=j)&&(i1[i]<=m)) {i=i+1;}
        while((j>=i)&&(i1[j]>m)) {j=j-1;}
        if(j>i)
        { for(isort=0;isort<nrear;isort++)
          { i1r=i2[isort];
            tmp=i1r[j];
            i1r[j]=i1r[i];
            i1r[i]=tmp;
      }}}
      ibig[iblock+2]=ibig[iblock+1];
      iend[iblock+2]=j;
      ibig[iblock+1]=i;
      iblock=iblock+2;
}}}}
```

Listing 10.22   Sorting a set of one-dimensional integer arrays – sorting procedure.

elements. All z-lists are at this stage considered to be 'new z-lists'. Many of these 'new' z-lists are empty, i.e. lists with no discrete elements assigned to them. By default, the head of each empty list is -1, which indicates non-existing discrete elements. All non-empty lists have heads greater than or equal to zero. This is because in C, numbering starts with zero.

To identify non-empty z-lists, a loop over all discrete elements is performed. As each discrete element belongs to one and only one list, the list to which a particular discrete element belongs is, by default, a non-empty list. This list is identified by the integerised z coordinate of the particular discrete element. However, the list may contain more than one discrete element, thus the same list would be visited as many times as there are discrete elements in a particular list. To avoid this, as soon as the 'new' list is detected, it is marked as an old list, as shown in Listing 10.29. In that way, when another discrete element from the same list points to this list, it will be found to be an 'old' list, and will therefore be ignored. Marking a list as an old list is done by setting i1cfz[iz]=i1cfz[iz]+nelemd;,

```
INT ncelx,ncely,ncelz; /* total number of x, y, z cells  */
INT nelemd;         /* twice total number of elements */
INT ielem;        /* element              */
INT ielemx;        /* element assigned to cell x  */
INT ielemy;        /* element assigned to cell y  */
INT ielemz;        /* element assigned to cell z  */
INT ihx,ihy,ihz;     /* x, y, z  head of a list */
INT iminx,iminy,iminz;   /* space boundaries         */
INT imaxx,imaxy,imaxz;    /* space boundaries         */
INT inod,jnod;       /* nodes              */
INT ix,iy,iz;     /* x, y, z cell     */
INT jelemx;        /* element assigned to cell x   */


INT *i1cnx;   /* contactor next  x      */
INT *i1cny;   /* contactor next  y      */
INT *i1cnz;   /* contactor next  z      */
INT *i1cfz;   /* contactor first z      */


INT *i1eccx;   /* element coordinate current x      */
INT *i1eccy;   /* element coordinate current y      */
INT *i1eccz;   /* element coordinate current z      */
INT  i1heax[5]; /* heads of 5 connected lists for x cells */
INT  i1heay[5]; /* heads of 5 connected lists for y cells */
INT  i1heaz[2]; /* heads of 2 connected lists for z cells */
INT *i2cfx[2]; /* contactor first x       */
INT *i2cfy[2]; /* contactor first y       */
```

Listing 10.23 A list of variables used in the C implementation of the NBS contact detection algorithm in 3D.


```
                 .........continued from previous listing.............
 /* find space boundaries */
      iminx=i1eccx[0];
      imaxx=i1eccx[0];
      iminy=i1eccy[0];
      imaxy=i1eccy[0];
      iminz=i1eccz[0];
      imaxz=i1eccz[0];
      for(ielem=1;ielem<nelem;ielem++)
      { iminx=MINIM(iminx,i1eccx[ielem]);
       imaxx=MAXIM(imaxx,i1eccx[ielem]);
       iminy=MINIM(iminy,i1eccy[ielem]);
       imaxy=MAXIM(imaxy,i1eccy[ielem]);
       iminz=MINIM(iminz,i1eccz[ielem]);
       imaxz=MAXIM(imaxz,i1eccz[ielem]);
      }
      iminx=iminx-1;
      iminy=iminy-1;
      iminz=iminz-1;
      imaxx=imaxx+2;
      imaxy=imaxy+2;
      imaxz=imaxz+2;
```

Listing 10.24 Space boundaries.

```
.........continued from previous listing.............
/* normalise coordinates */
for(ielem=0;ielem<nelem;ielem++)
{ i1eccx[ielem]=i1eccx[ielem]-iminx;
  i1eccy[ielem]=i1eccy[ielem]-iminy;
  i1eccz[ielem]=i1eccz[ielem]-iminz;
}
ncelx=imaxx-iminx;
ncely=imaxy-iminy;
ncelz=imaxz-iminz;
```

Listing 10.25   Total number of cells in the x, y and z direction.

```
.........continued from previous listing...........
/* allocate memory */
i1cfz=TallNT1(ncelz);   /* contactor first z        */
i2cfy[0]=TallNT1(ncely); /* contactor first y  (iz-1)  */
i2cfy[1]=TallNT1(ncely); /* contactor first y  (iz  )   */
i2cfx[0]=TallNT1(ncelx); /* contactor first x  (iz-1,iy-1)  */
i2cfx[1]=TallNT1(ncelx); /* contactor first x  (iz-1,iy )   */
i1cnz=TallNT1(nelem);     /* contactor next z        */
i1cny=TallNT1(nelem);     /* contactor next y        */
i1cnx=TallNT1(nelem);     /* contactor next x        */
```

Listing 10.26   Dynamic memory allocation.

```
.........continued from previous listing.............
/* assume no contactors at any cell */
for(iz=0;iz<ncelz;iz++)
{ i1cfz[iz]=-1;
}
for(ihy=0;ihy<2;ihy++)
{ for(iy=0;iy<ncely;iy++)
  { i2cfy[ihy][iy]=-1;
} }
for(ihx=0;ihx<2;ihx++)
{ for(ix=0;ix<ncelx;ix++)
  { i2cfx[ihx][ix]=-1;
} }
```

Listing 10.27   Initialisation.

where nelemd=2*nelem. Thus, even after marking the head of a list as an 'old' list, the list can still be accessed, and discrete elements from such a list are still available.

The criterion for a new list is therefore if(i1cfz[iz]<nelem). The new list is characterised by the specific integerised coordinate z, and therefore represents a specific layer iz of cells. This layer is called the central layer, as explained in Chapter 3.

In Listing 10.30, all the discrete elements from the central layer iz and the layer immediately below it (layer iz-1) are loaded onto y-lists representing rows of cells. Two separate

```
.........continued from previous listing.............
/* assign all contactors to z-cells */
for(ielem=0;ielem<nelem;ielem++)
{ if(d1erad[ielem]>R0)
  { i1cnz[ielem]=i1cfz[i1eccz[ielem]];
    i1cfz[i1eccz[ielem]]=ielem;
  } }
```

Listing 10.28 All elements are put onto simply connected z-lists, and all lists are marked new.

```
.........continued from previous listing.............
/* scan all loaded z cells */
for(ielem=0;ielem<nelem;ielem++)
{ iz=i1eccz[ielem];
  if(i1cfz[iz]<nelem)
  { i1heaz[0]=i1cfz[iz];
    i1heaz[1]=i1cfz[iz-1];
    if(i1heaz[1]>nelem)i1heaz[1]=i1heaz[1]-nelemd;
    i1cfz[iz]=i1cfz[iz]+nelemd;
```

Listing 10.29 All z-lists are searched for a new list, which is then marked as an old list by increasing the first element in the list by nelemd=2*nelem.

```
.........continued from previous listing.............
/* load elements from cells iz & iz-1  onto y cells */
for(ihz=0;ihz<2;ihz++)
{ ielemz=i1heaz[ihz];
  while(ielemz>=0)
  { i1cny[ielemz]=i2cfy[ihz][(i1eccy[ielemz])];
    i2cfy[ihz][(i1eccy[ielemz])]=ielemz;
    ielemz=i1cnz[ielemz];
  } }
```

Listing 10.30 Elements from a new z-list are loaded onto y-lists.

sets of y-lists are used: the first set i2cfy[0] contains all lists for discrete elements from the central layer iz; the second set i2cfy[1] contains all y-lists containing all discrete elements from the layer iz-1.

All y-lists are, at this stage, considered to be 'new y-lists'. Some of these y-lists are empty, thus a search for non-empty y-lists is performed, as shown in Listing 10.31. This is done by considering all discrete elements from iz-list, i.e. all discrete elements from the central layer of cells. First, a discrete element is therefore the head of this list i1heaz[0], which in Listing 10.29 was set to i1heaz[0]=i1cfz[iz]. Once a 'new' iy-list is found, row iy is called the central row, and the y-list corresponding to it is marked as 'old' by setting i2cfy[0][iy]=i2cfy[0][iy]+nelemd. The head of the y-list representing the central row is stored as i1heay[0], while the heads of y-lists representing neighbouring rows are stored as i1heay[1], i1heay[2], i1heay[3] and i1heay[4].

All discrete elements from the central row iy are loaded onto x-lists, as shown in Listing 10.32. This is done by looping over discrete elements from the y-lists

```
.........continued from previous listing...........
/* scan all loaded y cells */
ielemz=i1heaz[0];
while(ielemz>=0)
{ iy=i1eccy[ielemz];
  if(i2cfy[0][iy]<nelem)
  { i1heay[0]=i2cfy[0][iy];
    i1heay[1]=i2cfy[0][iy-1];
    i1heay[2]=i2cfy[1][iy+1];
    i1heay[3]=i2cfy[1][iy];
    i1heay[4]=i2cfy[1][iy-1];

    if(i1heay[1]>nelem)i1heay[1]=i1heay[1]-nelemd;
  i2cfy[0][iy]=i2cfy[0][iy]+nelemd;
```

Listing 10.31   All y-lists are searched to find a 'new' y-list, which is then marked as 'old' be setting i2cfy[0][iy]=i2cfy[0][iy]+nelemd;.

```
.........continued from previous listing........
/* load elements from y cells onto x cells */
ihx=0;
for(ihy=0;ihy<5;ihy++)
{ if(ihy>0)ihx=1;
  ielemy=i1heay[ihy];
  while(ielemy>=0)
  { i1cnx[ielemy]=i2cfx[ihx][(i1eccx[ielemy])];
    i2cfx[ihx][(i1eccx[ielemy])]=ielemy;
    ielemy=i1cny[ielemy];
} }
```

Listing 10.32   All elements from the 'new' y-list and neighbouring y-lists are loaded onto x-lists.

representing the central row and neighbouring rows according to the contact map. The heads of these lists are i1heay[0], i1heay[1], i1heay[2], i1heay[3], i1heay[4]; thus the loop for(ihy=0;ihy<5;ihy++). Discrete elements from the list i1heay[0] (central row) are placed onto the first set of x-lists (set of x-lists i2cfx[0]), while discrete elements from neighbouring rows are together placed onto the second set of x-lists (set of x-lists i2cfx[1]).

At this stage, all x-lists created are considered to be 'new x-lists'. However, all x-lists do not necessarily have any discrete elements assigned to them, i.e. some of the x-lists are empty lists. Thus, a search for non-empty x-lists is performed (Listing 10.33).

The search for new x-lists is done by looping over all discrete elements from the central row of cells iy, starting with the i1heay[0] discrete element. For a particular discrete element, the x-list to which that discrete element belongs is identified by the integerised coordinate x of that discrete element. For a particular ix, the corresponding ix-list is 'new' if (i2cfx[0][ix]<nelem). Such an ix represents the central cell, and the list i2cfx[0][ix] is therefore called the 'central x-list'. So as not to visit this list again with another discrete element from the central row, this list is immediately marked as an 'old x-list' by setting i2cfx[0][ix]= i2cfx[0][ix]+nelemd. The head of the central x-list is conveniently stored as

```
........continued from previous listing.............
/* scan all loaded x cells */
ielemy=i1heay[0];
while(ielemy>=0)
{ ix=i1eccx[ielemy];
  if(i2cfx[0][ix]<nelem)
  { i1heax[0]=i2cfx[0][ix];
    i1heax[1]=i2cfx[0][ix-1];
    i1heax[2]=i2cfx[1][ix+1];
    i1heax[3]=i2cfx[1][ix];
    i1heax[4]=i2cfx[1][ix-1];
    if(i1heax[1]>nelem)i1heax[1]=i1heax[1]-nelemd;
    i2cfx[0][ix]=i2cfx[0][ix]+nelemd;
```

Listing 10.33 All x-lists from the first set of lists are searched for an x-list marked 'new'.

i1heax[0], while the heads of x-lists corresponding to the neighbouring cells according to the contact map are temporarily stored as i1heax[1], i1heax[2], i1heax[3] and i1heax[4].

The cell (ix,iy,iz) is currently the central cell. All discrete elements mapped to this cell are on the i1heax[0] list. All discrete elements mapped to the neighbouring cells according to the contact mask are on the lists as i1heax[1], i1heax[2], i1heax[3] and i1heax[4]. Thus, each discrete element from the list i1heax[0] may be in contact with some of the discrete elements from lists i1heax[0], i1heax[1], i1heax[2], i1heax[3] and i1heax[4]. In other words, some contacts have been detected, and contact interaction processing needs to take place as shown in Listing 10.34. Once the contact interaction processing has been accomplished, the search for 'new x-lists' continues, thus ielemy=i1cny[ielemy]. In this way, another central (ix, iy, iz) cell is detected with its central x-list and neighbouring x-lists.

```
......continued from previous listing...........
/* detect contacts for cell (ix,iy,iz) */
        ielemx=i1heax[0];
        while(ielemx>=0)
        { for(ihx=0;ihx<5;ihx++)
          { jelemx=i1heax[ihx];
            while(jelemx>=0)
            { if((ihx!=0)||(ielemx>jelemx))
              {    CONTACT HAS BEEN DETECTED
                   RECORD IT, OR PROCESS CONTACT
                   INTERACTION
              }
              jelemx=i1cnx[jelemx];
            } }
            ielemx=i1cnx[ielemx];
        } }
        ielemy=i1cny[ielemy];
        }
```

Listing 10.34 All the elements from the 'new' x-list are checked against all elements from neighbouring x-lists according to the contact map.

At some point the ielemy=i1cny[ielemy]=-1 will be obtained, i.e. the end of the central y-list will have been reached. By that time, all x-lists will have been visited and will have become 'old x-lists'. There is no use for these lists any longer, and they have to be 'emptied', i.e. turned into empty lists. This is done by 'unloading' discrete elements from the x-lists, as shown in Listing 10.35.

A loop over discrete elements from the central row iy and neighbouring rows according to the contact mask is performed. For each discrete element, the corresponding x-list according to the integerised x coordinate is set to an empty list. This is done by simply setting i2cfx[ihx][(i1eccx[ielemy])]=-1.

After completion of these processes, no non-empty x-list is left. Thus the search for non-empty y-lists is continued by setting ielemz=i1cnz[ielemz], and another non-empty 'new y-list' is discovered and called the central y-list, while the row that list corresponds to is called the 'central row'. For this central row, x-lists are assembled, a search for new x-lists is performed, the central cell is located and contact detected, and contact interaction is performed. The process is repeated until ielemz becomes equal to -1. At that point, the end of the central z-list is reached and all iy-lists have become old y-lists, i.e. there is no 'new y-list' left. At this stage there is no use for these y-lists any longer, and they must be turned into empty y-lists. This is done by looping over all discrete elements from the central layer and neighbouring central layer, and for each discrete element setting the corresponding y-list (according to the integerised y coordinate of the discrete element) to an empty list by assigning -1 as the list head (Listing 10.35).

There is no need to empty z-lists, because by the time the last z-list has been processed, contact detection has reached completion. Thus, dynamically allocated memory is returned to the operating system, in opposite order of allocation, to avoid memory fragmentation (Listing 10.36).

```
.........continued from previous listing.............
/* unload elements from x cells */
ihx=0;
for(ihy=0;ihy<5;ihy++)
{ if(ihy>0)ihx=1;
  ielemy=i1heay[ihy];
  while(ielemy>=0)
  { i2cfx[ihx][(i1eccx[ielemy])]=-1;
    ielemy=i1cny[ielemy];
}}}
  ielemz=i1cnz[ielemz];
}
/* unload elements from y cells */
for(ihz=0;ihz<2;ihz++)
{ ielemz=i1heaz[ihz];
  while(ielemz>=0)
  { i2cfy[ihz][(i1eccy[ielemz])]=-1;
    ielemz=i1cnz[ielemz];
}}}}
```

Listing 10.35 Elements are removed from both the x-lists and y-lists.

.........continued from previous listing.............
```
                /* free memory */
        FREE(i1cnx);
        FREE(i1cny);
        FREE(i1cnz);
        FREE(i2cfx[1]);
        FREE(i2cfx[0]);
        FREE(i2cfy[1]);
        FREE(i2cfy[0]);
        FREE(i1cfz);
    } } }
```

Listing 10.36   Dynamically allocated memory is returned in reverse order to avoid any memory fragmentation.

10.7   DEFORMABILITY WITH FINITE ROTATIONS IN 3D

The basic concepts of finite displacement, finite rotation and finite strain deformability can be found in Chapter 4. Finite rotation elasticity using a constant strain, four-noded tetrahedron finite element is described in detail in Chapter 4. Implementation of the algorithmic procedures described in Chapter 4 is given in the form of code listings in this section. The notation employed is shown in Figure 10.4, while the local variables used to process this element are shown in Listing 10.37.

In Listing 10.38, the global components of the base vectors for the frames of reference are calculated. First the global components of the base vectors of the initial frame (see Figure 10.4) are calculated. These components are as follows:

$$\begin{bmatrix} \hat{i}_x & \hat{j}_x & \hat{k}_x \\ \hat{i}_y & \hat{j}_y & \hat{k}_y \\ \hat{i}_z & \hat{j}_z & \hat{k}_z \end{bmatrix} \tag{10.1}$$

and are represented by the two-dimensional array FO[3][3].

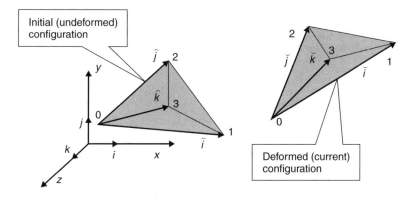

Figure 10.4   Global, initial and deformed initial frames of reference used in processing finite-rotation elasticity.

```
{ DBL nx,ny,nz,voli,volc;
  DBL B[3][3]; /* left Cauchy-Green strain tensor */
  DBL D[3][3]; /* rate of deformation (stretching) tensor */
  DBL E[3][3]; /* strain tensor (small strains) */
  DBL F[3][3];
  /* deformation gradient in global base delta ux/delta x */
  DBL F0[3][3]; /* initial local base */
  DBL FX[3][3]; /* current local base  also delta ux/delta X */
  DBL F0inv[3][3]; /* global base in initial local base */
  DBL FXinv[3][3]; /* global base in current local base */
  DBL L[3][3];
  /* velocity gradient in global base  delta vx/delta x   */
  DBL LX[3][3];
  /* velocity gradient in current local base = delta x/delta X */
  DBL T[3][3]; /* Cauchy stress */
  INT ielem;
  INT i,j,k,l;
```

Listing 10.37   Local variables used in processing deformability with finite rotation for a four-noded solid tetrahedron finite element.

```
              .........continued from previous listing.............
  for(ielem=0;ielem<nelem;ielem++)
  {   for(i=1;i<4;i++)
      { /* initial base */
        F0[0][i-1]=d1ncix[(i2elto[i][ielem])]-d1ncix[(i2elto[0][ielem])];
        F0[1][i-1]=d1nciy[(i2elto[i][ielem])]-d1nciy[(i2elto[0][ielem])];
        F0[2][i-1]=d1nciz[(i2elto[i][ielem])]-d1nciz[(i2elto[0][ielem])];
        /* current base */
        FX[0][i-1]=d1nccx[(i2elto[i][ielem])]-
                        d1nccx[(i2elto[0][ielem])];
        FX[1][i-1]=d1nccy[(i2elto[i][ielem])]-
                        d1nccy[(i2elto[0][ielem])];
        FX[2][i-1]=d1nccz[(i2elto[i][ielem])]-d1nccz[(i2elto[0][ielem])];
        /* velocity gradient */
        LX[0][i-1]=d1nvcx[(i2elto[i][ielem])]-
                        d1nvcx[(i2elto[0][ielem])];
        LX[1][i-1]=d1nvcy[(i2elto[i][ielem])]-d1nvcy[(i2elto[0][ielem])];
        LX[2][i-1]=d1nvcz[(i2elto[i][ielem])]-d1nvcz[(i2elto[0][ielem])];
        /* global base in initial local coordinates   */
        YMATINV3(F0,F0inv,voli);
        /* global base in current local coordinates   *
        /YMATINV3(FX,FXinv,volc);

      }
```

Listing 10.38   Global components of the base vectors of initial and deformed initial frames.

The global components of the base vectors of the deformed initial frame, as shown in Figure 10.4.

$$
\begin{bmatrix}
\check{i}_x & \check{j}_x & \check{k}_x \\
\check{i}_y & \check{j}_y & \check{k}_y \\
\check{i}_z & \check{j}_z & \check{k}_z
\end{bmatrix}
\tag{10.2}
$$

are represented by a two-dimensional array FX[3][3].

At the end of Listing 10.38, the matrix of the tensor of velocity gradient

$$
\mathbf{L} =
\begin{bmatrix}
\dfrac{\partial v_{xc}}{\partial \hat{x}_i} & \dfrac{\partial v_{xc}}{\partial \hat{y}_i} & \dfrac{\partial v_{xc}}{\partial \hat{z}_i} \\[2mm]
\dfrac{\partial v_{yc}}{\partial \hat{x}_i} & \dfrac{\partial v_{yc}}{\partial \hat{y}_i} & \dfrac{\partial v_{yc}}{\partial \hat{z}_i} \\[2mm]
\dfrac{\partial v_{zc}}{\partial \hat{x}_i} & \dfrac{\partial v_{zc}}{\partial \hat{y}_i} & \dfrac{\partial v_{zc}}{\partial \hat{z}_i}
\end{bmatrix}
=
\begin{bmatrix}
v_{1xc} - v_{0xc} & v_{2xc} - v_{0xc} & v_{3xc} - v_{0xc} \\
v_{1yc} - v_{0yc} & v_{2yc} - v_{0yc} & v_{3yc} - v_{0yc} \\
v_{1zc} - v_{0zc} & v_{2zc} - v_{0zc} & v_{3zc} - v_{0zc}
\end{bmatrix}
\tag{10.3}
$$

is also calculated and stored as the two-dimensional array L[3][3].

In Listing 10.38, the components of the global base vectors are also expressed in terms of the base vectors of the initial frame. These components are as follows:

$$
\begin{bmatrix} \mathbf{i} & \mathbf{j} & \mathbf{k} \end{bmatrix} =
\begin{bmatrix}
i_{\hat{x}} & j_{\hat{x}} & k_{\hat{x}} \\
i_{\hat{y}} & j_{\hat{y}} & k_{\hat{y}} \\
i_{\hat{z}} & j_{\hat{z}} & k_{\hat{z}}
\end{bmatrix}
=
\begin{bmatrix}
\hat{i}_x & \hat{j}_x & \hat{k}_x \\
\hat{i}_y & \hat{j}_y & \hat{k}_y \\
\hat{i}_z & \hat{j}_z & \hat{k}_z
\end{bmatrix}^{-1}
\tag{10.4}
$$

The matrix

$$
\begin{bmatrix}
i_{\hat{x}} & j_{\hat{x}} & k_{\hat{x}} \\
i_{\hat{y}} & j_{\hat{y}} & k_{\hat{y}} \\
i_{\hat{z}} & j_{\hat{z}} & k_{\hat{z}}
\end{bmatrix}
\tag{10.5}
$$

is stored as the two-dimensional array FOinv[3][3]. In a similar way, the components of the global base vectors are expressed in terms of the base vectors of the deformed initial frame. These components are as follows:

$$
\begin{bmatrix} \mathbf{i} & \mathbf{j} & \mathbf{k} \end{bmatrix} =
\begin{bmatrix}
i_{\check{x}} & j_{\check{x}} & k_{\check{x}} \\
i_{\check{y}} & j_{\check{y}} & k_{\check{y}} \\
i_{\check{z}} & j_{\check{z}} & k_{\check{z}}
\end{bmatrix}
=
\begin{bmatrix}
\check{i}_x & \check{j}_x & \check{k}_x \\
\check{i}_y & \check{j}_y & \check{k}_y \\
\check{i}_z & \check{j}_z & \check{k}_z
\end{bmatrix}^{-1}
\tag{10.6}
$$

The matrix

$$
\begin{bmatrix}
i_{\check{x}} & j_{\check{x}} & k_{\check{x}} \\
i_{\check{y}} & j_{\check{y}} & k_{\check{y}} \\
i_{\check{z}} & j_{\check{z}} & k_{\check{z}}
\end{bmatrix}
\tag{10.7}
$$

is represented by the two-dimensional array FXinv[3][3]. Both FOinv[3][3 and FXinv[3][3 are calculated using the MACRO YMATINV3, which is presented in the file frame.h. It takes matrix FO as input and returns the inverse matrix FOinv, together with the determinant voli of the matrix FO. In a similar way, this macro takes matrix FX and returns its determinant volc, together with the inverse matrix FXinv. It is worth mentioning that the ratio volc/voli represents volumetric stretch, i.e. the ratio between the volume of the deformed tetrahedron and the volume of the initial (nondeformed) tetrahedron.

The array FX[3][3] also represents the matrix of the deformation gradient:

$$
\mathbf{F} = \begin{bmatrix}
\dfrac{\partial x_c}{\partial \widehat{x}_i} & \dfrac{\partial x_c}{\partial \widehat{y}_i} & \dfrac{\partial x_c}{\partial \widehat{z}_i} \\[1.5em]
\dfrac{\partial y_c}{\partial \widehat{x}_i} & \dfrac{\partial y_c}{\partial \widehat{y}_i} & \dfrac{\partial y_c}{\partial \widehat{z}_i} \\[1.5em]
\dfrac{\partial z_c}{\partial \widehat{x}_i} & \dfrac{\partial z_c}{\partial \widehat{y}_i} & \dfrac{\partial z_c}{\partial \widehat{z}_i}
\end{bmatrix}
\tag{10.8}
$$

This matrix represents the deformation gradient tensor using the base vectors of the initial frame (undeformed tetrahedron). The matrix of the deformation gradient that uses the base vectors of the global frame is obtained as follows:

$$
\mathbf{F} = \begin{bmatrix}
\dfrac{\partial x_c}{\partial x_i} & \dfrac{\partial x_c}{\partial y_i} & \dfrac{\partial x_c}{\partial z_i} \\[1.5em]
\dfrac{\partial y_c}{\partial x_i} & \dfrac{\partial y_c}{\partial y_i} & \dfrac{\partial y_c}{\partial z_i} \\[1.5em]
\dfrac{\partial z_c}{\partial x_i} & \dfrac{\partial z_c}{\partial y_i} & \dfrac{\partial z_c}{\partial z_i}
\end{bmatrix}
= \begin{bmatrix}
\dfrac{\partial x_c}{\partial \widehat{x}_i} & \dfrac{\partial x_c}{\partial \widehat{y}_i} & \dfrac{\partial x_c}{\partial \widehat{z}_i} \\[1.5em]
\dfrac{\partial y_c}{\partial \widehat{x}_i} & \dfrac{\partial y_c}{\partial \widehat{y}_i} & \dfrac{\partial y_c}{\partial \widehat{z}_i} \\[1.5em]
\dfrac{\partial z_c}{\partial \widehat{x}_i} & \dfrac{\partial z_c}{\partial \widehat{y}_i} & \dfrac{\partial z_c}{\partial \widehat{z}_i}
\end{bmatrix}
\begin{bmatrix}
i_{\widehat{x}} & j_{\widehat{x}} & k_{\widehat{x}} \\
i_{\widehat{y}} & j_{\widehat{y}} & k_{\widehat{y}} \\
i_{\widehat{z}} & j_{\widehat{z}} & k_{\widehat{z}}
\end{bmatrix}
\tag{10.9}
$$

These calculations are performed in Listing 10.39, where the deformation gradient matrix

$$
\begin{bmatrix}
\dfrac{\partial x_c}{\partial x_i} & \dfrac{\partial x_c}{\partial y_i} & \dfrac{\partial x_c}{\partial z_i} \\[1.5em]
\dfrac{\partial y_c}{\partial x_i} & \dfrac{\partial y_c}{\partial y_i} & \dfrac{\partial y_c}{\partial z_i} \\[1.5em]
\dfrac{\partial z_c}{\partial x_i} & \dfrac{\partial z_c}{\partial y_i} & \dfrac{\partial z_c}{\partial z_i}
\end{bmatrix}
\tag{10.10}
$$

is represented by the array F[3][3].

In Listing 10.39, the matrix of the velocity gradient tensor is also calculated using the following formula:

$$
\mathbf{L} = \begin{bmatrix}
\dfrac{\partial v_{xc}}{\partial x_i} & \dfrac{\partial v_{xc}}{\partial y_i} & \dfrac{\partial v_{xc}}{\partial z_i} \\[1.5em]
\dfrac{\partial v_{yc}}{\partial x_i} & \dfrac{\partial v_{yc}}{\partial y_i} & \dfrac{\partial v_{yc}}{\partial z_i} \\[1.5em]
\dfrac{\partial v_{zc}}{\partial x_i} & \dfrac{\partial v_{zc}}{\partial y_i} & \dfrac{\partial v_{zc}}{\partial z_i}
\end{bmatrix}
= \begin{bmatrix}
\dfrac{\partial v_{xc}}{\partial \widehat{x}_i} & \dfrac{\partial v_{xc}}{\partial \widehat{y}_i} & \dfrac{\partial v_{xc}}{\partial \widehat{z}_i} \\[1.5em]
\dfrac{\partial v_{yc}}{\partial \widehat{x}_i} & \dfrac{\partial v_{yc}}{\partial \widehat{y}_i} & \dfrac{\partial v_{yc}}{\partial \widehat{z}_i} \\[1.5em]
\dfrac{\partial v_{zc}}{\partial \widehat{x}_i} & \dfrac{\partial v_{zc}}{\partial \widehat{y}_i} & \dfrac{\partial v_{zc}}{\partial \widehat{z}_i}
\end{bmatrix}
\begin{bmatrix}
i_{\widehat{x}} & j_{\widehat{x}} & k_{\widehat{x}} \\
i_{\widehat{y}} & j_{\widehat{y}} & k_{\widehat{y}} \\
i_{\widehat{z}} & j_{\widehat{z}} & k_{\widehat{z}}
\end{bmatrix}
\tag{10.11}
$$

```
.............continued from previous listing.............
for(i=0;i<3;i++)
{ for(j=0;j<3;j++)
 { F[i][j]=R0;
   L[i][j]=R0;
   for(k=0;k<3;k++)
   { F[i][j]=F[i][j]+FX[i][k]*F0inv[k][j];/* deform. gradient */
     L[i][j]=L[i][j]+LX[i][k]*FXinv[k][j];/* velocity gradient */
}}}
```

Listing 10.39 Deformation and velocity gradients.

The matrix of the velocity gradient tensor

$$
\begin{bmatrix}
\dfrac{\partial v_{xc}}{\partial x_i} & \dfrac{\partial v_{xc}}{\partial y_i} & \dfrac{\partial v_{xc}}{\partial z_i} \\[2mm]
\dfrac{\partial v_{yc}}{\partial x_i} & \dfrac{\partial v_{yc}}{\partial y_i} & \dfrac{\partial v_{yc}}{\partial z_i} \\[2mm]
\dfrac{\partial v_{zc}}{\partial x_i} & \dfrac{\partial v_{zc}}{\partial y_i} & \dfrac{\partial v_{zc}}{\partial z_i}
\end{bmatrix}
\tag{10.12}
$$

is stored as the two-dimensional array L[3][3].

In Listing 10.40, strain tensors are evaluated. These are as follows:

- Left Cauchy–Green strain tensor. The matrix of this tensor is stored as two-dimensional array B[3][3].
- Green–St.Venant strain tensor. The matrix of this tensor is stored as array E[3][3].
- Rate of deformation tensor. The matrix of the rate of deformation tensor is stored using array D[3][3].

In Listing 10.41, the Cauchy stress tensor is calculated. This is achieved by employing the appropriate constitutive law, depending on the material. At this point, a MACRO resolving the stress-strain relationship is called. A detailed description of a set of physical equations for homogeneous isotropic elastic material is given in Chapter 4. However, any constitutive law can be employed at this point, including material nonlinearity such as plastic, plastic hardening or softening material.

```
.............continued from previous listing.............
for(i=0;i<3;i++)
{ for(j=0;j<3;j++)
  { B[i][j]=R0;
    for(k=0;k<3;k++)
    { B[i][j]=B[i][j]+F[i][k]*F[j][k]; /* left Cauchy-Green strain */
    }
    D[i][j]=RP5*(L[i][j]+L[j][i]); /* rate of deformation     */
    if(i==j)
    { E[i][j]=RP5*(B[i][j]-R1); /* Green-St.Venant strain   */
    }
    else
    { E[i][j]=RP5*B[i][j];
}}}
```

Listing 10.40   Strain calculation.

```
.........continued from previous listing...........
/* Cauchy stress */
{ ConstitutiveLaw(E,T,volc/voli);
}
```

Listing 10.41   Cauchy stress tensor.

```
...........continued from previous listing..............
for(i=0;i<4;i++)/* Nodal Forces */
{ j=i+1; if(j>3)j=0;
  k=j+1; if(k>3)k=0;
  l=k+1; if(l>3)l=0;
  nx=((d1nccy[(i2elto[k][ielem])]-d1nccy[(i2elto[j][ielem])])*
     (d1nccz[(i2elto[l][ielem])]-d1nccz[(i2elto[j][ielem])])-
     (d1nccy[(i2elto[l][ielem])]-d1nccy[(i2elto[j][ielem])])*
     (d1nccz[(i2elto[k][ielem])]-d1nccz[(i2elto[j][ielem])]))/R6;
  ny=((d1nccz[(i2elto[k][ielem])]-d1nccz[(i2elto[j][ielem])])*
     (d1nccx[(i2elto[l][ielem])]-d1nccx[(i2elto[j][ielem])])-
     (d1nccx[(i2elto[k][ielem])]-d1nccx[(i2elto[j][ielem])])*
     (d1nccz[(i2elto[l][ielem])]-d1nccz[(i2elto[j][ielem])]))/R6;
  nz=((d1nccx[(i2elto[k][ielem])]-d1nccx[(i2elto[j][ielem])])*
     (d1nccy[(i2elto[l][ielem])]-d1nccy[(i2elto[j][ielem])])-
     (d1nccy[(i2elto[k][ielem])]-d1nccy[(i2elto[j][ielem])])*
     (d1nccx[(i2elto[l][ielem])]-d1nccx[(i2elto[j][ielem])]))/R6;
  d1nmct[(i2elto[i][ielem])]=d1nmct[(i2elto[i][ielem])]+
                 dpero*voli/R6;
  if((i==0)||(i==2))
  { d1nfcx[(i2elto[i][ielem])]=d1nfcx[(i2elto[i][ielem])]+
               (T[0][0]*nx+T[0][1]*ny+T[0][2]*nz);
  d1nfcy[(i2elto[i][ielem])]=d1nfcy[(i2elto[i][ielem])]+
               (T[1][0]*nx+T[1][1]*ny+T[1][2]*nz);
  d1nfcz[(i2elto[i][ielem])]=d1nfcz[(i2elto[i][ielem])]+
               (T[2][0]*nx+T[2][1]*ny+T[2][2]*nz);
  }
  else
  { d1nfcx[(i2elto[i][ielem])]=d1nfcx[(i2elto[i][ielem])]-
               (T[0][0]*nx+T[0][1]*ny+T[0][2]*nz);
  d1nfcy[(i2elto[i][ielem])]=d1nfcy[(i2elto[i][ielem])]-
               (T[1][0]*nx+T[1][1]*ny+T[1][2]*nz);
  d1nfcz[(i2elto[i][ielem])]=d1nfcz[(i2elto[i][ielem])]-
               (T[2][0]*nx+T[2][1]*ny+T[2][2]*nz);
}}}}}
```

Listing 10.42 Calculation of nodal forces.

In Listing 10.42, the nodal forces are calculated using surface tractions, as explained in Chapter 4. The surface traction force on each of the surfaces of the tetrahedron is replaced by equivalent forces on three nodes belonging to the particular surface. One-third of the traction force is applied to each node. Surface normals are obtained as a cross product of the corresponding edges (i.e. base vectors) in the deformed configuration. This cross product is divided by 2, because the surfaces of the tetrahedron are triangles (half of the area of the corresponding rectangle). Thus, division by 2 is followed by division by 3 (one-third of the traction force is assigned to each node), which is equivalent to division by 6. That is why normals are initially divided by R6, which is a constant defined in frame.h and is equal to 6.

Bibliography

1. Acharya, A. (1992) *Discrete element method for the simulation of ball mills*. Master's thesis, University of Utah.
2. Adams, M.J. and Edmondson, B. (1987) Forces between particles in continuous and discrete liquid media. In *Tribology in Particulate Technology*, edited by B.J. Briscoe and M.J. Adams, Adam Hilger.
3. Adams, M.J. and Perchard, V. (1985) The cohesive force between particles with interstitial liquid. *I. Chem. E. Symp. Series.* **91**. 147–160.
4. Aggson, J.R. (1979) *Stress Induced Failures in Mine Roof*, Bureau of Mines RI 8338.
5. Aggson, J.R. (1978) *Coal Mine Floor Heave in the Beckley Coalbed*. An Analysis,' Report of Investigations 8274, Denver Mining Research Center, Bureau of Mines, Denver, Colorado, USA.
6. Ahola, M. (1989) Application of the Discrete Element Method toward Roof Stability Problems in Underground Coal Mines, In *Proc. of 1st Conf. DEM*, Golden, CO, USA.
7. Aidanpää, J.O., Shen, H.H. and Gupta, R.B. (1996) Experimental and numerical studies of shear layers in granular shear cell. *ASCE Journal of Engineering Mechanics*, **122**(3), 187–196.
8. Aizawa, T., Iwai, T. and Kihara, J. (1992) Granular Modeling of Steel Powder Flow and Compaction in Injection Molding. In *Proc. of Powder Injection Molding Symposium* (American Powder Metallurgy Institute, California), 419–433.
9. Alshibli, K. and Sture, S. (2000) Shear band formation in plane strain experiments of sand. *ASCE Journal of Geotechnical and Geoenvironmental Engineering*, **126**(6), 495–503.
10. Anandarajah, A. (1994) Discrete Element Method for Simulating Behavior of Cohesive Soil, *J. Geotech. Eng., ASCE*, **120**(9), 1593–1613.
11. Anandarajah, A. (2000) Numerical simulation of one dimensional behaviour of kaolinite, *Geotechnique* **50**(5), 509–521.
12. Anandarajah, A. and Chen, J. (1997) Van der Waals Attractive Force Between Clay Particles in Water and Contaminant. *Soils and Foundations, Japanese Society of Soil Mechanics and Foundation Engineering*, **37**(2), 27–37.
13. Anderson, T. and Jackson, R. (1967) A fluid mechanical description of fluidized beds, *Ind. Eng. Chem. Fundam.* **6**(4), 527–539.
14. Antonellini, M.A. and Pollard, D.D. (1995) Distinct element modeling of deformation bands in sandstone. *J. Struct. Geol.* **17**, 1165–1182.
15. Babic, M., Shen, H.H. and Shen, H.T. (1990) The stress tensor in granular shear flows of uniform, deformable disks at high solids concentrations. *J. Fluid Mech.* **219**, 81–118.
16. Babuska, I. and Melenk, J.M. (1997) The partition of unity method. *Int. J. Numer. Meth. Engng*, **40**, 727–758.
17. Baerns, M. (1966) Effect of interparticle adhesive forces on fluidization of fine particles. *Industrial & Engineering Chemistry Fundamentals*, **5**, 508–516.
18. Bakhtar, K., Jones, A.H. and Reed, M.A. (1987) Physical Modeling of Complex Underground Structures, *28th US Symposium on Rock Mechanics*, 771–779.
19. Ban, A.H. (1981) Superquadrics and Angle-Preserving Transformations. *IEEE Computer Graphics and Applications*, **1**, 1–20.
20. Bangash, T., Munjiza, A. and John, N. (2001) Modelling of Reinforced Concrete Beam Failure using Combined Finite/Discrete Element Method. *Fracture Damage Mechanics Conference 2001*, Milan.

21. Barbosa, R. and Ghaboussi, J. (1988) Discrete Element Model for Granular Soils. In *Proceedings of Workshop on Fill Retention Structures*, Ottawa, Canada.
22. Bardet, J.P. and Proubet, J. (1991) Adaptive dynamic relaxation for statics of granular materials. *Computers and Structures*, **39**(3/4), 221–229.
23. Bardet, J.P. and Vardoulakis, I. (2001) The Asymmetry of Stress in Granular Media. *Int. J. Solids and Structures*, **38**, 353–367.
24. Bardet, J.E. and Proubet, J. (1991) A numerical investigation of the structure of persistent shear bands in granular media. *Geotechnique*, **41**, 599–613.
25. Bardet, J.P. and Scott, R.F. (1985) Seismic Stability of Fractured Rock Masses with the Distinct Element Method. *26th US Symposium on Rock Mechanics*, 139–149.
26. Barr, A. (1981) Superquadrics and Angle-Preserving Transformations. *IEEE Computer Graphics and Applications*, **1**, 1–20.
27. Barr, A. (1984) Global and local deformations of solid primitives. *Computer Graphics*, **18**(3), 21–30.
28. Bathe, K.J. (1996) *Finite Element Procedures*. Prentice-Hall, Englewood Cliffs, New Jersey.
29. Bathurst, R.J. and Rothenburg, L. (1989) Investigation of Micromechanical Features of Idealized Granular Assemblies using DEM. In *Proceedings of 1st U.S. Conference on Discrete Element Methods*, Golden, CO.
30. Bathurst, R.J. and Rothenburg, L. (1988) Micromechanical aspects of isotropic granular assemblies with linear contact interactions, *J. Appl. Mech.* **55**(1), 17–23.
31. Bathurst, R.J. and Rothenburg, L. (1988b) Note on a Random Isotropic Granular Material with Negative Poisson's Ratio, *Int. J. Eng. Sci.* **26**(4), 373.
32. Bathurst, R.J. and Rothenburg, L. (1990). Observations on stress-force-fabric relations in idealized granular materials. *Mechanics of Materials*, **9**, 65–80.
33. Bauer, A. and Fratzos, D. (1987) Finite element modelling of presplit blasting using measured pressure time curves. *Soc. of Explosive Engineers Annual Meeting*, Miami, FL.
34. Bazant, Z.P. and Pijaudier-Cabot, G. (1988) Non-local continuum damage, localization instability and convergence. *J. Appl. Mech.* **55**, 287–293.
35. Belytschko T., Krongauz Y., Organ D., Fleming M. and Krysl P. (1996) Meshless methods: an overview and recent developments. *Comput. Methods Appl. Mech. Eng.* **139**, 3–47.
36. Belytschko, T. and Hughes, T.J.R. (eds) (1983) *Computational Methods For Transient Analysis*, Vol. **I**, Computational Methods in Mechanics, Elsevier Science.
37. Belytschko, T.B., Yen, H.J. and Mullen, R. (1979) *Mixed Method for Time Integration, Computer Methods in Applied Mechanics and Engineering*, North-Holland, 259–275.
38. Bentley, J.L. (1975) Multidimensional binary search trees used for associative searching. *Commun. ACM*, **18**, 1.
39. Benziey, S.E. and Krieg, R.D. (1982) A Continuum Finite Element Approach for Rock Failure and Rubble Formation. *Int. J. Numer. Anal. Methods Geomech.* **6**, 277–286.
40. Berryman, J.G. (1983) Random Close Packing of Hard Spheres and Disks. *Phys. Rev. A*, **27**, 1053–1061.
41. Beus, M.J., Iverson, S. and Stewart, B. (1997) Application of physical modelling and particle flow analysis to evaluate ore pass design. *Trans. Inst. Min. Metal. (Sect. A: Mining Industry)*, **106**, 110–117.
42. Bicanic, N., William, K.J. and Pramono, E. (1985) Numerical Prediction of Concrete Fracture Localization, *Proceedings NUMETA '85*, Swansea, Balkema.
43. Bieniawski, Z.T. (1984) *Rock Mechanics Design in Mining and Tunneling*. Balkema, Rotterdam.
44. Bolton, M.D. (1986). The strength and dilatancy of sands. *Geotechnique*, **36**(1), 65–78.
45. Bonet, J. and Peraire, J. (1991) An alternating digital tree (ADT) algorithm for 3D geometric searching and intersection problems. *Int. J. Num. Meth. Eng.* **31**, 1–17.
46. Boutt, D.F. and McPherson, B.J. (2001) Discrete Element Models of the Micromechanics of Sedimentary Rock: The Role of Organization vs. Friction. *EOS Trans. AGU*, **81**(47), Abstract T32E-0913.
47. Brach, R.M. (1989) Rigid body collisions. *J. Appl. Mech.* **56**, 133–138.
48. Bray, J.W. (1987) Boundary Element and Linked Methods for Underground Excavation Design. In *Analytical and Computational Methods in Engineering Rock Mechanics*, ed. E.T. Brown, 164–202.
49. Brenner, H. (1980) A general theory of Taylor dispersion phenomena. *PhysicoChem. Hydrodyn.* **1**, 91–123.
50. Brenner, H. (1980) Dispersion resulting from flow through spatially periodic porous media. *Phil. Trans. Roy. Soc. Lond. A297.* **81**, 133.
51. Briscoe, B.J. and Rough, S.L. (1998) The effects of wall friction in powder compaction. *Colloids and Surfaces A*, **137**, 103–116.
52. Brown, R.L. and Richards, J.C. (1970) *Principles of Powder Mechanics*, Pergamon Press.

53. Buckingham, E. (1914) On physically similar systems; illustrations of the use of dimensional equations. *The Physical Review*, Vol. **IV**, Series II, 345–376.

54. Burchell, S.L. (1992) Analysis of High Speed Films at the B&LS Coal Mine. Internal Communication, ICI Explosives, USA.

55. Burington, R.S. (1965) *Handbook of Mathematical Tables and Formulas*. McGraw-Hill, 4th edition.

56. Butkovich, T.R., Walton, O.R. and Heuze, F.E. (1988) Insights in cratering phenomenology provided by discrete element modelling. In *Key Questions in Rock Mechanics*, Cundall, P.A. *et al.* (Eds.), Balkema, Rotterdam, 359–368.

57. Campbell, C.S., Cleary, P.W. and Hopkins, M.A. (1995) Large scale landslide simulations: Global deformation, velocities and basal friction. *J. Geophys. Res.* **100**, B5, 8267–8283.

58. Campbell, C.S. and Brennen, C.E. (1985) Computer simulation of granular shear flows. *J. Fluid Mech.* **151**, 167–188.

59. Campbell, C.S. and Brennen, C.E. (1983) Computer simulation of shear flows of granular material. In *Mechanics of Granular Materials: New Models and Constitutive Relations*, Jenkins, J.T. and Satake, M., eds., Elsevier, Amsterdam, 313–326.

60. Campbell, C.S. (1988) Boundary interactions for two-dimensional granular flows: asymmetric stress and couple stress. In *Micromechanics of Granular Materials*, Satake, M. and J.T. Jenkins (Eds.), Elsevier Science, Amsterdam, 163–173.

61. Chang, C.S. (1987) Micromechanical modelling of constitutive relations for granular material. In *Micromechanics of Granular Materials*, Satake, M. and J.T. Jenkins (Eds.), Elsevier, Amsterdam, 271–278.

62. Chang, C.S. and Misra, A. (1990) Application of uniform strain theory to heterogeneous granular solids. *J. Eng. Mech. ASCE*, **116** (10).

63. Chang, C.S. and Misra, A. (1989) Computer simulation and modelling of mechanical properties of particulates, *Computers and Geotech.* **7**(4), 269–287.

64. Chang, C.S., Misra, A. and Xue, J. (1989) Incremental stress-strain relationships for regular packings made of multi-sized particles. *Int. J. Solid and Structures*, **25**(6), 665–681.

65. Chappel, B.A. (1972) The Mechanics of Blocky Material, PHD thesis, Australian National University.

66. Chen, S. and Doolen, G. (1998) Lattice Boltzmann method for fluid-flows. *Ann. Rev. Fluid Mech.* **30**, 329–364.

67. Cheng, Y.M. (1998) Advancement and improvements in discontinuous deformation analysis, *Computational Geotechnics* **22**(2), 153–163.

68. Chung, S.H. and Katsabanis, P. (2001) An integrated approach for estimation of fragmentation. *Proceedings of the 27th Annual Conference on Explosives and Blasting Technique*, **1**, 247–256, Orlando, FL.

69. Cleary, P.W. (1998) Discrete element modelling of industrial granular flow applications, TASK. *Quarterly – Scientific Bulletin*, **2**, 385–416.

70. Cleary, P.W. (2000) DEM simulation of industrial particle flows: Case studies of dragline excavators, mixing in tumblers and centrifugal mills. *Powder Technology*, **109**, 83–104.

71. Cleary, P.W. and Campbell, C.S. (1993) Self-lubrication for long run-out landslides: Examination by computer simulation. *J. Geophys. Res.* **98**, No B12, 21911–21924.

72. Cleary, P.W. and Sawley, M.L. (1999) Three-dimensional modelling of industrial granular flows. *Second International Conference on CFD in the Minerals and Process Industries*, CSIRO, Melbourne, Australia, 95–100.

73. Cleary, P.W., Laurent, B.F.C. and Bridgwater, J. (2002) DEM prediction of flow patterns and mixing rates in a ploughshare mixer. *Proc. World Congress Particle Technology 4*.

74. Cleary, P.W. (1998) How well do discrete element granular flow models capture the essentials of mixing processes? *Appl. Math. Modelling*, **22**, 995–1008.

75. Cleary, P.W. and Hoyer, D. (2000) Centrifugal mill charge motion and power draw: comparison of DEM predictions with experiment. *Int. J. Miner. Process.* **59**(2), 131–148.

76. Cleary, P.W., Stokes, N. and Hurley J. (1997) Efficient collision detection for three dimensional super-ellipsoid particles. *Proc. 8th International Computational Techniques and Applications Conference*, World Scientific, Adelaide.

77. Cohen, J., Lin, M., Manocha, D. and Ponamgi, K. (1995) I-COLLIDE: An Interactive and Exact Collision Detection System for Large-Scaled Environments. *Proceedings of ACM Int. 3D Graphics Conference*, 189–196.

78. Cook, B., Noble, D., Preece, D. and Williams, J. (2000) Direct simulation of particle-laden fluids. In, *Pacific Rocks 2000*, Girard, Liebman, Breeds and Doe (Eds.), Balkema, Rotterdam, 279–286.

79. Cook, R.D., Malkus, D.S., Plesha, M.E. and Witt, R.J. (2001) *Concepts and Applications of Finite Element Analysis*. 4th ed., John Wiley & Sons, Chichester.

80. Cundall, P.A. (1971) A computer model for simulating progressive large scale movements in blocky rock systems. *Proc. Symp. Rock Fracture (ISRM)*, Nancy, Vol. I, paper 11–8.

81. Cundall, P.A. and Strack, O.D.L. (1983) Modeling of microscopic mechanism in granular material. In *Mechanics of Granular Materials; New Models and Constitutive Relations*, J.T. Jenkins and M. Satake (Eds.), Elsevier, Amsterdam.

82. Cundall, P.A. (1987) Distinct element models of rock and soil structure. In *Analytical and Computational Methods in Engineering Rock Mechanics*, E.T. Brown (Ed.), Allen and Unwin, London.

83. Cundall, P.A. (1988) Formulation of three-dimensional distinct element model-Part 1. A scheme to detect and represent contacts in system composed of many polyhedral blocks, *Int. J. Rock Mech. Min. Sci. 8 Geomech. Abstr.* **25**(3), 107–116.

84. Cundall, P.A. and Roger, D.H. (1989) Numerical modeling of discontinua. *Proc. of 1st U.S. Conference on Discrete Element Methods*, Golden, CO.

85. Cundall, P.A. (1976) Explicit finite-difference method in geomechanics. *Numerical Methods in Geomechanics, ASCE*, P13 2–150.

86. D'Addetta, G.A., Kun, F., Hemnaim, H.J. and Ramm, E. (2002) On the application of a discrete model to the fracture process of cohesive granular materials. *Granular Matter.* **4**(2).

87. De Borst, R. (2001) Some recent issues in computational failure mechanics, *Int. J. Numer. Meth. Engng.* **52**(1/2), 63–96.

88. Dialer, C. (1992) A distinct element approach for the defromation behavior of shear stressed masonry panels. *Proceedings of the 6th Canadian Masonry Symposium*, Saskatoon, 765–776.

89. Drake, T.G. (1990) Structural features in granular flows, *J. Geophys. Res.* **95**(B6), 8681–8695.

90. Eringen, A.C. (1968) Theory of micropolar elasticity. In *Fracture – An Advanced Treatise*, Chapter 7, Liebovitz (Ed.), Vol. II, Academic Press, New York, 621–693.

91. Feng, Y.Q. and Yu, A.B. (2002) Effect of bed thickness on fluidization behaviour of particles mixtures. *4th World Congress on Particle Technology*, Sydney, Australia (to appear).

92. Feng, Y.Q., Xu, B.H., Zhang, S.J., Yu, A.B. and Zulli, P. (2001) Size segregation of particle mixtures in a gas-fluidized bed. *7th Int. Conf. on Bulk Materials Storage, Handling and Transportation*, Newcastle, Australia, 377–385.

93. Feng, Y.T. and Owen, D.R.J. (2002) An augumented spatial digital tree algorithm for contact detection in computational mechanics. *Int. J. Num. Meth. Eng.* (in press).

94. Feng, Y.T. and Owen, D.R.J. (2002b) An energy based comer to corner contact algorithm. *3rd Int. Conf. Discrete Element Methods*, Santa Fe, NM, 23–25.

95. Feng, Y.T., Han, K. and Owen, D.R.J. (2002) Filling domains with disks: an advancing front approach. *Int. J. Numer. Meth. Eng.* (in press).

96. Ghaboussi, J. (1992) Some theoretical and computational aspects of large scale discrete element. Rock mechanics. In *Proceedings of the 33rd U.S. symposium*, Tillerson, J.R. and Wawersik, W.R. (Eds.), Balkema, **33**, 619–628.

97. Ghaboussi, J. and Barbosa, R. (1990) Three-dimensional discrete element method for granular materials. *Int. J. Numer. Anal. Meth. Geomech.* **14**, 451–472.

98. Goodman, R.E. and Shi, G.H. (1985) *Block Theory and its Application to Rock Engineering*. Prentice-Hall, New Jersey.

99. Goodman, R.E., Taylor, R. and Brekke, T.L. (1968) A model for the mechanics of jointed rock. *J. Soil Mech. Found. Div. ASCE*, **94**, 637–660.

100. Gregory, C.E. (1973) *Explosives for North American Engineers*. Trans Tech Publications, Cleveland, OH.

101. Hakuno, M. and Hirao, T. (1973) A trial related to random packing of particle assemblies. *Proc. JSCE.*, **219**, 55–63 (in Japanese).

102. Hazzard, J.F., Young, P.F. and Maxwell, S.C. (2000) Micromechanical modeling of cracking and failure in brittle rocks, *J. Geophys. Res.* **105**(B7), 16683–16697.

103. Herrmann, H.J. (1991) Patterns and scaling in fracture. In *Fracture Processes in Concrete, Rock and Ceramics*, Chapman & Hall, New York.

104. Hocking, G. (1993) Collision impact of a ship with multi-year sea ice. *2nd Int. Conf. on Discrete Element Methods*, MIT, Cambridge, MA.

105. Hocking, G., Mustoe, G.G.W. and Williams, J.R. (1988) Dynamic analysis for generalized three dimensional contact and fracturing of multiple bodies, INTERA Technologies, Inc.

106. Hocking, G., Mustoe, G.G.W. and Williams, J.R. (1987) Two and three dimensional contact and fracturing of multiple bodies. *NUMETA '87 Numerical Methods in Engineering, Theory and Application*, A.A. Balkema, Rotterdam.

107. Hocking, G., Mustoe, G.G.W. and Williams, J.R. (1985) Validation of the CICE discrete element code for ice ride-up and ice ridge cone interaction. *ASCE Speciality Conference*, ARCTIC '85, San Francisco.

108. Hocking, G. (1989) The discrete element method for analysis of fragmentation and discontinua. *Proc. 1st Conf. DEM*, Golden, CO.

109. Hocking, G. (1992) The discrete element method for analysis of fragmentation of discontinua. *Eng. Computations*, **2**, 145–155.

110. Hocking, G., Williams, J.R. and Mustoe, G.G.W. (1985) Validation of the CICE discrete element code for ice ride-up and ice ridge/cone interaction. *ARCTIC '85, ASCE*, San Francisco.

111. Hocking, G., Williams, J.R. and Mustoe, G.G.W. (1985) CICE Model Validation Project. AOGA Project No. 231.

112. Hocking, G. (1977) Development and Application of the Boundary Integral and Rigid Block Methods for Geotechnics, PHD thesis, Imperial College.

113. Hocking, G., Mustoe, G.G.W. and Williams, J.R. (1985) CICE discrete element analysis code–theoretical manual. Applied Mechanics Inc., Lakewood, CO.

114. Hocking, L.M. (1964) The behavior of clusters of spheres falling in a viscous fluid. Part2 Slow motion theory. *J. Fluid Mech.* **20**, 129–139.

115. Hopkins, M.A. and Louge, M.Y. (1991) Inelastic microstructure in rapid granular flows of smooth disks. *Phys. Fluids A*, **3**(1), 47–57.

116. Hopkins, M.A., Daly, S.F. and Lever, J.H. (1996) Three-dimensional simulation of river ice jams. *Proceedings of the 8th International Specialty Conference on Cold Regions Engineering*, Fairbanks, AK, 12–17.

117. Hughes, T.J.R. (1983) Analysis of transient algorithms with particular reference to stability behavior. In *Computational Methods for Transient Analysis*, Vol. **1**.

118. Iwashita, K. and Hakuno, M. (1988) Granular assembly simulation for dynamic cliff collapse due to earthquake. *Proc. 9th World Conf. on Earthquake Eng.*, **3**, 175–180, Tokyo-Kyoto.

119. Jacota, A. and Dawson, P.R. (1988) Micromechanical modeling of powder compacts – I. Unit problems for sintering and traction induced deformation, *Acta Metall.* **36**(9), 2551–2561.

120. Jayaweera, K.O.L.F., Mason, B.J. and Slack, G.W. (1964) The behavior of clusters of spheres falling in a viscous fluid. Part 1 Experiment. *J. Fluid Mech.* **20**, 121–128.

121. Jenkins, J.T. and Savage, S.B. (1983) A theory for the rapid flow of identical, smooth, nearly elastic, spherical particles. *J. Fluid Mech.* **130**, 187–202.

122. Jensen, R.P., Bosscher, P.J., Plesha, M.E. and Edil, T.B. (1999) DEM simulation of granular media – Structure interface: effects of surface roughness and particle shape. *Int. J. Num. Anal. Meth. Geomech.*, **23**, 531–547.

123. Johansson, C.H. and Persson, P.A. (1970) *Detonics of High Explosives*. Academic Press, London.

124. Jorgenson, G.K. and Chung, S.H. (1987) Blast simulation – surface and underground with SABREX model. *CIM Bulletin*, 37–41.

125. Kawaguchi, T., Tanaka, T. and Tsuji, Y. (1998) Numerical simulation of two-dimensional fluidised beds using discrete element method (comparison between two- and three-dimensional models). *Powder Technol.* **96**, 129–138.

126. Kelley, C.T. (1995) *Iterative Methods for Linear and Nonlinear Equations*. SIAM, Philadelphia.

127. Kernighan, B.W. and Ritchie, D.M. (1988) *The C Programming Language*. Prentice Hall, New Jersey, 2nd edition.

128. Kishino, Y. (1987) Disc model analysis of granular media. In *Micromechanics of Granular Materials*, Satake, M. and Jenkins, J.T. (Eds.), Elsevier, Amsterdam, 143–152.

129. Kitamura, R. (1981) Analysis of deformation mechanism of particulate material at particle scale. *Soils and Foundations*, **21**(2), 85–97.

130. Komodromos, P. (2002) On the simulation of deformable bodies using combined discrete and finite element methods. *3rd International Conference on Discrete Element Methods*, Santa Fe, NM.

131. Kreyszig, E. (1983) *Advanced Engineering Mathematics*. John Wiley & Sons, Chichester. 5th edition.

132. Lewis, R.W. and Schrefler, B.A. (1998) *The Finite Element Method in the Static and Dynamics Deformation and Consolidation of Porous Media*. John Wiley & Sons, Chichester, 2nd Ed.

133. Lian, J. and Shima, S. (1994) Powder assembly simulation by particle dynamics method. *Int. J. Num. Meth. Eng.* **37**, 763–775.

134. Lifshitz, E.M. (1956) The theory of molecular attractive forces between solids. *Soviet Physics*, **2**(1), 73–83.

135. Lin, J.S. (1995) Continuous and discontinuous analysis using the manifold method,' *Proceeding Working Forum on the Manifold Analysis*. Vol. 1, 1–20, Geotechnical Lab, US Army Engineers Waterways Experiment Station.

136. Livesley, R.K. (1978) Limit analysis of structures formed from rigid blocks. *Int. J. Num. Meth. in Eng.* **12**, 1853–1871.

137. Lloyd, S. (2000) Ultimate physical limits to computation. *Nature*, **406**, 1047–1054.

138. Londe, P. (1987) The Malpasset Dam Failure. *Engineering Geology*, **24**, 295–529.

139. Lorenz, A., Tuozzolo, C. and Louge, M.Y. (1995) Measurements of impact properties of small, nearly spherical particles. *Experimental Mechanics*, **37**(3), 292–298.

140. Mandel, J. (1963) Tests on reduced scale models in soil and rock mechanics – a study of the conditions of similitude. *Int. J. Rock Mech. Mining Sci.* **1**, 31–42.

141. Margolin, L.G. (1984) Generalized Griffith criteria for crack propagation. *Eng. Frac. Mech.* **19**, 539–543.

142. Mazzone, D.N., Tardos, G.I. and Pfeffer, R. (1986) The effect of gravity on the shape and strength of a liquid bridge between two spheres. *J. Colloid Interface Sci.* **113**, 544–556.

143. Meguro, K. and Hakuno, M. (1988) Fracture analysis of concrete structure by granular assembly simulation. *Bulletin of the Earthquake Research Institute*, **63**(4), 409–468 (in Japanese).

144. Metcalfe, G., Shinbrot, T., McCarthy, J.J. and Ottino, J.M. (1995) Avalanche mixing of granular solids. *Nature*, **374**, 39–41.

145. Minty, E.J. and Kearns, O.K. (1983) Rock mass workability. In *Collected case studies in Engineering Geology, Hydrogeology, Environmental Geology*, Knight, M.J., Minty, E.J. and Smith, R.B. (Eds.), 59–81.

146. Mishra, B.K. and Murty, C.V.R. (2001) On the determination of contact parameters for realistic simulation of tumbling mills. *Powder Technology*, **115**, 290–297.

147. Mroz, Z. and Zubelewicz, A. (1982) On initiation of flow of granular materials from hoppers. In *Deformation and Failure of Granular Materials*, Vermeer, P.A. and Lucrer, H.J. (Eds.), Balkema, 569–577.

148. Muhlhaus, H.B. and Vardoulakis, I. (1987) The thickness of shear bands in granular materials. *Geotechnique*, **37**(3), 271–283.

149. Muhlhaus, H.B. (1989) Application of Cosserat theory in numerical solutions of limit load problems. *Ing.-Archiv.*, **59**(2), 124–137.

150. Munjiza, A. (1999) Fracture, fragmentation and rock blasting models in the combined finite-discrete element method. In *Fracture of Rock*. Computational Mechanics Publications.

151. Munjiza, A., Latham, J.P. and John, N.W.M. (2003) 3D dynamics of discrete element systems comprising irregular discrete elements. *Int. J. Num. Methods Eng.* **56**, 35–55.

152. Munjiza, A. and John, N.W.M. (2001) Mesh size sensitivity of the combined FEM/DEM fracture and fragmentation algorithms, *Eng. Fract. Mech.* **69**(2), 281–295.

153. Munjiza, A. and Andrews, K.R.F. (2000) Discretised penalty function method in combined finite-discrete element analysis. *Int. J. Num. Meth. Eng.* **49**, 1495–1520.

154. Munjiza, A. and Andrews, K.R.F. (2000) Detonation gas model for combined finite-discrete element modelling of fracture and fragmentation. *Int. J. Num. Meth. Eng.* **49**, 1377–1396.

155. Munjiza, A., Andrews, K.R.F. and White, J.K. (1999) Combined single and smeared crack model in combined finite-discrete element method. *Int. J. Num. Meth. Eng.* **44**, 41–57.

156. Munjiza, A., Latham, J.P. and Andrews, K.R.F. (1999) Challenges of a coupled combined finite-discrete element approach to explosive induced rock fragmentation. *FRAGBLAST – Int. J. Fragmentation and Blasting*, **3**, 237–250.

157. Munjiza, A. and Andrews, K.R.F. (1998) NBS contact detection algorithm for bodies of similar size. *Int. J. Num. Meth. Eng.*, **43**, 131–149.

158. Munjiza, A. and Owen, D.R.J. (1998) A K^m proportional damping in explicit integration of dynamic structural systems. *Int. J. Num. Meth. Eng.* **41**, 1277–1296.

159. Munjiza, A., Owen, D.R.J. and Bicanic, N. (1995) A combined finite-discrete element method in transient dynamics of fracturing solids. *Int. J. Eng. Computations*, **12**, 145–174.

160. Munjiza, A. and Latham, J.P. (2002) Grand challenge of discontinuous deformation analysis, plenary lecture. *5th Int. Conf. on Analysis of Discontinuous Deformation*, Israel.

161. Munjiza, A. and Latham, J.P. (2002) Computational and algorithmic challenge of modelling discontinua, keynote lecture. *3rd Int. Conf. On Discrete Element Methods*, Santa Fe, CA.

162. Munjiza, A. and Latham, J.P. (2002) Challenge of modelling particulate and fracturing solids, keynote lecture. *5th World Congress on Computational Mechanics*, Vienna.

163. Munjiza, A. and Andrews, K.R.F. (1999) A FEM/DEM model for flow through cracked solids. *7th ACME Conference on Computational Mechanics in the UK*, Durham, UK.

164. Munjiza, A. and Andrews, K.R.F. (1998) Improved fracture solutions for the combined finite-discrete element method. *6th ACME Conference on Computational Mechanics in the UK*, Exeter.

165. Munjiza, A. (1996) Combined finite-discrete element models for blasting and mining operations. In *Discontinuous Deformation Analysis (DDA) and Simulations of Discontinuous Media*, Reza Salami, M. and Banks, D. (Eds.), Berkeley, CA, 518–525.

166. Munjiza, A., Owen, D.R.J. and Crook, A.J.L. (1995) Energy and momentum preserving contact algorithm for general 2D and 3D contact problems. *Proceedings of the Third International Conference on Computational Plasticity: Fundamentals and Applications*, Barcelona, Spain, 829–841.

167. Munjiza, A., Bicanic, N. and Owen, D.R.J. (1992) Object oriented programming concepts in discrete element analysis of fracturing media. *Proceedings of the Third International Conference on Computational Plasticity: Fundamentals and Applications*, Barcelona, Spain, 1949–1966.

168. Munjiza, A., Owen, D.R.J., Bicanic, N. and Xian, L. (1991) A concept of contact element in the discrete element method. In *Proceedings NEC-91, Int. Conf. on Nonlinear Engineering Computations*, Bicanic *et al.* (Eds.), Pineridge Press, 435–448.

169. Munjiza, A., Andrews, K.R.F. and White, J.R. (1997) Discretized contact solution for combined finite-discrete method. *5th ACME Conf.*, London, UK, 96–100.

170. Munjiza, A., Owen, D.R.J. and Bicanic, N. (1995) A combined finite-discrete element method in transient dynamics of fracturing solids. *Int. J. Eng. Computation*, **12**, 145–174.

171. Mustoe, G.G.W. (2000) A numerical and experimental study of the performance and safety issues for ore pass system. (RP-5). WMRC (Western Mining Resources Center) annual progress.

172. Mustoe, G.G.W. (1989) Special elements in discrete element analysis. *1st U.S. Conference on Discrete Elements*, Golden, CO.

173. Mustoe, G.G.W., Henriksen, M. and Huttelmaier, H.P. (Eds.) (1989) *Proceedings of the 1st U.S. Conf. on Discrete Element Methods*, Golden, CO.

174. Mustoe, G.G.W., Williams, J.R., Hocking, G. and Worgan, K. (1988) Penetration and fracturing of brittle plates under dynamic impact. INTERA Technologies, Inc.

175. Mustoe, G.G.W., Williams, J.R. and Hocking, G. (1987) The discrete element method in geotechnical engineering. In *Developments in Soil Mechanics and Foundation Engineering–3*, Banerjee, P.K. and Butterfield, R. (Eds.), New York; Elsevier, 233–263.

176. Mustoe, G.G.W., Williams, J.R. and Hocking, G. (1977) The discrete element method in geotechnical engineering. In *Developments in Soil Mechanics and Foundation Engineering*, (Ch. 7), Elsevier, Barking, U.K.

177. Nedderman, R.M. (1992) *Statics and Kinematics of Granular Materials*. Cambridge University Press, Cambridge.

178. Nemat-Nasser, S. (1990) Certain basic issues in finite-deformation continuum plasticity. *Mechanica*, **25**, 223–229.

179. Obert, L. and Duvall, W.I. (1968) *Rock Mechanics and the Design of Structures in Rock*. John Wiley & Sons, New York.

180. Oda, M. and Konishi, J. (1974) Microscopic deformation mechanism of granular material in simple shear. *Soils and Foundations*, **14**(4), 25–38.

181. Ogawa, H. and Takeuchi, M. (1969) Dispersion of dumped sand from hopper-barges. *Proceedings of the JSCE*, **161**, 39–49.

182. Ohnishi, Y., Mimuro, T., Hakevakl, N. and Yoshida, J. (1985) Verification of Input parameters for distinct element analysis of jointed rock mass. *Proc. Int. Symp. on Fundamentals of Rock Joints*, Bjorkllden.

183. Ouyang, J., Yu, A.B. and Pan, R.H. (2001) Simulations of plug flow in vertical pipe by hard sphere model. *7th Int. Conf. on Bulk Materials Storage, Handling and Transportation*, Newcastle Australia, 801–815.

184. Owen, D.R.J., Munjiza, A. and Bicanic, N. (1992) A finite element–discrete element approach to the simulation of rode blasting problems. *Proceedings FEMSA-92, 11th Symposium on Finite Element methods in South Africa*, Cape Town, 39–59.

185. Pande, G.N., Beer, G. and Williams, J.R. (1990) *Numerical Methods in Rock Mechanics*. John Willey & Sons, England.

186. Papradakakis, M. (1981) A method for the automatic evaluation of the dynamic relaxation parameters. *Computer Methods in Applied Mechanics and Engineering*, **25**, 35–48.

187. Pentland, A. and Williams, J.R. (1989) Fast simulations on small computers: modal dynamics applied, to volumetric nodes. *Proc. 20th Annual Modeling and Simulation Conference*, Pittsburg.

188. Pentland, A.P. and Williams, J.R. (1988) Virtual construction. *Construction*, **3**(4).

189. Pentland, A.P. and Williams, J.R. (1989) Good vibrations: modal dynamics for graphics and animation. *SIGGRAPG '89, ACM Computer Graphics*, **23**(3).

190. Perkins, E. and Williams, J. (2001) A fast contact detection algorithm insensitive to object size. *Engineering Computations*, **18**(1/2), 48–61.

191. Perkins, P. and Williams, J.R. (2001) C-grid: Neighbor searching for many body simulation. *ICADD-4*, 427–438.

192. Preece, D.S. (1992) The influence of damping on computer simulations of rock motion. In *Proceedings of the 25th Annual Oil Shale Symposium*, Golden, CO.

193. Preece, D.S. and Chung, S.H. (2002) Rock blasting 3-D discrete element heave predictions for surface coal mines and rock quarries. In *Proceedings of NARMS-TAC 2002*, Toronto, Ontario.

194. Preece, D.S., Burchell, S.L. and Scovira, D.S. (1993) Coupled explosive gas flow and rock motion modeling with comparison to bench blast field data. In *Proceedings of the Fourth International Symposium on Rock Fragmentation by Blasting*, Vienna, Austria.

195. Preece, D., Jensen, R., Perkins, E. and Williams, J. (1999) Sand production modeling using superquadric discrete elements and coupling of fluid flow and particle motion. *Proceedings of the 37th U.S. Rock Mechanics Symposiu*, Amadei, Kranz, Scott and Smeallie (Eds.), Balkema, Amsterdam.

196. Preece, D.S. and Taylor, L.M. (1989) Complete computer simulation of crater blasting including fragmentation and rock motion. *Proceedings of Research Symposium*, Society of Explosive Engineers Spring Meeting.

197. Preece, D.S. and Taylor, L.M. (1990) Spherical element bulking mechanisms for modeling blasting induced rock motion. *Proceedings of the Third International Symposium on Rock Fragmentation by Blasting*, Brisbane, Queensland, Australia.

198. Preece, D.S. and Knudsen, S.D. (1992) Computer modeling of gas flow and gas loading of rock in a bench blasting environment. *Proceedings of the 33rd U.S. Symposium on Rock Mechanics*, Santa Fe, NM.

199. Preece, D.S. (1990) Rock motion simulation of confined volume blasting. *Proceedings of the 31st U.S. Symposium on Rock Mechanics*, Golden, CO.

200. Reddy, J.N. (1984) *Energy and Variational Methods in Applied Mechanics*. John Wiley & Sons, New York.

201. Richman, M.W. (1988) Homogeneous shear flow of highly inelastic disks: the full range of solid fraction. *Proc. 7th Conf. of Engineering Mechanics*, Blacksburg, Virginia.

202. Ristow, G.H. (1998) *Flow Properties of Granular Materials in Three-dimensional Geometries*. Verlag Gorich & Weiershäuser, Marburg, Germany.

203. Rothenburg, L. and Bathurst, R.J. (1991) Numerical simulation of idealized granular assemblies with plane elliptical particles. *Comput. and Geotech.* **11**, 315–329.

204. Sakaguchi, H. and Muhlhaus, H.B. (1997) Mesh free modelling of failure and localization in brittle rock. In *Deformation and Progressive Failure in Geomechanics*, Asoaka, Adachi and Oka (Eds.), Pergamon, 15–21.

205. Sawamoto, Y., Tsubota, H., Kasai, Y., Koshika, N. and Morikawa, H. (1998) Analytical studies on local damage to reinforced concrete structures under impact loading by discrete element method. *Nuclear Eng. Des.* **179**, 157–177.

206. Schamaun, J.T. (1984) Methods for predicting rubble motion during blasting. *Proceedings of 25th U.S. Symposium on Rock Mechanics*, Northwestern University.

207. Schubert H., Herrmann, W. and Rumpf, H. (1975) Deformation behavior of agglomerates under tensile stress. *Powder Technology*, **2**, 121.

208. Serrano, A.A. and Rodriguez-Ordz, J.M. (1973) A contribution to the mechanics of heterogeneous granular media. *Proc. Symp. on the Role of Plasticity in Soil Mechanics*, Cambridge, 215–228.

209. Shahinpoor, M. and Sharpass, A. (1982) Frequency distribution of voids in monolayers of randomly packed equal spheres. *Bulk Solids Handling*, **2**, 825–838.

210. Shi, G. and Goodman, R.E. (1989) Generalization of two-dimensional discontinuous deformation analysis for forward modelling. *Int. J. Num. Analy. Meth. Geomech.* **13**, 359–380.

211. Shi, G. (1988) Discontinuous deformation analysis: a new method for computing stress, strain and sliding of block systems. PhD Thesis, Department Civil Engineering, University of California, Berkeley.

212. Shi, G. (1993) *Block System Modeling by Discontinuous Deformation Analysis*. Computational Mechanics Publications, Boston.

213. Shi, G.H. (1989) Block System Modeling by Discontinuous Deformation Analysis. Doctoral Dissertation, Department of Civil Engineering, University of California, Berkeley.

214. Shi, G.H. (1992) Modeling rock joints and blocks by manifold method. *Proceedings of the 33rd U. S. Rock Mechanics Symposium*, Santa Fe, New Mexico, 639–648.

215. Shi, G.H. (1991) *Block System Modelling by Discontinuous Deformation Analysis*. Computational Mechanics, London.

216. Shi, G.H. and Goodman, R.E. (1985) Two dimensional discontinuous deformation analysis. *Int. J. Numer. Anal. Meth. Geomech.* **9**, 541–556.

217. Shi, G.H and Goodman, R.E. (1988) Discontinuous deformation analysis, a new method For computing stress, strain and sliding of block systems. *Proc. 29th U. S. Symposium on Rock Mechanics*, Minneapolis, 381–393.

218. Shi, G.H. (1989) Discontinuous deformation analysis – a new numerical model for the statics and dynamics of block structures. *Proc. 1st Conf. DEM*, Golden, CO.

219. Shimizu, Y. and Cundall, P.A. (2001) Three-dimensional DEM simulations of bulk handling by screw conveyors. *J. Eng. Mech. ASCE*, **127**(9), 285–292.

220. Shodja, H.M. and Nezami, E.G. (2002) Stress-induced anisotropy in random assemblies of oval granules. *8th International Symposium on Numerical Models in Geomechanics*, Rome, Italy.

221. Sims, K. (1990) Particle animation and rendering using data parallel computation. *Comput. Graphics*, **24**(4), 405–413.

222. Skinner, A.E. (1969) A note on the influence of interparticle friction on the shearing strength of a random assembly of spherical particles. *Geotechnique*, **19**, 150–157.

223. Southwell, R.V. (1940) *Relaxation Methods in Engineering Science*. Oxford University Press, London.

224. Sowizral, H., Rushforth, K. and Deering, M. (1998) *The Java 3D Specification*. The Java Series, Addison-Wesley, Reading, MA.

225. Stark, C.P. (1991) An invasion percolation model of drainage network evolution. *Nature*, **352**, 423–425.

226. Stewart, R., Bridgwater, J., Zhou, Y.C. and Yu, A.B. (2001) Simulated and measured flow of granules in a bladed mixer – a detailed comparison. *Chem. Eng. Sci.* **56**, 5457–5471.

227. Stroustrup, B. (1986) *The C++ Programming Language*. Addison-Wesley, Reading, MA.

228. Sun, Y. and Vinogradov, O. (1998) Numerical simulation of jamming of solid particles transported by fluid in planar channels. *Comput. Modeling & Simulation in Eng.* **3**, 27–32.

229. Swan, G. (1983) Determination of stiffness and other joint properties from roughness measurements. *Rock Mech. Rock Eng.* **16**, 19–38.

230. Takagi, Y., Mizutani, H. and Kawakami, S. (1984) Impact fragmentation experiments of basalts and pyrophyllites. *Icarus*, **59**, 462–477.

231. Tarumi, Y. and Hakuno, M. (1989) A DEM simulation for sand liquefaction. *Proc. 1st U.S. Conf. on Discrete Element Methods*, Golden, CO.

232. Taylor L.M. and Preece D.S. (1989) Simulation of blasting induced rock motion using spherical element models. *Proceedings of the 1st U.S. Conf. on Discrete Element Methods*, Golden, CO.

233. Taylor, L.M. and Preece, D.S. (1992) Simuladon of blasting induced rock motion using spherical element models, *Eng. Computations*, **9**(2).

234. Terzaghi, K. (1936) Stress distribution in dry and in saturated sand above a yielding trap-door. *Proceedings First International Conference on Soil Mechanics and Foundation Engineering*, Cambridge, MA, 307–311.

235. Thomas, P.A. and Bray, J.D. (1999) Capturing nonspherical shape of granular media with disk clusters, *ASCE J. Geotech. Geoenvironmental Eng.* **125**(3), 169–178.

236. Thornton, C. and Barnes, D.J. (1982) On the mechanics of granular materials. *IUTAM Conf. on Deformation and Failure of Granular Materials*, Delft, 69–77.

237. Thornton, C. and Barnes, D.J. (1986) Computer simulated deformation of compact granular assemblies. *Acta Mechanica*, **64**, 45–61.

238. Thornton, C. (1989) Applications of DEM to process engineering problems. *Proceedings 1st U.S. Conference on Discrete Element Methods*, Golden, CO.

239. Thornton, C. (Ed.) (2000) Numerical simulations of discrete particle systems. *Powder Technology*, Special Issue, **109**(1–3), 1–298.

240. Thornton, C. and Randall, C.W. (1988) Applications of theoretical contact mechanics to solid particle system simulation. In *Micromechanics of Granular Materials*, Satake, M. and Jenkins, J.T. (Eds.), Elsevier, 133–142.

241. Thornton, C. and Yin, K.K. (1991) Impact of elastic spheres with and without adhesion. *Powder Technology*, **65**, 153–166.

242. Thornton, C. (1991) Interparticle sliding in the presence of adhesion. *J. Phys. D: Appl. Phys.* **24**, 1942–1946.

243. Thornton, C. and Antony, S.J. (2000) Quasi-static shear deformation of a soft particle B; system. *Powder Tech.*, **109**, 179–191.

244. Throop, G.J. and Bearman, R.J. (1965) Numerical solutions of the Percus–Yevick equation for the hard-sphere potential. *J. Chem. Phys.* **42**, 2408–2411.

245. Tillemann, H.J. and Herrmann, H.J. (1995) Simulating deformations of granular solids under shear. *Physica A*, **217**, 261–288.

246. Ting, J.M. and Corkum, B.T. (1988) Strength behavior of granular materials using discrete numerical modeling. *Proc. 6th Int. Conf. Numerical Models in Geomechanics*, Innsbruck, Austria, 305–310.

247. Ting, J.M. (1992) A robust algorithm for ellipse-based discrete element modelling of granular materials. *Comput. and Geotech.*, **13**(3), 175–186.

248. Toffoli, T. and Margolus, N. (1987) *Cellular Automata Machines: A New Environment for Modeling.* MIT Press, USA.

249. Tsuji, Y., Kawaguchi, T. and Tanaka, T. (1993) Discrete particle simulation of two-dimensional fluidized bed. *Powder Technology*, **77**, 79–87.

250. Tuzun, U., Houlsby, G.T., Nedderman, R.M. and Savage, S.B. (1982) The flow of granular materials II: Velocity distributions in slow flow. *Chem. Eng. Sci.*, **37**(12), 1691–1709.

251. Uchida, Y. and Hakuno, M. (1989) A DEM simulation of debris flow. *Proc. 1st U.S. Conf. on Discrete Element Methods*, Golden, CO.

252. Van Nierop, M.A., Glover, G., Hinde, A.L. and Moys, M.H. (2001) A discrete element method investigation of the charge motion and power draw of an experimental two-dimensional mill. *Int. J. Miner. Process.* **61**, 77–92.

253. Walker, D.M. (1966) An approximate theory for pressures and arching in hoppers. *Chem. Eng. Sci.* **21**, 975.

254. Walton, O.R. and Braun, R.L. (1988) Viscosity and temperature calculations for assemblies of inelastic frictional disks. *J. Rheology*, **30**(5), 949–980.

255. Walton, O.R., Braun, R.L., Mallon, R.G. and Cervelli, D.M. (1988) Particle-dynamics calculations of gravity flows of inelastic, frictional spheres. In *Micromechanics of granular material*, Satake, M. and Jenkins, J.T. (Eds.), Elsevier, Amsterdam, 153–161.

256. Walton. O.R. and Braun, R.L. (1986) Viscosity, granular-temperature and stress calculations for shearing assemblies of Inelastic, frictional disks. *J. Rheology*, **30**, 949–980.

257. Wang, C.Y., Wang, C.F. and Sheng, J.A. (1999) Packing generation scheme for the granular assemblies with 3D ellipsoidal particles. *Int. J. Numer. Anal. Meth. Geomech.* **23**, 815–828.

258. Warburton, P.M. (1981) Vector stability analysis of an arbitrary polyhedral rock block with any number of free faces. *Int. J. Rock Mech.* **18**, 415–427.

259. Watanabe, H. (1999) Critical rotation speed for ball-milling. *Powder Technology*, **104**, 95–99.

260. Weaver, J.M. (1975) Geological factors significant in the assessment of rippability. *The Civil Eng. in Sth. Africa*, **17**(12), 313–316.

261. Wei, Q., Cheng, X.H. and Liu, G.T. (1991) The elliptic discrete element method as a new approach to simulating granular media. *Proc. Asian Pacific Conf. Computational Mechanics*, Hong Kong.

262. Wen, C. and Yu, Y. (1966) Mechanics of fluidization. *Chem. Eng. Prog. Symp. Ser.*, **62**(62), 100.

263. William, K.J., Bicanic, N., Pramono, N. and Sture, S. (1985) Composite fracture mode for strain softening computations of concrete. *Int. Conf. Fracture Mechanics of Concrete*, Lausanne, Switzerland.

264. Williams, J.R. and Mustoe, G. (1987) Nodal methods for the analysis of discrete systems. *Int. J. Comput. and Geotech.* **4**, 1–19.

265. Williams, J.R. and Pentland, A.P. (1989) Superquadrics and modal dynamics for discrete elements in concurrent design. In *Proceedings of the 1st U.S. Conf. on Discrete Element Methods*, Mustoe, G.G.W., Henriksen, M. and Huttelmaier, H.P. (Eds.), Golden, CO.

266. Williams, J.R. and O'Connor, R. (1995) A linear complexity intersection algorithm for discrete element simulations of arbitrary geometrics. *Int. J. CAE-Eng. Computations*, **12**(2), 185–201.

267. Williams, J.R., O'Coimor, R. and Rege, N. (1996) Discrete element analysis and granular vortex formation. *Electronic J. Geotech. Eng.* Available at http://139.78.66.61/ejge/.

268. Williams, J. and Mustoe, G. (Eds.) (1993) *Proceedings of the 2nd International Conference on Discrete Element Methods (DEM)*, IESL Publications.

269. Williams, J.R. and Mustoe, G.G.W. (1987) Modal methods for the analysis of discrete systems. *Comput. and Geotech.* **4**, 1–19.

270. Williams, J.R. and Pentland, A. (1991) Superquadrics and modal dynamics for discrete elements in concurrent design. Technical Report Order No. IESL91-12, Intelligent Engineering Systems Laboratory, Massachusetts Institute of Technology.

271. Williams, J.R. and Pentland, A.P. (1989) Good vibrations: modal dynamics for graphics and animation. *ACM Comput. Graphics*, **23**(3).

272. Williams, J.R. (1987) Contact analysis of large numbers of interacting bodies using discrete modal methods for simulating material failure on the microscopic scale. Technical Report Order No. IESL91-12, Intelligent Engineering Systems Laboratory, Massachusetts Institute of Technology, 1991.

273. Williams, J.R. (1988) Contact analysis of large numbers of interacting bodies using discrete modal methods for simulating material failure on the microscopic scale. *Eng. Comput.* **5**, 198–209.

274. Williams, J.R. and Mustoe, G.G.W. (1987) Modal methods for the analysis of discrete systems. *Comput. and Geotech.* **4**, 1–19.

275. Williams, J.R., Hocking, G. and Mustoe, G.G.W. (1985) The theoretical basis of the discrete element method. *Proceedings of the International Conference on Numerical Methods in Engineering: Theory and Applications*, Swansea, 897–906.

276. Williams, J.R. and Pentland, A.P. (1989) Interactive, integrated design – object representation and modal analysis. *Human Computer Interface International '89*, Boston, MA.

277. Wilson, E.L, Farhoomand, I. and Bathe, K.J. (1973) Nonlinear dynamic analysis of complex structures. *Earthquake Eng. Struct. Dyn.* **1**, 241–252.

278. Wittke, W. and Leonards, G.A. (1987) Modified hypothesis for failure of Malpasset Dam. *Eng. Geology*, **24**, 567–594.

279. Worgan, K.J and Mustoe, G.G.W. (1989) Application of the discrete element method to modeling the subsurface penetration of a uniform cover. *Proc. 1st Conf. DEM*, Golden, CO.

280. Wriggers and Simo, J.C. (1985) A note on tangent stiffness for fully nonlinear contact problems. *Commun. Appl. Numer. Meth.* **1**, 199–203.

281. Wright, T.W. (1987) Steady shearing in a viscoplastic solid. *J. Mech. Phys. of Solids*, **35**, 269–282.

282. Wylen, G.J.V. and Sonntao, R.E. (1985) *Fundamentals of Classical Thermodynamics*. 3rd edition, John Wiley & Sons, New York.

283. Xu, B.H., Yu, A.B., Chew, S.J. and Zulli, P. (2000) Numerical simulation of the gas-solid flow in a bed with lateral gas blasting. *Powder Technol.* **109**, 13–26.

284. Yamamoto, T. and Hakuno, M. (1989) A DEM simulation for tunnel excavation. *Proc. 1st U.S. Conf. on Discrete Element Methods*, Golden, CO.

285. Yang, R.Y., Zou, R.P. and Yu, A.B. (2001) Microdynamic analysis of the flow of particles in horizontal rotating drum. *7th International Symposium on Agglomeration*, Aibi, France, 569–588.

286. Zhang, D.Z. and Rauenzahn, R.M. (1997) A viscoelastic model for dense granular flows. *J. Rheol.* **41**, 425–453.

287. Zhang, D.Z. and Rauenzahn, R.M. (2000) Stress relaxation in dense and slow granular flows. *J. Rheol.* **44**, 1019–1041.

288. Zhang, Y. and Cundall, P.A. (1986) Numerical simulation of slow deformations. *Proc. Symp. on the Mechanics of Particulate Media, Tenth U.S. National Congress of Applied Mechanics*, Austin, TX.

289. Zhang, Z.P., Yu, A.B. and Oakeshott, R.B.S. (1996) Effect of packing method on the randomness of disk packing. *J. Phys. A: Mathematical & General*, **29**, 2671–2685.

290. Zhang, Z.P., Liu, L.F., Yuan, Y.D. and Yu, A.B. (2001) Numerical study of the effects of dynamic factors on the packing of particles. *Powder Technol.* **116**, 23–32.

291. Zhong, Z.H. and Nilsson, L. (1990) A contact searching algorithm for general 3D contact-impact problems. *Comp. Struct.* **34**, 327–335.

292. Zhou, Y.C., Xu, B.H., Yu, A.B. and Zulli, P. (2002) An experimental and numerical study of the angle of repose of coarse spheres, *Powder Technol.* **125**(1), 45–54.

293. Zhou, Y.C., Xu, B.H., Yu, A.B. and Zulli, P. (2001) Numerical study of sandpile formation and force evolution. *Powders and Grains*, Sendai, Japan, 495–498.

294. Zhou, Y.C., Xu, B.H., Yu, A.B. and Zulli, P. (2002) A numerical and experimental study of the angle of repose of granular particles. *Powder Technol.* **125**, 45–54.

295. Zhou, Y.C., Wright, W.D., Yang, R.Y., Xu, B.H. and Yu, A.B. (1999) Rolling friction in the dynamic simulation of sandpile formation. *Physica A*, **269**, 536–553.

296. Zhu, H.P. and Yu, A.B. (2001) Weighting ftinction in the averaging theory of granular materials. *Bulk Solids Handling*, **21**, 53–57.

297. Zhu, H.P. and Yu, A.B. (2001) Stress distribution of hopper flow. *7th Int. Conf. on Bulk Materials Storage, Handling and Transportation*, Newcastle, Australia, 283–290.

298. Zienkiewicz, O.C. and Taylor, R.L. (1989) *The Finite Element Method; Basic Formulation and Linear Problems. 1*, 4th edition. McGraw-Hill, London.

299. Zietlow, W.K. and Labuz, J.F. (1998) Measurement of the intrinsic process zone in rock using acoustic emission. *Int. J. Rock Mech. Min. Sci. & Geomech. Abstr.*, **35**(3), 291–299.

300. Zubelewicz, A. and Mroz, Z. (1983) Numerical simulation of rockburst processes treated as problems of dynamic instability. *Rock Mech. And Eng.* **16**, 253–274.

Index

The Combined Finite-Discrete Element Method   A. Munjiza
© 2004 John Wiley & Sons, Ltd   ISBN: 0-470-84199-0